GOVERNMENT LABORATORY TECHNOLOGY TRANSFER

Government Laboratory Technology Transfer

Process and impact

SALLY A. ROOD
Federal Laboratory Consortium, Washington, USA

LONDON AND NEW YORK

First published 2000 by Ashgate Publishing

Reissued 2018 by Routledge
2 Park Square, Milton Park, Abingdon, Oxon OX14 4RN
711 Third Avenue, New York, NY 10017, USA

Routledge is an imprint of the Taylor & Francis Group, an informa business

Publisher's Note
The publisher has gone to great lengths to ensure the quality of this reprint but points out that some imperfections in the original copies may be apparent.

Disclaimer
The publisher has made every effort to trace copyright holders and welcomes correspondence from those they have been unable to contact.

A Library of Congress record exists under LC control number: 99076157

ISBN 13: 978-1-138-70404-6 (hbk)
ISBN 13: 978-1-138-70402-2 (pbk)
ISBN 13: 978-1-315-20288-4 (ebk)

Contents

Tables

Foreword
by C. Dan Brand, Chair
Federal Laboratory Consortium

Government Laboratory Technology Transfer is technology transfer *evolved*, the vital area that federal laboratory research and development has become today—and its growing economic and entrepreneurial impact.

Acknowledging the important ongoing role of the federal laboratories in basic research, author Sally A. Rood goes much beyond the old stereotypes of lab activity to point out what the general public does not always realize: Businesses, large and small, apply that government research—often breakthrough information in a number of scientific areas—to create innovative emerging technologies, which, in turn, may develop into entire new industries.

With more than 20 years government service in specialized areas of technology transfer, including her longtime association and current work with the Federal Laboratory Consortium (FLC), Sally Rood employs her experience to chart the growth and dramatic changes in tech transfer in the last two decades and define how our system works—or does not work. Along the way, the traditional paradigms of lab R&D are expanded, especially in the last few years. And inspiring success stories demonstrate the results from the resources in our government laboratories—products developed from lab technologies that continue to improve our quality of life, as well as contribute to the U.S. economy and enterprise development, in unexpected ways and places.

As a veteran in government service and tech transfer myself, I'm pleased that *Government Laboratory Technology Transfer* has documented its history and development—we can see how far we've come in a short period of time. Federal workers, as well as those in industry and the university scientific communities, will welcome the close examination of the government system, with an element of surprise—at the number and kinds of technologies available in the labs and their expertise in specialized areas, for example—and an increasing interest in the potential for future collaborations.

I also endorse Dr. Rood's forecast for continued faster-paced procedures and the evident changes and challenges that the government is sure to encounter in the future in technology transfer, especially in those labs with fewer resources and personnel. In our field, we too are becoming part of an increasingly global economy, and that international angle will drive much of the activity in the technology transfer arena. Therefore, our laboratories must learn to be flexible in their processes, and be able to identify and respond to the critical needs of society—as well as to the needs of users in industry, universities and state and local governments.

Rood acknowledges the government's inconsistencies, however, in setting performance measures to determine the outcomes of technology transfer activities. That status, of course, is to change, as the Government Performance Results Act of 1993 requires. As the competition increases each year, we would do well to look to our entrepreneurial industrial partners as guides on tracking these important metrics and in other ways. With their proven success record and tactics, the private sector provides examples to follow—primarily serving our customers and increasing program outreach and technology marketing. In addition, as a representative of the FLC, I can see, as many labs and agencies have discovered, that our forum to provide strategies and opportunities will continue to enable expanded and increased technology transfer activities on a number of levels.

In chronicling the recent successes and ongoing, positive change in the labs and agencies, *Government Laboratory Technology Transfer* lays the groundwork for continued growth—and inspires us to meet the increasing needs of our growing national and international technology community. Therein lies the great challenge in technology transfer, as we go into the new century.

Preface and Acknowledgments

As is evident from the References section of this book, organized topically, there are many government reports and policy studies on technology policy and transfer, but few books available to increase our understanding of this timely subject. I anticipate this book will be useful to all levels of academic programs focused on technology partnerships and commercialization. It may also prove useful to government-oriented training programs. Furthermore, it may be helpful to potential users and collaborators in government technology programs such as company execs, university researchers, and state and local governments.

I am pleased that the Federal Laboratory Consortium (FLC) gave its blessing for this publication since I now work for the organization! More explicitly, the FLC Executive Committee, headed by FLC Chair C. Dan Brand, agreed that publication of this data is a positive step for technology transfer. But going back further in time, I would not have been able to do the research in the first place if it were not for Dr. Loren C. Schmid, Former Chair of the FLC. With concurrence of the (then) FLC Executive Committee a number of years ago, he gave me permission to pursue the research and shared helpful background materials.

I would also like to acknowledge Joe Allen and the late Lee Rivers, the current and former heads of the National Technology Transfer Center in Wheeling, West Virginia. Having spent a number of years working with the NTTC, my understanding of the laboratory technology transfer system in this country has been enhanced by them and my other colleagues there.

I am grateful to Professor John W. Dickey at Virginia Tech for recommending this work to be included in a published series on public affairs. The original work in this volume was performed as part of the requirement in earning a Ph.D. in Public Affairs and Administration through Virginia Tech's Center for Public Administration and Policy. In that program, known for its outstanding educators, Dr. Jim Wolf, Dr. John W. Dickey, Dr. Larkin S. Dudley, and Dr. Orion F. White all were inspirations—each in their own way. I feel fortunate to have had them as teachers. Dr. Alistair M. Brett and Dr. Richard L. Chapman also contributed greatly to the original doctoral dissertation.

Each of the laboratory scientists and company officials who spent numerous hours being interviewed is appreciated. It is evident that they are dedicated to their work because, in many cases, they graciously forwarded articles, packages of information, additional contacts and helpful documents. In some cases, they even forwarded product samples. They were also willing to respond to follow-up questions to verify the accuracy of information.

Finally, Susan W. Gates and Linda D. Voss provided editorial assistance which is immensely appreciated.

Abbreviations

AF	Air Force
AFB	Air Force Base
ARS	Agricultural Research Service
ATP	Advanced Technology Program
AUTM	Association of University Technology Managers
BNL	Brookhaven National Laboratory
©	copyright
CATI	Colorado Advanced Technology Institute
CPC	Competitiveness Policy Council
CRADAs	cooperative research and development agreements
CRS	Congressional Research Service
DOC	Department of Commerce
DOD	Department of Defense
DOE	Department of Energy
EDA	Economic Development Administration
FETC	Federal Energy Technology Center
FLC	Federal Laboratory Consortium
FOIA	Freedom of Information Act
FPL	Forest Products Laboratory
FTTA	Federal Technology Transfer Act
GAO	General Accounting Office
GATT	General Agreements on Tariffs and Trade
GOCO	government-owned contractor-operated (laboratory)
GOGO	government-owned government-operated (laboratory)
GPRA	Government Performance and Results Act
IRI	Industrial Research Institute
JILA	Joint Institute for Laboratory Astrophysics
LaRC	Langley Research Center
LBL	Lawrence Berkeley Laboratory
LLNL	Lawrence Livermore National Laboratory
MEP	Manufacturing Extension Partnership
MIT	Massachusetts Institute of Technology
MSFC	Marshall Space Flight Center

NAS	National Academy of Sciences
NASA	National Aeronautics and Space Administration
NCAUR	National Center for Agricultural Utilization Research
NIST	National Institute of Standards and Technology
NREL	National Renewable Energy Laboratory
NRL	Naval Research Laboratory
NSF	National Science Foundation
NTTC	National Technology Transfer Center
NWC	Naval Weapons Center
OTT	(DOD) Office of Technology Transition
ORNL	Oak Ridge National Laboratory
ORTAs	offices of research and technology applications
OSTP	Office of Science and Technology Policy
OTA	Office of Technology Assessment
PETC	Pittsburgh Energy Technology Center
PNNL	Pacific Northwest National Laboratory
PTO	Patent and Trademark Office
R	registered trademark
R&D	research and development
RD&E	research, development and engineering
RTTCs	regional technology transfer centers
SBIR	Small Business Innovation Research (Program)
SNL	Sandia National Laboratories
SRRC	Southern Regional Research Center
STC	Southern Technology Council
STTR	Small Business Technology Transfer (Program)
TM	trademark
TRP	Technology Reinvestment Project
TU	technology utilization
USDA	Department of Agriculture

1 Technology Transfer Introduction

This book examines the process and impact of government laboratory technology transfer. This introductory chapter provides background on government technology transfer, briefly highlighting recent trends in government technology programs. It then explains the core elements involved in the government technology transfer process in more detail. Next, it discusses the purpose of the book, which presents a qualitative comparative analysis of successful pre- and post-legislation technology transfer cases. Lastly, it outlines the research design for the study, including the philosophical basis, methodology, and potential obstacles.

Technology Transfer Legislation

Government laboratories as well as corporate and university laboratories perform federally-funded research and development (R&D). Because taxpayers support this research, the federal government has gone to great lengths to make the results of this R&D publicly available. In recent times, society has experienced rapid technological progress along with a rise in the availability of scientific and technical information. As a result, research discoveries have been applied to a wide variety of commercial products.

Much of the technology transfer-related legislation of the 1980s was enacted to provide industry and other users greater access to federally-funded R&D. The following is a listing of the major legislation in this area with a brief summary of the focus:

- Stevenson-Wydler Technology Innovation Act of 1980 – Required the creation of offices within many federal laboratories to facilitate access to the laboratories.
- Bayh-Dole University and Small Business Patent Procedures Act of 1980 – Waives ownership of federally-funded R&D to not-for-profits (universities, small businesses) performing the research.
- National Cooperative Research Act of 1984 – Stipulates that antitrust

criteria do not apply to consortia of companies (eg., SEMATECH) registered as such with the Justice Department.

- Federal Technology Transfer Act of 1986 – Gives federal and defense laboratory directors the right to enter into cooperative research and development agreements (CRADAs) with private parties.
- National Competitiveness Technology Transfer Act of 1989 – Gives contractor-operated national laboratories the right to enter into cooperative research.

In terms of the role of government laboratories, the key pieces of legislation occurred in 1986 and 1989 with passage of the Federal Technology Transfer Act and the National Competitiveness Technology Transfer Act.[1] These two laws dramatically changed the basic nature of both federal laboratories and national (contractor-operated) laboratories by encouraging cooperative research between the public and private sectors. Previous to the 1986 and 1989 acts, technology transfer from government laboratories was most often accomplished through informal collaboration and technical assistance or through the more traditional licensing fashion. It has been shown that technology transfer is most effectively accomplished, not by "passing off a baton" as in technology licensing, but by joint cooperative technology development between the laboratory providing the technology and the technology user or commercializer.

Brief Background

In the past two decades, three interrelated trends or series of events have profoundly affected federal science and technology, ultimately impacting government technology transfer. The first is the end of the Cold War, resulting in defense conversion and the movement toward dual use technologies. The second is globalization and the rise of the international "economic wars," resulting in a wider understanding of how technology can contribute to improving the economy. The third is attention to the federal budget deficit at a time of increased emphasis on high-technology industrial sectors by government policies and programs, including an emphasis on public-private partnering. This resulted in partisan "budget wars" over science and technology.

End of the Cold War Brought Defense Conversion, Dual Use

During the period from late 1980s to the early 1990s, the end of the Cold War unfolded. Defense conversion resulted not only in the closing of military bases and converting facilities from military to civilian purposes, but also in funding cutbacks for R&D, testing, and evaluation of defense systems by large defense contractors. As a result, the military-industrial complex is now turning towards the commercial marketplace and dual uses for its technologies,[2] and is teaming with many buyers and suppliers in contrast to having a single customer in the past (the Defense Department). This is contributing to a convergence of interests by a number of communities, including this military-industrial complex, academia and universities, and certain elements of the scientific community like the biomedical establishment which is ripe for commercialization.

Competitiveness Brought International Economic Wars[3]

In terms of their contribution to the economy, high-technology sectors tend to have certain advantages over more traditional ones.[4] First, high-technology companies tend to conduct more R&D, which is associated with greater innovation, resulting in more new products and processes for the marketplace. Also, high-technology companies tend to pay higher wages.[5] Therefore, high-technology industries and companies contribute more to the economy than traditional industries, and have actually become the engine driving economic growth.[6] This makes technology the basis for international competitiveness and for the rise in international economic "wars."[7] Early-on in the 1980s, the United States' economy was faced with Japanese and German competition, and the threat of the European single market. More recently, Pacific Rim nations have been rising to prominence in technological areas.

Consequently, government policies and programs have been putting increased emphasis on high-technology industries at the national level. And because high-technology firms tend to cluster regionally, this makes them amenable to state and local government programs and economic development.[8] In addition to direct subsidies to certain sectors, government incentives also include R&D tax credits, loan guarantees, and other financial measures.

On a related note, both industry and government are becoming more "global" due to communications technologies and international mobility.[9]

Companies are exporting more to overseas countries and/or partnering with foreign firms.[10] Government technology programs and policies are now interrelated with globally-oriented measures such as trade and investment practices, international patent regimes, international industry standards certification, and international environmental regulations and standards.[11] In addition, with the high cost of big science projects, nations are co-funding these projects with other nations where possible. Examples are the international space station and the major particle physics facility in Switzerland known as CERN.[12]

Budget Wars Caused Science/Technology Dichotomy

In order to understand the issues underlying the budget problems related to science and technology, it is important to understand the longer-term historical context and more details on the recent government programs. The build-up of the nation's R&D infrastructure was first proposed by Presidential Science Advisor Vannevar Bush in the mid-1940s.[13] The post World War II period became known for "big science" characterized by mega-projects like the space missions and many others. During this 50-year era, the science community was well-endowed.

The big science era came to an end in the late 1980s and early 1990s with the growing attention to the federal budget deficit and its effect on the economy in the long run. At this time, in terms of the budget, science began being treated differently from technology. Federal funding for fundamental science began to decrease,[14] particularly funds for large expensive facilities conducting basic research. For example, the superconducting supercollider in Texas was abandoned. At the same time that federal funding for basic science was decreasing, federal funding for technology programs increased. This includes (1) industry consortia, (2) technology development funds, and (3) technical assistance programs.

The technology funding trend began with the creation of R&D consortia allowed by the 1984 National Cooperative Research Act. In addition to anti-trust exemptions for joint R&D, several of these private-sector consortia were heavily funded by the U.S. Department of Defense, including SEMATECH and Microelectronics Computer Corporation (MCC), both located in Austin, Texas.

The government also has several new high-visibility government technology funding programs with cooperative research as their underlying theme. They are high-visibility because they involve large amounts (multimillions, and in some cases, billions of dollars) of government funding. For example, the 1991 American Technology Preeminence Act established the Advanced Technology Program (ATP) during the Bush Administration, which has since been expanded by President Clinton. ATP supports "pre-competitive" technologies, and has recently focused on specific technology areas. As another example, President Clinton's DOD Technology Reinvestment Project (TRP) was enacted by the 1992 Defense Conversion, Reinvestment and Transition Assistance Act. Although this program was subsequently killed by a new Republican Congress, it was replaced by a Dual Use Applications program. Another example is the Partnership for a New Generation Vehicle program for fostering new technologies in the automobile industry, a collaborative project between the big three auto makers. Both of the latter two programs have encouraged public-private teaming and interagency co-sponsorship.

Companies are also partnering with the federal government through cooperative R&D and other direct technology transfer activities with government laboratories. By entering into such agreements with laboratories, companies gain new technologies and intellectual property, thereby leveraging their resources to better compete in both domestic and global markets.

Another area of cooperation on technology programs has involved federal-state cooperation. Several decades ago, only a small percentage of the larger states had some form of technology development programs. Now all fifty states have not only technology funding programs, but also well-funded technology-based economic development and technical assistance programs that are often regionally-oriented within states.[15] These state programs are often partially or jointly funded by the federal government. For example, the Manufacturing Extension Partnership now has hundreds of sites across the country supported by a combination of federal, state and local funding, and corporate fees. Like the ATP program, this program was created during the Bush Administration (as the Manufacturing Technologies Centers program by 1989 legislation[16]) and vastly expanded by the first Clinton Administration.

In the 1990s, support for federal technology programs erupted into a highly-partisan political battle between the new Democratic Administration and new Republican Congress. Funding for basic research was pitted against funding for the industry R&D consortia, technology development programs, and technical assistance programs. This caused the programs in these areas to become controversial and contributed to a strong call for program evaluation at the same time that program evaluation was becoming more important government-wide.

So, the encouragement of government technology transfer is one policy area among a variety of interrelated policy and program areas. This will become more apparent in this book as the topics of funding and financial incentives, dual use, state economic development programs, and others are addressed in the context of technology transfer. The broad context for technology programs is discussed in more detail in Chapter 2.

Study Purpose

This study most directly relates to the effects of the federal technology transfer legislation. In enacting key legislation in 1986 and 1989, Congress presumed that federal technology transfer activities would benefit the economy in the long run by creating new spinoff companies, jobs, products, and other beneficial outcomes. Lawmakers also assumed the legislation would impact government laboratory activities, resulting in, for example, an increase in the number of public-private cooperative R&D agreements. It is important to measure these resulting outcomes in order to understand the impact of the legislation.

This study secondarily examines the process of technology transfer and how it is implemented by the laboratory personnel. For example, in earlier decades, government-funded science and technology were viewed as "public goods" not to be promoted for private gain. The problem is that the private sector won't become involved in many forms of technology transfer unless there is some economic benefit to be gained via private rights to the technology. Thus, there is a "catch 22" inherent to public sector technology transfer, and government personnel have been struggling to adjust to the new paradigm for public action in this area. This is illustrated by the fact that agencies and laboratories at first responded weakly to implementing the technology transfer laws.[17]

One aspect of this study is to provide an understanding of technology transfer from the perspective of bench scientists and researchers in the government laboratories, as well as their industry counterparts. In this sense, the study enhances our level of understanding from the *actor's* frame of reference. This approach works to uncover powerful social forces operating within the laboratories (powerful in the sense of information sharing). This effect, in turn, works to create a dynamic for action.

The findings point out the policy implications in terms of the need for new or different lenses to evaluate government technology transfer. For example, what sort of time frame is required for technology transfer and commercialization? Particular emphasis is placed on policy recommendations or perspectives proffered by the scientists and industry partners directly involved with technology transfer, so that these recommendations can serve as a point of comparison—if distinct—from existing policies.

The study findings also point to factors that are relevant for agencies in evaluating government technology transfer. Some federal agencies have established data collection guidelines and evaluation methodologies, yet there are still many questions related to the line of study on evaluation.

The next section provides activity-level background for understanding this study. It presents the core elements of the government technology transfer process.

Government Technology Transfer Process – Core Elements

The following core elements of technology transfer will be useful for understanding the broad context presented in the next chapter, as well as the particulars of the study described in this book. Many of the key words and definitions noted here relate to interview topics described later in this chapter.

Government Laboratory

The phrase "government laboratory" refers to the spectrum of federal (not state) government laboratories—both civilian and defense—including DOE's national laboratories, NASA's field centers, and the military services' R&D laboratories. Most of these are government-owned and government-operated, with the exception of the national laboratories which are often contractor-operated. Some laboratories have been transferring technologies since the early 1900s, although not under the auspices of the 1980 and 1986 legislation discussed in this book. For example, the Department of Commerce's National Institute of Standards and Technology (NIST) has been working with industry since 1905. The Department of Agriculture's (USDA) centers have been disseminating technologies to outside users since 1914. Today, some 700 federal laboratories and research centers from a variety of departments and agencies across the country, with approximately 100,000 researchers, perform approximately $24 billion in R&D each year.[18]

Research and Development

The National Science Foundation uses the following definitions for basic and applied research and development (R&D)[19]: *Basic research* is for gaining an understanding of the subject under study. *Applied research* is aimed at gaining knowledge to determine the means by which a specific recognized need may be met. *Development* is the systematic use of the knowledge gained from research directed toward the production of useful materials, devices, systems, or methods, including the design and development of prototypes and processes.

Technology Transfer

The Federal Laboratory Consortium for Technology Transfer defines federal technology transfer as "the process by which existing knowledge, facilities or capabilities developed under federal R&D are utilized to fulfill public or private domestic needs."[20]

Roles of Laboratory Researchers and Other Personnel

Government scientists perform everything from basic and applied research to technology and prototype development, including experiments and test and evaluation work of all kinds. There are many types of prototypes with varying degrees of sophistication, ranging from laboratory prototypes and pre-production prototypes to commercially-oriented production prototypes developed by outside engineering groups. Laboratory scientists conduct experiments and demonstrations for a variety of purposes, such as to determine feasibility or proof-of-concept or for technology marketing.

The laboratory technology transfer staff serves as a point of contact for outside users, thereby acting as a buffer for the laboratory researchers if necessary. The laboratory technology transfer function is generically called an "office of research and technology applications" (ORTA) by the 1980 Stevenson-Wydler Technology Innovation Act which mandated creation of such an office or function.

Technology transfer is a responsibility of both laboratory technology transfer officers *and* researchers,[21] so both researchers and technology transfer personnel perform technology marketing. In the literature, this role of "technology champion" has proven to be an important success factor in technology transfer. Technology transfer is often described as being a "body contact" sport, and it is generally acknowledged that effective technology transfer requires good communication skills. Proactive marketing includes packaging the technology, and planning and implementing marketing strategies such as demonstrations or technology transfer conferences. Somewhat more passive approaches may include, for example, making technology-related databases available for access. Laboratory efforts to promote technologies to outside users are commonly referred to as "technology push." User-initiated interest in a technology is often referred to as "market pull."

Other than legislation and formal incentives, other factors influencing technology transfer success include laboratory-level aspects such as laboratory mission statements, management support, promotion criteria, informal culture, awards, and other forms of acknowledgment. These factors vary from laboratory to laboratory. Government laboratories run by maintenance and operation contractors have both for-profit and not-for-profit operators, so internal policies vary accordingly. Similarly, there is no overarching, federally-mandated conflict-of-interest policy that applies to all the laboratories across the board so those laboratory policies also vary. For example, at some laboratories, the researchers may leave their jobs to start new businesses without losing some of their government benefits. Pacific Northwest National Laboratory (PNNL) and Sandia National Laboratories are two laboratories with "entrepreneurial" leave-of-absence policies.[22]

Technologies and Applications

The cases cover a variety of technology areas because a wide range of highly-visible technologies have originated from government laboratories. Examples include: penicillin (from a Department of Agriculture center), global positioning and night vision (from Defense laboratories), and insulin and the hepatitis vaccine (from the National Institutes of Health). Products that are often mistakenly attributed to federal laboratories include Tang, Teflon and velcro. The technologies that went into these products actually emanated from university and corporate laboratories and from independent inventors.

In terms of the applications process, some technologies gradually become more sophisticated and eventually result in improved commercial products or processes; or, the technologies slowly branch out and spread from one application area to another. These are known as incremental improvements or applications. Other technologies develop within one discipline, and are then suddenly applied to an entirely different area. These are known as revolutionary applications.

University Involvement

Government laboratories partner with universities for technology transfer purposes, just as they do with companies. For example, universities can be CRADA partners, or they can team with laboratories under the Small Business Technology Transfer (STTR) program. While the 1980 Stevenson-Wydler Act made technology transfer a mission of the federal laboratories, the comparable legislation for universities is the 1980 Bayh-Dole University and Small Business Patent and Trademark Amendments Act. This law gives universities and small businesses title to inventions they produce using federal funds. So, 1980 was the first time that universities or federal laboratories could profit commercially by licensing their technologies.

It wasn't until 1986 (and passage of the Federal Technology Transfer Act) that GOGO laboratories could manage their inventions as universities do, or until 1989 for GOCO laboratories. Consequently, government laboratories are generally said to be about ten years behind universities in terms of related management and cultural changes.

Funding, Financing

Government technology transfer activities enabled by the legislation do not involve the exchange of government funds. In government laboratories, the technology transfer office is funded as a percentage of each laboratory's budget (per the Stevenson-Wydler Act). Each laboratory is allowed to keep its share of license revenues after sharing with the laboratory inventor(s).[23] For example, Lawrence Livermore National Laboratory, with an annual budget of about $1 billion, brought in about $1 million in revenues from licensing royalties in one recent year. It is generally up to each laboratory director as to how licensing income is used. Regarding CRADAs, companies often contribute funds and the laboratories do not. Many federal agencies *require* company funds in order to implement CRADAs. Most agencies support laboratory CRADA activities through existing programs.

Special Technology Transfer Funds   The only other source of funding for laboratory technology transfer activities has involved "special" agency funding. In 1991, Congress created two funds for DOE CRADAs called the "Technology Transfer Initiative"—one fund for the DOE defense program laboratories and the other for the DOE civilian laboratories. However, this special funding was drastically cut back when the new Republican Congress came in a couple of years later. For example, at the program's peak, the University of California system, which operates various DOE laboratories, signed 185 CRADAS worth $218 million in 1994. In 1995, only 22 new CRADAs were signed by the same university operating contractor. In 1997, only $59 million in special funding was available for DOE defense laboratory CRADAs, and there were no special funds for the DOE civilian laboratories. Certain DOE headquarters programs now provide some CRADA support, but these programs are very restrictive. For example, the civilian part of DOE funds only high-risk CRADAs at five specific laboratories in three technology areas: intelligent manufacturing, tailored materials, and sustainable environments.

Follow-on Funding   R&D and technology transfer comprise only a small part of the total cost associated with bringing a technology to market. Ramping up to full-scale production (including manufacturing prototype development, scale-up engineering, pilot plant implementation, and product testing and marketing) involves an investment of money, time, and personnel by the partnering company. A rule-of-thumb indicates commercialization costs five to ten times as much as the transferring laboratory's investment in the initial technology, or roughly $5 to $15 million per project. Therefore, follow-on funding or financing is usually needed by laboratory partners. Undercapitalization is a particular problem with small-firm partners.

Other than in-house corporate funds, additional sources of support include private venture and angel capital, government funding programs, and dedicated laboratory venture capital. Trends in the mid-1990s indicate the availability of more private venture capital funds and somewhat more early-stage seed capital. With venture financing, a company is generally expected to come up with a product within a couple of years. The Small Business Innovation Research (SBIR) program, implemented by eleven government agencies,[24] has been a competitively-bid source of R&D funding for small firms since the early 1980s.[25]   The Small Business

Technology Transfer (STTR) program was established ten years later, and is being implemented by five agencies. It is devoted to cooperative R&D projects between small firms and either universities, GOCO laboratories, or non-profit research institutions such as a Federally-Funded R&D Centers (FFRDCs). Several government laboratory contracting operators have initiated dedicated non-profit venture capital funds for their laboratories. Examples are ARCH Development Corporation established by the University of Chicago for Argonne National Laboratory, and Technology Ventures started by Lockheed Martin for Sandia National Laboratories.

Intellectual Property[26]

Intellectual property rights for technologies most commonly involve patents, but they also involve trade secrets, trademarks, and copyrights.[27]

Researchers publicly disclose their inventions by formally submitting invention disclosures to their laboratories, publishing articles, and similar means. Inventions are protected in the United States for a year after disclosure. That is, once an article is published, the researcher still has a year-long grace period when a patent can be filed with the U.S. Patent and Trademark Office (PTO) without losing the opportunity. Beating the deadline for filing a patent application is now a more exact timing issue with the advent of real-time publishing on the world wide web.

Theoretically, each year, more invention disclosures are filed than patent applications. Similarly, more patents are filed each year than licenses are signed. Technology licenses are often, but not always, based upon patents.[28] For example, in 1995, DOE's Sandia National Laboratories received 275 invention disclosures from its researchers, filed 100 patent applications, and issued 40 licenses.[29]

The patent application process involves patent searches of prior art and the preparation of application papers by attorneys which can be a complex and costly process. The cost of applying for a patent with PTO varies by the type of patent (utility, design, etc.) and form of organization (e.g., there is a "small entity" fee category). Once the patent is granted, there is a recording fee, issue fee, and recurring maintenance fees. The fees for a basic utility patent total from about $8,000 to as much as $40,000 for the term of the patent. In the biomedical field, one of the most complex in terms of intellectual property issues, the average total cost of a patent can be as high as $100,000 or more.

Until recently, patents provided seventeen years of protection from the date of issue. The 1985 General Agreement on Trade and Tariffs (GATT) and the 1986 North American Free Trade Agreement (NAFTA) changed the U.S. patent system. Patents now provide twenty years of protection from the date of filing. The GATT and NAFTA agreements also offered the new option of filing a "provisional" patent application, a low-cost ($150) informal way to establish a patent filing date for a year.

PTO recently simplified its patent filing procedures and shortened the patent-pending period.[30] This is in spite of the fact that patent applications have been increasing over time. Federal agencies are pursuing patents more aggressively, as exhibited by the expansion of patent and legal support personnel, for one thing.[31]

It is not unusual for non-agency personnel (eg., graduate researchers, post-doctoral fellows, contractors, or Intergovernmental Personnel Act assignments) to be involved in government laboratory research. In these situations, the university personnel are entitled to their portion of any intellectual property they help to create.[32]

Any company claiming rights in a product name may use the "TM" (trademark) designation with that name on the package to alert the public to the claim and exclude others from its use. It is not necessary for the name to be registered with PTO in order to use it; however, when the name is registered with PTO, the "R" (registered) designation is used.[33]

GOCO laboratories often have different intellectual property and technology transfer procedures than GOGO laboratories. For example, GOCO laboratories can copyright software and other written material because the copyright is in the name of the operating contractor. Federal employees cannot copyright their work. Similarly, with GOCOs, license agreements are signed with the contractor rather than with the government laboratory it is operating. Most of the DOE laboratories are GOCO laboratories, which explains why the 1989 law was passed (in order to apply provisions of the 1986 law to the DOE system).

Technology Transfer Mechanisms

The Interagency Technology Transfer Committee's Working Group on Measurement and Evaluation[34] identified the *major* "technology transfer mechanisms" as: collegial interchange, cooperative R&D, exchange programs, licensing, reimbursable work, technical assistance, and use of facilities. Other mechanisms for transferring technology include standards-setting, procurement contracts, and cooperative research agreements.

Various types of demonstrations and prototypes are sometimes viewed as technology transfer mechanisms when they serve a marketing purpose. Other mechanisms used less commonly include consulting to a laboratory, consulting by laboratory personnel, cost-shared contracts, educational grants and awards, and informal mentor-protege relationships. Some consider technology funding programs as technology transfer mechanisms (such as the SBIR program, STTR program, or Defense-funded dual use programs). Also, certain agencies use special mechanisms unique to them (e.g., the National Institutes of Health's Clinical Trial Agreements). The more common mechanisms are explained in more detail below.

Collegial Interchange   This involves free and informal exchange of information among colleagues or through publications.[35]   Journals and databases work to document and/or archive an R&D project's findings. In addition to publications, this category includes workshops and conferences, preparation and distribution of other written materials, and computerized databases and bulletin boards.

Cooperative R&D Agreements   These agreements are often referred to as "CRADAs;" they also include NASA's similar Space Act Agreements. CRADAs allow joint laboratory-industry research to be conducted at the laboratories while minimizing the possibility of conflicts of interest for the laboratory personnel.

Personnel Exchange Programs   Exchange programs involve the exchange of technical personnel or equipment between government laboratories and universities or industry laboratories.

Technology Licenses   Licenses are usually royalty-bearing. That is, when a government technology is licensed by a commercial firm, the company pays the government an up-front fee and running royalty based upon product sales. There is usually a minimum annual royalty stream requirement, as well as a requirement for the licensee to submit a development or marketing plan, milestone dates, and projected sales in order to ensure commercial success.

Laboratory inventors receive the first $2,000 and at least fifteen percent of royalties per year (with an annual cap of $150,000), according to the 1995 National Technology Transfer and Advancement Act. Since DOE is taking the position that the 1995 law doesn't apply to its contractor-operated laboratories, they have different minimums.[36]

Licenses are generally negotiated based upon patents, but they can also be based upon other legal mechanisms such as trade secrets[37] or can include software which may be patented or copyrighted.[38] It is not uncommon for a technology license to integrate multiple patents or patent applications. Technology licenses are often limited to particular fields of use or geographical areas. Furthermore, they can be exclusive,[39] partially exclusive, non-exclusive, or co-exclusive.

Licensing can be considered as a stand-alone mechanism or in conjunction with a CRADA. The 1995 National Technology Transfer and Advancement Act guarantees exclusive licenses to CRADA partners in applicable fields of use. Although the principles of any licensing provisions are pre-negotiated along with CRADA negotiation, once a CRADA-related patent is filed, a laboratory has a year to complete the licensing before the technology is available to other potential licensees. In other words, the act gives company partners rights to their own inventions resulting from a CRADA along with the option to exclusively license the laboratory's portion of rights. All of this is intended to expedite CRADA negotiations. Before now, a typical license took one year to make it through laboratory/industry negotiations, and CRADA approval sometimes took even longer.

Reimbursable Work   This type of technology transfer mechanism is assistance provided to non-federal partners and outside users performed under the auspices of a laboratory. An example is the U.S. Department of Energy's "Work for Others" program.

Technical Assistance   This type of assistance, such as via telephone calls or face-to-face visits, is more informal and short-term than reimbursable work. Isolated statistics show that the laboratories are doing more non-reimbursed technical assistance than in the past. For example, Sandia National Laboratories provided some 300 instances of technical assistance to small businesses in 1995, up from sixteen instances in 1991.

Laboratory User Facilities   Laboratory or scientific user facilities are those unique and often large-scale machines in laboratories available to industry for a fee. The availability of these facilities is touted as a technology transfer mechanism and marketed by laboratories housing these types of facilities (many DOE laboratories are in this category). With user facility agreements, the outside party comes to the laboratory.

User Groups

Technology users can be found in industry, academia, government, or even other countries. In other words, they are generally outside the developing laboratory. Users may represent large businesses, small[40] and minority-owned firms, universities, state and local agencies, other federal laboratories and programs, or foreign-owned businesses.

Technology users become laboratory partners once a technology transfer agreement is signed. Agreements can be implemented through a variety of configurations: individual users and companies, consortia of companies performing collaborative R&D,[41] joint ventures, universities, research foundations, private technology management groups, trade associations reflecting industrial sectors, or other configurations. In the case of other government users, agreements can be arranged on an inter-agency or multi-agency basis.

Barriers to Commercialization

In terms of the study described in this book, a description of "commercialization barriers" does not lend itself to up-front definitions, as it was desirable to have actual barriers evolve from the interview discussions. However, as an example, it can be stated that an obstructive management culture is often cited in the literature as a major obstacle to transferring technologies. For instance, in the military laboratories, officers sometimes inaccurately associate technology transfer with the security breach of exporting technologies to overseas enemies. Also, the DOD and DOE weapons laboratories have been described as having a structured, seniority-oriented, "follow the rules" mentality, which stifles innovation and is a product of their weapons era culture.

Other Factors Involved in Technology Transfer

This topic serves as a catch-all category. Examples can be cited since there are some miscellaneous unresolved issues related to government technology transfer. For one thing, government laboratories are viewed by some as competing with the private sector, rather than helping it.[42]   Another unresolved technology transfer issue is whether to require a preference for partners that are U.S.-owned or -based.[43]

User Benefits/ Economic Impact/ Outcomes

User benefits and outcomes are considered to be broader than economic impact measures. Outcomes are "the longer-term results to which a program contributes, like military effectiveness or environmental improvement."[44] For example, in the case of a Defense-oriented technology, "outcomes" could include: breadth and criticality of military applications, cost savings achieved through dual-use approaches, increases in DOD use of commercial technologies, impact of spinoffs on defense costs, or preservation of defense industrial base, skills, processes and facilities. Impacts are "the total consequences of [a] program, including both intended benefits and unintended positive and negative results"[45] (see "Government Gains," below).

Traditionally, government laboratories have measured the success of their R&D projects in terms of technical performance, schedule and cost, while aiming for "better, faster and cheaper" results. With technology transfer projects, laboratories measure themselves in terms of the above criteria, plus their contribution to the economy and return on the taxpayers' investment. For example, DOE and its laboratories generally evaluate technology transfer performance according to four measurement categories: (1) overall performance measures – involves counting the number of: CRADAs, patent licenses, royalties, new products (and processes), start-up businesses, and workers hired or trained; (2) value to the government – involves measuring the amount of economic stimulation, U.S. competitiveness, jobs and business, sales and exports, and leveraged research; (3) customer satisfaction – ascertains a partner's level of satisfaction with the technology program or technical assistance, measured through repeat business, referral partnerships, favorable feedback, and support for technology transfer; and (4) medical and environmental benefits. Other agencies may measure additional indicators such as the amount of private capital leveraged or other outside investments and grants made in a transferred technology.

Measuring industry impact was more difficult for agencies before 1989 because partnering companies were leery of providing company sales and cost data due to the Freedom of Information Act (FOIA). However, the 1989 National Competitiveness Technology Transfer Act requires that CRADA-related information (such as partnering company names, and confidential and proprietary company information) must be protected from disclosure under FOIA stipulations for up to five years.

The term "metrics" is often informally used to refer to general measurement and evaluation. Formally, however, it refers to "standards of measurement that rely on counts of discrete entities to infer levels of accomplishment; e.g., improved health status, increased production, bibliometrics (publications and references), or degrees awarded."[46]

International Activity

When an agency chooses not to file for a patent in another country, the option is open to the laboratory inventor to do so on his or her own. The 1986 Federal Technology Transfer Act automatically grants inventors foreign rights if their agency didn't file within six months after the patent application. However, in 1996, the U.S. Department of Commerce issued an interim rule so that inventors must now wait eight months rather than six, and can only foreign file under certain conditions. The proposed provisions are intended to bring foreign rights into closer conformity with the domestic rights, but the final rule is still being worked out.[47]

As noted earlier, technology licenses can be limited according to geographical areas covered (eg., United States, another country, worldwide).

Government Gains

Government gains are anticipated or unanticipated. Unanticipated government gains are also referred to as "spinbacks." Dual use technology, an anticipated government gain, is intentionally developed for military or federal use but also planned to have civilian uses (or the reverse, where defense is the secondary use). Defense conversion, often confused with dual use, involves reductions in resources committed to national security and redirecting those resources to other purposes.

Economic Development, Technical Assistance

Business assistance services are provided by state or local economic development agencies, university-based organizations, and even government laboratory-based organizations. The Commerce Clearinghouse on State and Local Productivity, Technology and Innovation[48] categorized business assistance services as: feasibility studies, market development, finance, location, business planning, patent searches, database searching, training, technical/engineering consulting and testing, legal services, and accounting.

Elapsed Time

In the commercial world, who develops a technology first is not as important as who makes it to the market first. An early window of market penetration can mean the gain or loss of millions of dollars to a company because when competing products hit the market, the first-to-market firm does not necessarily maintain its market share. The issue of timing becomes even more crucial when a new product is based upon technology transferred from a government laboratory, which may involve taking the time to get a license negotiated or CRADA approved.

Research Design Involved Qualitative Survey

Methodologically, this study involved a survey research approach to perform a qualitative comparative analysis examining government technology transfer from two different time periods. Interviews were conducted with award-winning laboratory scientists and industry partners who successfully transferred technology in the mid-1980s (pre-enactment). The responses were compared to those involving similarly successful transfers in the early 1990s (post-enactment).

Philosophical Basis – Fourth Generation Evaluation

The epistemological framework for this study was based upon Guba and Lincoln's concept of "fourth generation evaluation."[49] Fourth generation-type evaluations are very different from traditional evaluations based upon scientific methods which employ random samples, objective survey instruments and controlled settings, and cause-effect relationships. The problem, according to Guba and Lincoln, is that the "value-free" approach associated with scientific methodologies is logically inconsistent with evaluation's goal of making value judgments. The following are some key features of a fourth generation evaluation:

- Fourth generation evaluations involve a more subjectively-based view in that the stakeholders' views are the basis for exploring needed information, as opposed to traditional before-the-fact postulates.

- With fourth generation evaluation, qualitative methods such as case studies are preferred over quantitative methods because they are more adaptable in dealing with complex multiple realities (versus generalizations) and with exploratory discovery.
- The sampling procedures used in fourth generation evaluations differ from conventional research samples which provide statistical representativeness and randomness. Qualitative sampling is "purposive" because it expands information such as by sampling extreme cases, deviant cases, typical cases, critical cases, politically-important cases, or sensitive cases. In this study, the sample of cases were all successful award-winning cases.
- With qualitative research, the criterion used to determine when to *stop* sampling is informational redundancy, not a pre-determined statistical significance or confidence level. The research design "unfolds" as data is collected and as the analysis proceeds, and there is continuous interplay between data collection and analysis. Therefore, the data collection instrument must be adaptable and flexible, as opposed to controlled, because information from early analyses is used to structure the data collection that follows. This "emergent design" process continues until a consensus appears.
- The data analysis techniques used are inductive rather than deductive. Deductive analysis involves reaching a conclusion through reasoning, and shows whether cause-effect statements were true or whether initial conditions were met. Inductive analysis involves deriving general principles from actual facts or cases and, in this study, the cases were assembled without a preconceived notion of how the results would turn out. The findings evolved as the research progressed.

Implementation Framework – Case Survey Method

While Guba and Lincoln's concept of fourth generation evaluation was used as the underlying *philosophical* framework for the methodology of this study, case-study expert Robert Yin's[50] approach to analyzing evidence was used as the underlying framework for implementing the methodology. This study used Yin's "qualitative case survey approach"[51] to examine government laboratory technology transfer. This approach involved analyzing a number of cases by applying the same survey instrument to each case. The answers are tallied and analyzed in the same way as those of a regular survey."[52] This study is an example of the desirable situation[53] in which multiple individual cases are developed as part of the same study, whereas most case survey research expands upon existing cases which represent the "literature" of numerous unrelated studies.

Level I Analysis – The Sample   The cases for this study were drawn from a listing of awards and nomination sheets on actual technology transfers compiled by the Federal Laboratory Consortium for Technology Transfer (FLC). The FLC, a Congressionally-created network of federal and national laboratory technology transfer officials, has held an annual awards program since 1984. As part of its overall awards program, the FLC Special Awards for Excellence in Technology Transfer recognize laboratory scientists and teams of researchers in government laboratories (as opposed to laboratory technology transfer officials) who did outstanding work in transferring technology. About twenty to thirty such "special awards" are presented each year, totaling 333 awards from 1984 until 1996.

The award criteria are three-fold: uncommon creativity and initiative must be demonstrated in the transfer of technology; the benefits to industry or state and local government must be significant; and, the achievements must be recent. The laboratories nominate individuals and teams of individuals for the FLC Special Awards. The nominations are judged by a panel of technology transfer experts, including representatives from industry, state and local government, academia, and the laboratories.

Level II Analysis – Time Frame Delineation   The FLC cases identified for this study spanned two different time frames. The two years chosen to represent the pre-legislation period were 1985 and 1986; the two years chosen to represent the post-legislation period were 1992 and 1993. A total of fifty FLC Awards of Excellence were made in the 1985/1986 period (21 in 1985, and 29 in 1986). A total of 57 awards were made by the FLC in the 1992/1993 period (29 in 1992, and 28 in 1993).

Of those 107 awards from pre-legislation and post-legislation times, a subset of awards was chosen for further information gathering through in-depth telephone interviews. The subset was derived from a series of cases updated by the FLC in 1993. Based upon updated information from the laboratories, the FLC created one-page summaries for its 1994 publication, *Winners in Technology Transfer: Success Stories from the Federal Laboratory Consortium.*[54]   For the identified pre-legislation and post-legislation periods, there were a total of 25 cases in the "Winners" document; eight of those were from 1985-86, and 17 were from 1992-93.

The information contained in the 25 one-page cases in the Winners document were analyzed from the perspective of the interview questions (see Appendix A). Briefly, the topics of the questions were:

- Role of the Laboratory Researchers and Other Personnel
- Technology and Applications
- University Involvement
- Funding, Financing
- Intellectual Property
- Technology Transfer Mechanisms
- User Groups
- Barriers to Commercialization
- Other Factors
- User Benefits/Economic Impact/Outcomes
- International Activity
- Government Gains
- Economic Development, Technical Assistance
- Elapsed Time

Background on the laboratories was also included in the cases, as compiled from available materials on the laboratories.[55]   The one-page cases in the Winners document did not go into a level of detail that provided answers to all of the topics; regardless, the information was only current up to 1993.

Level III Analysis – Case Development   In preparing for the interviews, since the FLC award nomination forms did not necessarily contain phone numbers, a first step was to locate the laboratory contacts. A variety of sources were used to determine the telephone numbers, addresses, and other contact information.[56]   With some cases, it was necessary to go through the laboratory technology transfer office for which a phone number was more readily available through the FLC. Other sources of information for locating the interviewees included:

- National Technical Information Service's "Directory of Federal Laboratory and Technology Resources: A Guide to Services, Facilities, and Expertise" (5th Edition);
- Individual agency telephone books and technology transfer office listings, such as "Technology Transfer 1994: U.S. Department of Energy"; and,
- Six FLC-sponsored regional directories of laboratories, as follows: "Federal Laboratory Consortium for Technology Transfer Far West

Region Directory," "Federal Laboratory Consortium Mid-Continent Region Directory: A Guide to Technology Transfer Resources," "Federal Laboratory Consortium for Technology Transfer Northeast Region Directory," "Industry Guide to Federal Laboratories in the Mid-Atlantic Region," "Technology Resources: Federal Laboratory Consortium Midwest Region."

Once the laboratory scientist was contacted, that individual provided leads to the outside sources in companies or universities. As noted, information about each laboratory was also gathered in conjunction with this step.

Telephone interviews with the laboratory awardees and industry personnel updated and expanded upon the data in the original FLC nomination forms, and filled in missing information. The interviews covered a variety of dimensions and factors related to technology transfer, and the interviewees were asked a number of open-ended questions about their work related to receiving the FLC award. The complete list of interview questions is shown in Appendix A.

At this point in the research, ethnographic interviewing techniques outlined by James Spradley[57] were employed. Ethnographic interviewing differs from other types of interviewing or from ordinary conversations in that the researcher/interviewer slowly introduces new elements to assist the interviewee. Spradley says that ethnographic interviewing is one strategy for getting people to talk about what they know. The point is to avoid having the interview seem like a formal interrogation so that rapport is not achieved. For example, the interviewees were asked whether they were aware of any benefits to the users resulting from the technology being developed and transferred. If there was a pause in response to this question, some options were suggested like, ". . .product sales, or jobs created or saved?"

A list of the individuals interviewed is contained in Appendix C. Some forty people, or an average of two to three people per case, were interviewed over five months. This necessitated over 85 calls out (to locate sources and interview them) and countless returned calls, averaging 30 to 45 minutes per interview, and in some cases an hour or more. Callbacks tended to be longer because the calls were at their convenience. A ledger of names, telephone numbers, fax numbers, and e-mail addresses was used to track calls out, dates, availability, and other data.

Due to the inherent nature of ethnographic interviewing, the analytical technique for this research involved content analysis of the interview notes organized by interview questions. Summary notes were constructed after each interview so that the data and emerging themes could be analyzed throughout the research process.[58] This way, themes emerged cumulatively as the interviews unfolded. Glaser and Strauss[59] call this a "constant comparison technique," where the researcher identifies differences between the groups of cases, and trends within the groups, as the cases proceed. This way, each interview could be re-tailored appropriately as the trends emerged. "Saturation" occurred after six to eight interviews for each of the two-year time periods. Saturation refers to that point in data collection where "nothing new" emerges in terms of data, patterns, and themes.[60]

Once the private partners were located, as it turns out, their information did not always correlate with the basic information provided by the federal laboratories up front (nomination forms, etc). In some of the cases, the information conflicted simply because the information gained directly from the partners was more up to date. Similarly, information contained in government databases was incomplete or not up to date. In some of the cases, the private partner had a different perspective from what was included on the laboratory's nomination form. This made it apparent, for example, that in conducting a technology transfer impact study, one must go directly to the outside user as the source for information.

Addressing, Overcoming Research Obstacles

As with any such study, obstacles cropped up while the research was being conducted, but they were addressed and overcome.

Implementing Fourth Generation Evaluations Using Software

The first challenge with this study was the sheer difficulty in implementing a fourth generation-type evaluation. Guba and Lincoln's comments on conducting this type of research are as follows:

> "It may be a difficult and cumbersome process, and may take much longer than we are accustomed to thinking evaluations should take. More resources may be required. But in the end, it works . . ."[61]

Keeping track of the interview notes and accumulated documents for the study was a major undertaking.[62] Organizing the data traditionally involved manual cutting and pasting which is time-consuming and cumbersome. Therefore, a computerized software package[63] was used so more time and attention could be devoted to the interpretive aspects of the analysis. The software allowed creation of a document system whereby text was stored, and words and phrases were automatically indexed. Information about the cases, as well as emerging thoughts (in the form of "memos"), were organized according to a "tree-like" structure via the software.

Making Contact with Laboratory Researchers, Companies

A second obstacle was that the interview process hinged largely on the availability of the original laboratory researchers and industry partners. Some of these people moved on to new jobs and had to be tracked down. Fortunately, many of the cases involved teams of researchers, making it possible to locate at least one person involved in each laboratory transfer.

Most everybody contacted was helpful and willing to cooperate. There were no refusals to cooperate. However, a certain level of trust and respect had to be built with the interviewees, so they would be willing to open up and share their information. Many were eager to talk about their technology before too many questions were asked, and in many cases, it turned out that most of the questions were answered before being asked.

At least two laboratory attitudes toward their company partners surfaced. One was protective, and the other encouraged outside interaction. In certain cases, the laboratory response depended upon the type of company involved. For example, one case involved a workers' cooperative building customized machines with a reputation for ultra-high quality and detail. Until a level of trust was developed through several interviews, the laboratory wanted to serve as the intermediary in providing information from the company. In another case, on the other hand, the laboratory researchers recommended that a new company executive be interviewed so he could have his chance at doing some marketing.

Contacts with the private sector were more difficult to make than those with the government personnel. Sometimes, smaller companies had been bought out by larger companies. Where larger companies had changed hands, it was particularly difficult to track down company personnel. Not everybody contacted was still associated with the technology being tracked.

Addressing Questions of Validity, Reliability with Consistency, Redundancy

As with any research methodology, the chosen method had advantages as well as limitations with regard to validity and reliability.[64] For example, bias is inherent to any study that uses inductive analytical techniques, and uses interviews as the primary data source. Since bias was purposefully built into the methodology, there was no attempt to eliminate it, but rather to keep it consistent. To address this, the interviews were largely transcribed verbatim.

In addition, a process called triangulation was used. Triangulation involves multiple data collection methods to study the same thing. The multiple methods included: (1) a review of the legislation, Congressional testimony, regulations, agency documents, policy reports, background studies, and scholarly papers, and (2) compilation and analysis of the case survey interviews. Step one provided an in-depth understanding of the underlying philosophies and assumptions on a government-wide and agency-by-agency basis. The Literature Review chapter brings together the policy studies and other materials, and subsequent chapters present the cases. The cases are factual descriptions, but they show the complexity of the relationships among the actors involved in technology transfer activities. They also capture the "contexts, emotions, and webs of relationships" described by Denzin[65] and show how technology transfer is accomplished in a real-life context.[66]

Maintaining Survey Instrument Confidence, Adaptability

An early issue related to ensuring a reasonable amount of confidence in the survey instrument. It was impossible to predict what would come from the one-on-one interviews, so the use of a questionnaire was intended to provide structure to guide them (see Appendix A). It was difficult to develop an interview questionnaire that was comprehensive, yet also adjustable enough to fit the adaptable research format. Therefore, the questionnaire was tested and refined throughout the study. Questions were tailored for each interview, depending upon what other information had already been gathered. Also, the use of "probe" questions promoted flexibility in data collection. Even more probe-type questions became apparent as the interviews proceeded. When it appeared that a case could be analyzed in terms of additional dimensions, then those factors were integrated.

Generalizing to "Award-Winning" Situations

Finally, the generalizability of research findings is an issue regularly debated by scientists, and the issue can become complicated because there are different types and levels of generalizations. For example, this study's findings only relate to the award-winning series of cases analyzed. It is not assumed that the findings are generalizable to *all* conceivable technology transfer situations because it is possible that a series of non-award-winning cases might generate entirely different findings.[67]

Table 1.1 summarizes the sources of each aspect of this study's research design: epistemological framework, methodology, data collection and analysis techniques, and software.

Table 1.1   Research Design Sources

	Primary Basis	**Secondary Basis**
Epistemological Framework	Guba and Lincoln (1989)	Miles and Huberman (1994)
Methodology	Yin (1994)	Denzin and Lincoln (1994)
Data Collection and Analysis*	Spradley (1979) Denzin (1996)	Glaser and Strauss (1967) Guba and Lincoln (1989)

*Software: QSR NUD-IST 3.0 (1996)

Summary and Book Organization

This chapter highlighted the technology transfer legislation, the core elements involved in technology transfer, and the purpose and methodology for this study.

The book is divided into six chapters. The next chapter, Chapter 2, examines the broader context for technology transfer, paralleling the historical backdrop and recent trends briefly highlighted in this introductory chapter: defense conversion, international competitiveness, and attention to the budget deficit. Chapter 2 also explains how this study relates to existing studies in this area. Chapters 3 and 4 discuss the pre-legislation (1980s) and post-legislation (1990s) cases, respectively. Chapter 5 presents the findings of the comparative analysis. Chapter 6 provides a summary and conclusions.

Notes

1   The National Technology Transfer and Advancement Act of 1995 is another piece of major technology transfer legislation. However, it was not passed before the events in the cases, so it is noted only where relevant. It gives CRADA partners the right to an exclusive license to any technologies developed. It also sets a minimum royalty return for government inventors.

2   U.S. Economic Development Administration, History of Defense Conversion, 1995.

3   Industrial policy is not mentioned here because, to an extent, "competitiveness" is the politically-correct way of referring to "industrial policy" which is not regarded highly by those in the corporate world and certain political parties.

4   Until the 1990s, technology was thought to cause unemployment as new technologies displace workers. Lately, this view has been replaced. See National Research Council et al. *Conflict and Cooperation in National Competition for High-Technology Industry,* Washington, DC: National Academy Press, 1996, pp. 33-35.

5   U.S. Department of Commerce, Economics and Statistics Administration.

6   One-third to 49 percent of economic growth and productivity is technology-based, according to the White House Office of Science and Technology Policy and the Commerce Department.

7   Economic Strategy Institute.

8   Annalee Saxenian, *Regional Advantage: Culture and Competition in Silicon Valley and Route 128,* Cambridge, Massachusetts: Harvard University Press, 1994. See also Michael Porter, *Competitive Advantage of Nations*, Cambridge: Harvard University Business Press, 1992.

9   Council of Academies of Engineering and Technological Sciences, *Globalization of Technology: International Perspectives*, Washington, DC: National Academy Press, 1988.

10 National Academy of Engineering, *National Interests in an Age of Global Technology: Prospering in a Global Economy*, Washington, DC: National Academy Press, 1991.

11 National Research Council et al, *Conflict and Cooperation in National Competition for High-Technology Industry,* Washington, DC: National Academy Press, 1996.

12 European Laboratory for Particle Physics.

13 Vannevar Bush, *Science: The Endless Frontier*, 1945.

14 R&D funding reached its lowest levels in FY1995, although funding for the National Science Foundation remained steady. As this study is being completed, Congress is considering increasing R&D budgets again.

15 See, for example, *State Funding for Cooperative Technology Programs*, State Science and Technology Institute, June 1996, Columbus, Ohio; Christopher Coburn and Dan Berglund, editors, *Partnerships: A Compendium of State and Federal Cooperative Technology Programs,* Columbus, Ohio: Battelle Memorial Institute, 1995; Paul B. Phelps and Paul R. Brockman, *Science and Technology Programs in the States*, Alexandria, Virginia: Advanced Development Distribution, Inc., February 1992; Marianne K. Clarke and Eric N. Dobson, *Increasing the Competitiveness of America's Manufacturers: A Review of Industrial Extension Programs*, Washington, DC: National Governors' Association, 1991; Charles Bartsch, *Enhancing Competitiveness: Selected State Technology Transfer Initiatives*, Washington, DC: Northeast-Midwest Institute,

January 1994.
16   The 1989 Authorization Act for the National Institute of Standards and Technology.
17   Only laboratories that had active Federal Laboratory Consortium representatives implemented the provisions of the act early-on; few agencies bothered issuing regulations for implementing that act. It was not until the 1986 amendment to the act that agencies started to pay attention to technology transfer, yet the regulations for the 1986 law were also slow in coming. This is documented in various reports described in Chapter II.
18   National Science Foundation (Intramural R&D plus Federally-Funded R&D Centers).
19   National Science Board, *Science and Engineering Indicators - 1993*, Washington, D.C.: U.S. Government Printing Office, NSB 93-1, 1993.
20   As noted, the Federal Laboratory Consortium is a Congressional-chartered organization of U.S. government laboratory technology transfer offices.
21   The 1986 Federal Technology Transfer Act makes this clear, and points out that it is consistent with laboratory missions.
22   Sandia's policy allows researchers to leave up to two years to form companies or help existing companies apply laboratory-developed technology. PNNL allows its staff to leave up to three years or work part-time and still receive medical, dental and other benefits; they can access PNNL or Battelle technologies for commercialization purposes, and enter into agreements to use laboratory equipment to further develop technologies. See Pacific Northwest National Laboratory, *Exploring Entrepreneurial Frontiers: The First Year*, 1997.
23   License revenues used to be distributed back into the U.S. Treasury.
24   With SBIR-funded projects, government patents can be transferred (or title waived), the same as the government does for university and small firms contractors under the Bayh-Dole Act.
25   By the 1982 Small Business Innovation Development Act.
26   For more details on technology transfer procedures and mechanisms, see *Mining the Nation's Brain Trust: How to Put Federally-Funded Research to Work For You*, by Fred E. Grissom, Jr. and Richard L. Chapman, Reading, Massachusetts: Addison-Wesley Publishing Company, Inc., 1992.
27   Patents are protected by federal laws; trade secrets are protected by state laws.
28   In 1995, PTO granted 1,867 patents to the U.S. government.
29   As a matter of interest, this can be compared to 129 invention disclosures, sixteen patent applications, and only two licenses just four years earlier at the same laboratory [presentation by Kevin W. Bieg, Sandia National Laboratories, Association of Federal Technology Transfer Executives Winter Conference, March 2, 1995].
30   Regardless of PTO's streamlining effort, there are still problems with the system in certain overloaded technical areas (like genetic patenting).
31   For example, the Agricultural Research Service system of more than a hundred laboratories and centers grew from a couple of patent advisors to seven patent advisors and five technology transfer coordinators marketing a portfolio of over 800 technology patents.
32   If the "outside" individual elects to exercise their rights rather than assigning them over to the government, the individual's organization (eg., the university) may agree to file the patent on the individual's behalf so that the technology is owned 50-50 with the

government. The individual does not even have to allow their organization or "agent" to receive a portion of their half of any royalty revenues.

33   Filing a trademark application with PTO costs $245.

34   This working group met for about three years in the late 1980s and early 1990s, and ultimately produced *Collective Reporting and Common Measures: Draft for Comment* in November 1994. The Interagency Committee and its working groups have been largely defunct during the Clinton Administration.

35   Working with government technologies, classified information causes special challenges with regard to technology transfer. As of two years ago, about 6,000 U.S. invention disclosures (some dating from the 1940s) remained under government secrecy orders. This prohibits their discussion or publication and prevents their conversion into patents, all of which discourages technology transfer. Any related papers are "sanitized" and therefore are of lower quality and usefulness, and the technologies discussed are often less advanced. See Dr. David C. Sayles, "Anticipated Barriers Confronting the Implementation of Technology Transfer," *Technology Transfer Partnerships Proceedings, 19th Annual Meeting - June 22-24, 1994, Huntsville, Alabama*, Kenneth E. Harwell, Kathy Wagner, Carl Ziemke, editors, Technology Transfer Society, 1994.

36   For example, through Battelle, Pacific Northwest National Laboratory's "Recognition and Reward" program provides: (1) a $300 cash award on the issuance of a U.S. patent, for software development, or for development of trademarked material, and a $500 cash award for receipt of a national award like an FLC award or "R&D 100" award; and (2) a cash award of $100 to $1,000 based upon the (up-front) license transaction fee, and ten percent of the gross royalty revenue stream up to $1 million over time. Also, Battelle can take equity ownership on behalf of PNNL (government laboratories can't do this). In these cases, transactions involves stock or partnership shares (2.5 percent, up to $1 million).

37   Unpatented "knowledge-based" or "proprietary" technologies are generally treated as trade secrets.

38   Federal agencies cannot copyright, but they can patent software. Software protection is not an issue at contractor-operated laboratories because the contractors can copyright or patent software.

39   Intent to grant an exclusive license must be published in the *Federal Register*.

40   The Small Business Administration defines a small firm as having 500 or fewer employees.

41   The 1984 National Cooperative Research Act allows firms to perform joint research without breaking anti-trust laws; later amendments allow firms to perform certain types of joint production, as well.

42   One of the reasons behind the creation of CRADAs was to avoid such conflicts of interest and competition with the private sector.

43   At present, each agency has its own definition/interpretation of a "U.S. corporation."

44   Ibid.

45   Susan E. Cozzens, *Evaluation of Fundamental Research Programs: A Review of the Issues*, A Report on Discussions in the Practitioners' Working Group on Research Evaluation, August 15, 1994.

46   Office of Science and Technology Policy, National Science and Technology Counicl, Committee on Fundamental Science, Subcommittee on Research, *Assessing*

Fundamental Science, Washington, D.C., July 1996.

47   Also, the GATT and NAFTA agreements produced changes to the U.S. patent system whereby U.S. inventors do not have certain advantages and filing privileges over foreign inventors that they previously had.

48   The Commerce Clearinghouse is now defunct and other organizations are fulfilling this role.

49   Egon G. Guba and Yvonna S. Lincoln, *Fourth Generation Evaluation*, Newbury Park, California: Sage Publications, Inc., 1989.

50   Robert K. Yin, *Case Study Research: Design and Methods*, 2nd edition, Newbury Park, California: Sage Publications, Inc., 1994.

51   Ibid.

52   Ibid, p. 124.

53   Ibid.

54   The "Winners" document contained a total of 65 cases (of the total awards) covering the years 1984 through 1993.

55   Several publications describing the laboratories are listed in the next subhead.

56   Contact information other than telephone numbers were obtained or verified because in a couple of cases, the sources wanted to review the pieces of information so they were sent copies.

57   James P. Spradley, *The Ethnographic Interview*, Ft. Worth, Texas: Harcourt Brace Jovanovich College Publishers, 1979. See also, Norman K. Denzin, *Interpretive Ethnography: Ethnographic Practices for the 21st Century*, Newbury Park, California: Sage Publications, Inc., 1996.

58   Miles and Huberman call this documentation "contact summary sheets" because they contain the rich narrative answers to each question posed to the "field contacts." Matthew B. Miles and A. Michael Huberman, *Qualitative Data Analysis: An Expanded Sourcebook*, 2nd edition, Newbury Park, California: Sage Publications, Inc., 1994.

59   Barney G. Glaser and Anselm L. Strauss, *The Discovery of Grounded Theory*, Chicago, Illinois: Aldine, 1967.

60   Glaser and Strauss, 1967; this is comparable to what Guba and Lincoln refer to as "consensus."

61   Guba and Lincoln, pp. 226, 227.

62   In addition to papers and descriptive brochures, the laboratory researchers and company partners even forwarded laboratory and product samples.

63   Qualitative Solutions and Research Pty, Ltd., *QSR NUD-IST: Non-numerical Unstructured Data Indexing Searching and Theorizing*, Version 3.0, Newbury Park, California: Scolari, Software Division of Sage Publications, Inc., 1996. Also useful in this effort was: Eben A. Weitzman and Matthew B. Miles, *Computer Programs for Qualitative Data Analysis: A Software Sourcebook*, Newbury Park, California: Sage Publications, Inc., 1995.

64   Inter-researcher reliability was not a problem, because there was only one researcher (see Yin, 1994).

65   Norman K. Denzin and Yvonna S. Lincoln, editors, *Handbook of Qualitative Research*, Thousand Oaks, California: Sage Publications, Inc., 1994.

66   Yin, 1994.

67   However, it is conceivable that another sample of *successful* cases could be generated

from the same FLC master database of over 300 cases identified since 1984, making the sample logically replicable (if one were using traditional scientific methods).

2   Policy and Evaluation Background

In order to understand this study of government laboratory technology transfer, it is necessary to have a broad background regarding science and technology policy. Therefore, this chapter is divided into two sections. The first section deals with the context for technology transfer activities and their evaluation, from technology policy to government laboratory policy. It covers how the field of government technology transfer has evolved from the trends of the 1980s and 1990s introduced in Chapter 1.

The second section of this chapter deals with the policy, practice, and evaluation of technology transfer. This work is not a program evaluation, it is an assessment of the effectiveness of the technology transfer legislation and policies to date. However, performing a policy assessment of this type draws upon techniques used in program evaluation. The section begins by discussing the evaluation of technology transfer as part of broader science and technology programs, such as efforts to evaluate the impacts of R&D. It proceeds to discussing the evaluation of technology transfer explicitly. Table 2.1 summarizes evaluation efforts by or for the government, proceeding from the legislative branch to the executive departments and independent agencies. It also covers inter-agency and multi-agency analyses, some conducted by outside evaluators. A section at the end of the chapter points out how this study fits into the existing evaluation efforts.

As the chapter progresses, it becomes apparent that most of our knowledge base and the literature in this area is in the form of reports, such as from policy organizations and committees. Relatively little of the literature is in the form of refereed journals or books. Related literature not noted in this chapter is listed in the References in the back of the book.

Section One – Science and Technology Policy Context

Three trends impacting science and technology are discussed in the following order: defense conversion, international competitiveness, and attention to the budget deficit.[1]

End of the Cold War Called for New Missions

The impact of defense conversion on the U.S. science and technology system is evident. The National Academy of Sciences, which provides science-related advice to Congress, produced a series of science and technology policy reports in the 1980s and 1990s including a landmark piece in 1993 which showed how, instead of being driven by the Cold War, science and technology now require new national objectives.[2] In addition to military security, these new goals include such as industrial performance, health care, and environmental protection.

The Carnegie Commission on Science, Technology, and Government is another group which examined technology policy from the angle of changing global relations. Twenty reports from the Carnegie Commission during the 1980s and early 1990s[3] culminated in a "concluding report" in 1993. It called for a "transformation" in the way science and technology policy making is organized in all branches of government in order to meet new challenges.[4] The "challenges of the human future" include encouraging long-term economic growth, sustaining the environment, and creating and maintaining peaceful relations among nations in the post-Cold War world.

Leading up to this, Congress had established funding through the Defense Department for consortia working in certain technology areas, such as SEMATECH, so that the government would not be dependent upon potentially unfriendly foreign sources for key technologies such as semiconductors and microchips.[5] In the early 1990s, the Office of Science and Technology Policy (OSTP)[6] and the Departments of Commerce[7] and Defense[8] all analyzed specific technology and industry sectors from a public policy perspective, producing summaries of "critical" or "emerging" technologies considered essential to the nation's well-being whether economic well-being or national security. The Defense Department in conjunction with the Commerce Department's Export Administration had been doing this for years; the new element was the economic aspect. The technology consortia, which were originally intended for a defense purpose, were urged to take on a commercial purpose.[9] Subsequently, several technology-related trade associations produced comparisons of the key technology lists and began producing "road maps" of industry and technology sectors.[10] Similarly, the private-sector Council on

Competitiveness specified critical technologies in 1991 and recommended ways to ensure U.S. competitiveness in those areas.[11] The Congressional Research Service followed suit with a 1993 summary on legislative and executive branch activities regarding critical technology policy-making.[12]

In terms of planning and implementing the defense conversion, the Congressional Office of Technology Assessment addressed strategies for converting the military-industrial complex to civilian uses.[13] The Economic Development Administration at the Commerce Department provided a historical context for defense conversion, outlining how the Cold War defense conversion differed from previous conversions of this century, particularly World War II, the Korean War, and the Viet Nam War.[14]

Some in the science and technology policy circles would say it is a good thing the Cold War conversion unfolded, because science and technology policy was "adrift" at that point, and defense conversion gave this area a new focus.[15] One result of all this was the Technology Reinvestment Project, touted at times as a defense conversion project and at other times as a dual use technology funding program before being cut from the budget.

The defense conversion of the 1980s occured at a time when there were concerns that technology had the capability to displace masses of workers, an issue addressed by several groups.[16] On this issue, in particular, defense conversion merged with the competitiveness trend, discussed next.

Competitiveness Spawned Interest in Technology Policy

During the 1980s, when it appeared that certain high-technology sectors of Japan's economy had overtaken the United States, a series of organizations sponsored studies comparing the United States, Japan, and other industrialized nations. A National Academy of Sciences project resulted in books of readings,[17] including a report on an economic study which found that the contribution of technological change was the most important source of growth to five industrialized nations.[18] Similarly, the American Enterprise Institute sponsored a multi-year project, "Competing in a Changing World Economy," and a resulting publication offered a five-nation comparison of high-technology policies.[19] Also, the National Science Foundation sponsored a number of studies on the high-tech sectors of individual nations, as well as global comparisons.[20] Even the General Accounting Office examined how the United States compared to Germany and Japan with regard to government policies and corporate activities as they impact competitiveness.[21]

It became evident that technology is a driver of economic growth and a major contributor to international competitiveness. This theme spanned several political administrations at the national level. Beyond the international comparisons, the early domestic literature included a 1987 report by the Conference Board and the National Governors Association on the role of science and technology in economic competitiveness.[22]

In the mid-1980s, the National Academy of Sciences initiated an ongoing dialogue between engineers and economists in a project focused on harnessing technology for economic growth.[23] This dialogue was continued in a 1991 report, *Technology and Economics*,[24] from the Academy's sister organization, the National Academy of Engineering. Since first being written in the 1980s, Congressional Research Service (CRS) reports on this topic have been periodically updated, as well.[25]

Towards the end of the Bush Administration in the late 1980s, the White House Office of Science and Technology Policy (OSTP) developed the short document, *U.S. Technology Policy*.[26] Looking back, it appears somewhat one-dimensional, yet it was the first official government technology policy.[27] President Bush also created an interagency Council on Competitiveness, another first, which was chaired by the vice president.[28]

As the relationship between technology and the economy became more universally-accepted, the subsequent presidential administration made this link more explicit. The Clinton Administration presented its technology policy in 1993[29] and, within the year, produced a brief status report on implementing it.[30] A Commerce Department report noted that the Clinton Administration marked the first time the various bureaus and agencies of that department had developed an integrated comprehensive strategy for technology policy.[31] Because technology programs have been a cornerstone of the Clinton Administration, the White House put into place a structure to produce a number of policy and position papers early in the Administration. President Clinton created the National Science and Technology Council (NSTC) within OSTP in 1993 which subsumed the activities of several previous groups within OSTP's purvue.[32] The NSTC is a virtual agency for coordinating R&D across agencies, chaired by the Under Secretary of Commerce for Technology. In 1994, the Clinton Administration released the first major statement on science policy since the Carter Administration which called for the government to move beyond its Cold War focus on military-driven research.[33] Perhaps more importantly, this was also the first time the government linked basic research with competitiveness. About the time the Commerce Department was criticized[34] for producing "public relations" reports for the Administration,[35] it also produced some analytical reports. For example, the Commerce Department's Economics and Statistics Administration (ESA) showed that the use of advanced technologies was associated with enhanced plant survival and faster employment growth.[36] More specifically, ESA found that companies investing in technologies paid 14.4 percent higher wages than other companies. The National Institute of Standards and Technology's (NIST) 1995 report to Congress, prepared by the NIST Senior Economist, provided economic analyses justifying government investments in technological infrastructure.[37]

During the years leading up to passage of the 1986 and 1989 technology transfer acts, Congress held a series of hearings on science and technology issues. A House Technology Policy Task Force produced the report, *Technology Policy and its Effect on the National Economy,* based upon testimony presented by expert witnesses.[38] This was during the time when the relationship between technology and competitiveness was just beginning to be understood. Eventually, the themes in the literature progressed to recommendations on *how* the relationship between technology and the economy should be influenced (such as through further defense conversion, workforce training and education, technology funding programs, technical assistance and manufacturing extension programs, industry consortia, or technology transfer).

The Competitiveness Policy Council (CPC) is an independent federal advisory committee created in the late 1980s,[39] encompassing a number of "subcouncils" including one for technology policy. The subcouncil on technology, headed by the former director of the National Science Foundation, produced a series of technology policy reports to accompany CPC's annual reports to the President and Congress. The most recent technology policy report,[40] accompanying CPC's 1994 report,[41] addressed a number of issue areas impacting competitiveness broadly including: financial markets and capital formation, trade policy, manufacturing issues, and transportation, telecommunications and other public infrastructure. The technology policy report specifically documented new concerns in the technology area, such as the need for a larger federal role in technology financing, and the need to broaden existing preferences for U.S. firms.[42]

The Center for Strategic and International Studies (CSIS) is an example of an independent public policy research institution focusing on science and technology policy, among other issue areas. In 1996, CSIS contributed a major report on international competitiveness which noted that future U.S. competitiveness hinged on the ability to integrate the elements of innovation into a national strategy, including: corporate competitiveness, a world-class science and technology system, a strong education system, and an investment-friendly environment.[43]

The private-sector Council on Competitiveness has summarized the technology policy themes every few years.[44] To do so, the Council generally surveys its hundred or more members who are CEOs, and compares U.S. data on investments, productivity, trade, savings and standard of living to data of other industrialized nations. The Council's 1993 policy assessment[45] examined eleven recommendations covering the following topics: federal R&D, coordination and cooperation, tax policy, and U.S. manufacturing.[46] It noted that a consensus on the issues had been evolving since the mid-1980s, and determined that, in areas such as refocusing the national laboratories from defense to industry needs, "moderate progress" was made.

Just as the defense conversion trend brought about the Technology Reinvestment Project, a result of the international competitiveness trend was increased funding for the Manufacturing Technology Centers program. This federal-state initiative was later re-named the Manufacturing Extension Partnership.

Big Science Era Ended, Attention Turned to the Budget Deficit

Vannevar Bush's landmark 1945 book, Science: The Endless Frontier, marked the beginning of the "big science" era.[47] After fifty years, several works marked the end of big science, such as the Council on Competitiveness 1996 report, Endless Frontier, Limited Resources: U.S. R&D Policy for Competitiveness.[48] Given that expectations were increasing and resources were decreasing, the report's message was that priorities and mechanisms must be established so that industry, academia and government can share.[49] It also called for an end to the politicized debate over the federal role in R&D, and provided guidelines for industry, government and academia to make the paradigm shift necessary in the post-Cold War era.

As budget issues came to the forefront in the 1990s, several reports were of significance. The National Academy of Sciences issued a 1995 report, known as the "Press Report" after its chair (former presidential science advisor) Frank Press, which recommended new criteria for judging and funding R&D programs.[50] Because the science community has traditionally been conservative in dealing with federal budget matters (probably due to its favorable treatment over the decades), this community did not greet this report favorably.[51]

A 1996 Congressional Research Service[52] report is another example of the attention among policy circles to the science and technology funding issue.[53] This report on budget constraints suggested creating a Department of Science and Technology to better coordinate federal R&D funding priorities, given funding reductions by the federal and state governments and industry.

As defense conversion, the international economic wars, and the budget debate all geared up, these three trends converged in the mid-1990s. Conservative Republicans accused the government of "picking winners" with its funding of technology programs.[54] This resulted in partisan wars over science and technology, and a new round of policy analyses in the mid-1990s focused on the *role of government* in science and technology. In response to the more conservative views of the new Congress, some of the reports offered alternative approaches to heavy government involvement such as a 1995 Congressional Research Service report which outlined potential indirect inducements to fostering commercial technology development in lieu of direct government spending.[55] Other reports were supportive of the Administration's more liberal spending approach, such as a 1995 Council of Economic Advisors report which advocated more R&D funding.[56] On this issue, the Competitiveness Policy Council managed to appear bipartisan with a white paper by the former Commerce Department Under Secretary for Technology which provided a historical view of the government's role in technology policy.[57]

Trends Affected Government Laboratory Policy

The trends outlined above caused particular attention to be focused on the laboratories producing and stockpiling nuclear weapons, the Department of Energy's (DOE) defense laboratories. The issue of the relevancy of these laboratories came to a head during highly-politicized discussions about downsizing the DOE bureaucracy or disbanding it altogether.

In 1992, an advisory board to the Secretary of Energy submitted a report which emphasized the changing environment and new national challenges, and presented "guiding principles" for defining each laboratory's role.[58] An advisory board to the next Energy Secretary submitted another review of the DOE laboratories and their missions in 1995.[59] This report, known as the "Galvin Report" because the advisory board was headed by former Motorola chair Robert Galvin, received a great deal of fanfare with its release.

However, the report's controversial recommendation to re-establish the laboratories as non-profit public corporations was not heeded by DOE top management. The Galvin Report even specified laboratory "metrics" for general laboratory work[60] and for technology transfer activities[61] stating, "The degree to which a laboratory engages in the process of renewal would be a significant measurement." The main theme was change in the laboratories, their missions, and their science.

Meanwhile, Congressional analysts also studied the DOE laboratories. The now-defunct[62] Congressional Office of Technology Assessment produced a duo of reports on defense conversion,[63] with the second report focusing on DOE's three major nuclear weapons laboratories and the issue of redirecting their R&D to civilian missions. Similarly, the Congressional Research Service produced a 1993 report with the same theme, and has revisited the topic since then.[64]

From 1994 to 1996, the General Accounting Office produced a series of reports on the DOE system, beginning with testimony about what a challenge it would be to convert the laboratory missions.[65] GAO developed a baseline inventory of human and capital resources within the national laboratories, and determined their activities were not related to commercial product development.[66] GAO called for clearer missions and better management[67] and urged a restructuring of the system, but did not make more specific recommendations because it assumed that Congress would come up with various approaches.[68] In 1996, GAO raised a ruckus with a report calling for DOE to recover from private sector partners its investment in technology development and commercialization.[69] DOE officials knew from their experience trying to come up with technology project metrics that the record-keeping requirements would be onerous.

Section Two – Evaluating Technology Transfer

Introduction – Brief History, Current Issues

In the 1970s, 1980s and 1990s, researchers produced countless reports and articles on all aspects of technology transfer, much of it supported by the National Science Foundation and NASA.[70] Early-on, the study of technology transfer was most closely related to the theory and process of technological innovation and diffusion. Everett Rogers, a professor of communications, is considered a classic theorist in this area since he wrote the seminal text, *Diffusion of Innovations*, first published in 1962.[71] The early network of innovation theorists included social scientists from a variety of academic disciplines focusing on how R&D produces new knowledge, resulting in new products, and how such technological innovations impact users and adopting organizations. This aspect of technological innovation overlaps the study of behavioral science and organizational behavior.[72] Modeling the innovation process is also related to another field, the management of R&D and product development cycles within organizations.[73] The early innovation theorists informally networked around a National Science Foundation (NSF) program on the innovation process. Several researchers in this network originated from the University of Michigan's Center for Research on the Utilization of Scientific Knowledge. This team, including Louis Tornatzky and J. D. Eveland, compiled a comprehensive literature review of the field of technological innovation.[74] Later, the team summarized more than a decade of research into an updated book, the lead authors being Tornatzky and Mitchell Fleischer.[75]

Early-on, the theory and practice of technology transfer related to the dissemination and communication of scientific and technical information.[76] This perspective viewed technology as a "baton" to be passed off. However, over time, research showed that technology transfer is more effectively accomplished in an integrative fashion involving joint development between a laboratory and outside user. Therefore, more recently, the practice of technology transfer relates to cooperative research, with commercialization generally being the intended outcome of such joint development.[77] Most recently, the practice of technology transfer also involves the complicated intellectual property and other legal aspects of partnerships.[78] A variety of how-to handbooks in this area are targeted to both industry and laboratory technology transfer officers.[79]

Viewing the process of technology transfer from another angle, it also relates to studies of economic development and regional clustering.[80] This focus often involves technologies transferred from universities and the related entrepreneurial development in technology corridors such as Silicon Valley near Stanford University and Massachusetts' Route 128 near the Massachusetts Institute of Technology.[81] Studies of entrepreneurial behavior, in turn, are related to small business development and the contributions of small firms to the economy.[82] Viewing technology transfer from this angle, government laboratories are relatively new actors in the process.

The entrance of government laboratories into a scenario first dominated (commercially) by universities has brought more interest in program evaluation. And this interest has shifted from performance monitoring and process evaluation to impact evaluation. In any case, Congress passed the Government Performance and Results Act (GPRA) in 1993 which called for general programmatic accountability on the part of federal agencies. It required agencies to set quantitative performance targets and to report annually on their progress. Nevertheless, it was really the partisan budget wars' focus on science and technology that caused both Congress and the Administration to become increasingly concerned with evaluating the effectiveness of federal investments in science and technology, including technology transfer. A number of reports by Congressional committees and Administration entities either called for such measurement activity or actually attempted to assess the outcomes of scientific research and technology.

At the same time that there has been this increased need to justify the direct investment of taxpayer dollars, technology transfer policy has been changing. Philosophically, technology transfer policy has progressed towards a balancing of public and private concerns in order to provide motivation to commercialize (ie., intellectual property rights for private-sector partners and royalties for government personnel). As noted, the early years of the big science era involved proactive dissemination of government science and technology to the public. Technology transfer provisions somewhat restrict broad dissemination, but the assumption is that American society and societies worldwide will ultimately benefit from resulting contributions to the economy.

Consequently, the studies analyzing government laboratory missions eventually focussed on the laboratories' role in technology transfer, and laboratory policy converged with technology transfer policy. While technology transfer policy has been in state of flux, many groups have offered their policy recommendations in this area. The Commerce Department's Technology Administration summarized technology policy recommendations made by forty industry associations and private organizations contained in nearly a hundred published reports.[83] In the area of technology transfer, the analysis found the two points having the greatest agreement related to: (1) orienting federal laboratories to industry needs, and (2) ensuring private sector input. Several examples are highlighted here.

An advisory committee[84] comprised of individuals from industry, academia, and federal laboratories oversaw a Council on Competitiveness study[85] which made general recommendations on laboratory funding, management, and industry initiatives. The study offered two recommendations regarding technology transfer: (1) authority to sign cooperative R&D agreements should rest with the laboratories, themselves, not their federal agencies, and (2) technology transfer does not require new funds, but a reprioritization of existing funds.

Another steering committee of representatives from industry, academia, national laboratories, and "key government observers" guided a study by CSIS.[86] This project initially focused on technology transfer from DOE's national laboratories. However, the committee found that this ignored the laboratories' multiprogrammatic character and capabilities extending beyond weapons research. It also assumed there were no more challenges for the laboratories, or that industry could address the challenges more effectively. The report outlined three requirements for enhancing benefits from the laboratories ("benefits that go well beyond technology transfer"): (1) identify national missions to meet the nation's grand challenges; (2) create strategic partnerships with industry and academia; and (3) ensure that the agencies of government coordinate an R&D investment strategy. Regarding technology transfer specifically, the report also contained proposals for streamlining the administrative process for cooperative R&D agreements.

Even another group, the Atlantic Council, urged stronger laboratory-industry relationships (through liaison programs, industry briefings, industry assistance in priority-setting, and advisory committees).[87]

It now appears confirmed that the remaining large issues in federal technology transfer policy center around the public-private interface. For example, how do CRADAs fit into the trend to eliminate "corporate welfare"?[88] What is the private sectors' responsibility to provide feedback on technology transfer results? As a Congressional Research Service report succinctly stated, "At issue is whether additional legislative initiatives are necessary to encourage increased technology transfer or if the responsibility now rests with the private sector to use the available resources."[89]

Furthermore, it should be clear by now that evaluating technology transfer differs from evaluating R&D in that there are identifiable activities labeled "technology transfer" beyond the specific research activities, themselves. Nevertheless, some approaches to evaluating technology transfer have grown out of approaches to evaluating R&D. The next section begins by describing R&D evaluation that may (or may not) include explicit technology transfer activities. Table 2.1 provides a summary sketch of evaluations described in the upcoming sections.

Evaluating Technology Transfer as Part of Broader R&D Programs

When government laboratory R&D produces economic impacts, this implies the technology is transferred to an outside user. Yet the transfer of the technology may be implicit to the R&D effort. In the mid-1980s, the Office of Technology Assessment investigated whether the benefits of investments in science programs could be predicted and measured.[90] In its 1986 background report, OTA concluded, "While there are some quantitative techniques that may be of use to Congress in evaluating specific areas of research, basic science is not amenable to the type of economic analysis that might be used for applied research or product development."[91] OTA also said, "even in the business community, decisions about research are much more the result of open communication followed by judgment than the result of quantification."[92]

In 1993, Congress appropriated money for the Office of Science and Technology Policy to explore performance measures for basic research. The project produced a set of nine principles for assessing fundamental science, such as the use of multiple sources of measures, and a mix of quantitative and qualitative indicators and narrative text.[93] It also touted three assessment methods: (1) qualitatively-based peer reviews – either retrospectively, prospectively, or in-process reviews; (2) customer satisfaction ratings – with users of the research products; and (3) quantitative metrics, where possible.

Evaluation by peer review is one area where the field of technology transfer has learned from evaluation of basic science. DOE laboratory researchers were among those pioneering the concept of evaluating technology transfer by peer review. This concept was discussed in a 1994 paper by a research team associated with DOE's High-Temperature Superconductivity program.[94] The procedures and evaluation criteria are those used in peer review for conventional R&D projects, but they were adapted to the characteristics of technology transfer projects.

Econometric Studies of R&D   In order to produce quantitative metrics, one way to analyze R&D outcomes (with technology transfer implicit within the process) involves using econometric techniques. Economists have been retrospectively estimating the impact of R&D on society since the 1950s.[95] The early econometric studies focused on R&D performed in corporate[96] and academic[97] settings, and by industry sectors, particularly agriculture.[98] Some of these studies explicitly examined technology transfer, but not government laboratory technology transfer.[99] As an analogy, examining corporate or academic impact is roughly comparable to estimating the economic impact of a laboratory (regardless of its technology transfer activities).[100]

Econometric studies fall into two categories: (1) studies of changes in national or regional productivity to measure the role of R&D and technological innovation; and (2) studies of social or private benefits, such as jobs created or consumer savings, based upon technological innovation. The first category involves statistical techniques that determine relationships between dependent variables (output, productivity, etc.) and independent variables (technology, patent data, skills, capital, labor, etc.).[101] The second category involves cost-benefit techniques that assess societal gains not attributable to profits in order to determine a rate of return. The costs and benefits are calculated from changes in costs, prices, and sales through certain steps.[102] The late Edwin Mansfield is generally credited with developing this approach,[103] and it is closely related to another technique for analyzing R&D outcomes called bibliometrics.[104] Bibliometric tools involve counting scientific publications, technology patents, and citations. They show that 5,000 new scientific papers are published and 2,000 patents are issued each day, indicators of both R&D success *and* technology transfer. More intricate methods analyze science papers citing papers, technology patents citing patents, or linkage patents citing papers.

A 1995 Council of Economic Advisors paper[105] and other papers[106] compared key studies by Mansfield and others. These reviews found that the mean private rate of return for a technological innovation is twenty to thirty percent, and the social rate of return is almost fifty percent.[107] Over a series of his own studies, Mansfield found a median rate of private return of 25 percent, and a median rate of social return of seventy percent.[108] Because econometric studies always produce these strong positive correlations between technology and the economy,[109] their use can be justified to retrospectively measure an R&D program or to develop related general policies. However, they produce estimates rather than precise impacts, so they are not necessarily appropriate for making R&D funding decisions.[110]

Benefit-cost econometric models have been used successfully by certain government laboratories to measure the impact of their R&D and technology transfer activities. The agencies and programs that have done the most work in terms of using econometric and benefit-cost analyses to evaluate

Table 2.1   Summary of R&D and Technology Transfer Evaluations

Program, Scope	Sponsor/Evaluator	Approach	Result
Basic research	OTA	Analysis	Inconclusive
Fundamental science	OSTP	Analysis	Guidelines
Superconductivity-related Tech Transfer projects	DOE In-house	Peer Review	
R&D by sectors, firms	NSF/ Mansfield & other economists	Cost-Benefit Cases, IRR	Positive
R&D	NSF/ CHI Research	Bibliometrics	Positive
NASA TU Program	NASA TU/ (1) Mathematica, (2) DRI	Cost-Benefit Cases	Positive
NASA Space Program, R&D	NASA/ (1) Chase Econometrics, (2) MRI	Cost-Benefit	Positive
NASA *Spinoff*	NASA TU/ Chapman	Survey	Positive
NASA Tech Transfer	NASA LaRC/ Bush	Benchmarking w/universities, Surveys, Three Cases	Mixed
NASA LaRC Spinbacks	NASA LaRC/ Chapman	Case Studies	Positive
NASA Southeast Alliance	NASA SE centers, RTTC	Surveys	Positive
NASA MSFC	NASA MSFC In-house	Surveys, Extrapolations	Positive
NIST Laboratories	NIST In-house (Tassey)	Cost-Benefit, IRR	Positive
NIST Laboratory Programs	NIST/ Link	Cost-Benefit	Positive
National Lab Tech Transfer	DOE Tech Partnerships Program	(1) Data analysis, (2) Cases	Positive
Tech Transfer, NREL	NREL/ Chapman, Moran	Survey, Extrapolations	Positive

Manufacturing Assistance, ORNL	ORNL/ CMT In-house	Survey, Extrapolations	Positive
Manufacturing extension	(1) GAO, (2) Census, (3) Individual centers	Surveys	(1) Negative (2) Mixed (3) Mostly positive
DOD Laboratory Tech Transfer	DOD/OTT In-house	Survey	Process measures
Air Force Tech Transfer	Air Force/ Battelle, ESI	Lit search, Survey, Workshops, Interviews	Best practices
Govt. Lab Tech Transfer, Multi-agency	House of Rep.– Small Business Committee	Survey	Negative
Govt. Lab Tech Transfer, Multi-agency	GAO	Survey, Input/ Output	Inconclusive
Govt. Lab Tech Transfer, Multi-agency	DOC (3 "Biennial" reports)	Surveys, Data analysis	Process measures
Govt. Lab Tech Transfer, Multi-agency	Interagency Committee	Establish Framework	Inconclusive
Govt. Lab Tech Transfer, Mid-Continent Region Labs	FLC/ Chapman	Survey	Positive
Govt. Lab Tech Transfer, Multi-agency	DOE, NSF, etc./ Bozeman et al	Surveys, Input-Output	Mixed
Govt. Lab Tech Transfer, Multi-agency	Papadakis		Mixed
Govt. Lab Tech Transfer, Multi-agency	NSF/ Geisler	Survey	Best practices
(continued)	(continued)	(continued)	(continued)

Govt. Lab, University Tech Transfer	NSF, DOE/ (IRI) Roessner, Bean	Surveys	Mostly positive
Govt. Lab, University, Intermediary Tech Transfer	DOC, CATI/ CITTI (Anderson)	Survey	Best Practices
University Tech Transfer	(1) AUTM, (2) Individual universities	Surveys, Public benefit Analysis	(1) Positive growth, (2) Positive
University Tech Transfer, Southeastern Region	STC	Benchmarking	Benchmark-ing data, Best practices
Govt., University Tech Transfer	Carr	Lit Review, Interviews	Best Practices
Laboratory Tech-based Economic Development	NASA MSFC, USASDC / University of Alabama	Survey	Positive
Laboratory, University Incubators	EDA/ NBIA, STC, ILGARD, UMich	Surveys, Benchmarking	Positive impacts, Best practices

laboratory impact include: the laboratories comprising the National Institute of Standards and Technology (NIST), the NASA headquarters technology transfer program, and certain NASA R&D programs such as the Space Program. As will be pointed out, some of this work has explicitly focused on technology transfer activities, and some of the evaluation work has implicitly included technology transfer. Therefore, the next section begins with NIST and NASA and proceeds to other agencies and approaches.

Before proceeding to the next section, it should be noted that the only multi-agency R&D funding program that involves explicit technology transfer from government laboratories (or other R&D institutions) is the Small Business Technology Transfer Program (STTR) which was first implemented in the mid-1990s. But it has been too soon to conduct serious STTR economic impact evaluations. The Defense Department's Technology Reinvestment Project (TRP) also funded some government-industry partnerships, thereby explicitly involving technology transfer. However, DOD has not evaluated the early TRP projects since this program was cut from the budget. It is also noted that benefit-cost approaches have been used to evaluate the R&D performed through technology funding programs at the state level. However, as with federally-funded R&D, state-funded R&D programs may or may not involve government laboratory technology transfer.[111] In any case, what state R&D programs *do* have in common with government laboratory technology transfer is the difficulty in measuring their ultimate economic impact.

Evaluating Government Technology Transfer

NIST Programs, Laboratories Focused on Impact Evaluation A 1994 report from the National Institute of Standards and Technology (NIST) discusses how each of NIST's four program areas set priorities, evaluate performance, and measure economic impact.[112] NIST's four major program areas are: (1) the Advanced Technology Program, (2) Manufacturing Extension Partnership, (3) the Malcolm Baldrige National Quality Award Program (not a technology transfer program), and (4) the various NIST laboratories that perform R&D and technology transfer. The report stated that metrics could not be reduced to simple formulas yielding unambiguous, quantitative answers because judgments are inherent to the process and must be guided by both qualitative and quantitative data. Consequently, a companion report presented case studies describing the industrial impacts of these four program areas.[113]

NIST Laboratory Research and Tassey Econometrics: Laboratory research is the NIST area with the most depth as far as impact studies. NIST's Economist, Gregory Tassey, produced a 1996 report which examined the impact of the NIST laboratories.[114] In this report, as well as a book by Tassey,[115] the laboratory infrastructure was labeled "infratechnology." The report stated that measuring the economic impact of laboratory research projects requires both quantitative (eg., sales, reduced time to market) and qualitative metrics. He identified NIST's qualitative metrics as: effects on standardization, R&D results, and improved collaboration with industry. NIST's most common methods for measuring economic impacts are: (1) benefit-cost ratios and (2) internal or social rates of return. The report noted that in comparing NIST with similar private-sector research and technology investments, the NIST research projects produced "estimated social rates of return above estimates for private-sector innovations."[116] A follow-up book by Tassey carried this theme further, covering infratechnology as well as standardization and other mechanisms, and provided R&D policy impact assessment methods.[117]

NIST Laboratory Research and Link Econometrics: Throughout the 1990s, another economist, Albert Link produced a series of economic impact studies for NIST in various technology and industrial sectors related to NIST laboratory research such as optical fibers, semiconductors, and electromagnetic interference.[118] A 1996 book by Link on evaluating public-sector R&D brings together seven NIST benefit-cost cases, although he emphasized several times that the purpose was not to compare across cases or projects because benefit-cost methodologies are not useful for that purpose.[119] The benefit-cost evaluations documented ratios ranging from 7-to-1 up to 1,041-to-1 for the laboratory research, as well as some preliminary findings for NIST's ATP program.

There are certain steps involved in calculating cost-benefit ratios. Computing the cost portion generally involves obtaining program budgets over a period of time, and many of the Link analyses were calculated or projected back to the early 1980s. Computing benefits involves questioning industry users according to a survey instrument of open-ended questions which may uncover both tangible and intangible responses. Where the benefits (eg., sales revenue or other economic gains) or savings are not clear-cut, economists empirically estimate the value of non-market-type goods and services[120] based upon survey responses. These are, in other words, social benefits not attributable to profits or other monetary amounts. Often this involves extrapolating into the future or the past to estimate non-market values. Sometimes it must involve measuring benefits indirectly based upon an attribute of the output (such as a citation or patent count). These subjective or speculative aspects of an economist's work are said to be determined through "informed opinion" or by developing a consensus among the respondents ("peer evaluation").[121]

The Link method incorporated technology transfer, in the sense that technology transfer activities were included under costs. Costs are categorized in various ways (e.g., push, industry pull). Where relevant, the technology transfer costs were included under the category of tangible "push" costs. However, not all of the evaluated NIST research projects involved explicit technology transfer. In most of the NIST laboratory cases, technology transfer was subsumed under R&D costs.

NASA Technology Utilization Assessment Began Early-on   The National Aeronautics and Space Administration (NASA) has a long history of program evaluation because the 1958 Space Act charged NASA with performing technology transfer. The NASA headquarters Technology

Utilization office sponsors publications and other technology transfer-related activities such as publications and intermediaries.[122] The NASA "field centers" (laboratories) each have technology transfer offices which implement "Space Act Agreements" (or CRADAs).

NASA TU Program Econometrics: In the late 1970s, Congress required a cost-benefit study of NASA technology transfer.[123] In order to assess the feasibility of conducting such a study and to compare alternative analysis methods, the NASA Technology Utilization (TU) office initiated several studies through the Denver Research Institute and Mathematica, Inc. Mathematica's Mathtech Division studied the value added by the technology transfer activities of the NASA TU office by quantifying the economic benefits of secondary applications of NASA-related R&D.[124] Mathematica selected four specific technology areas[125] to analyze because: (1) data were available, (2) NASA's role was widely acknowledged, and (3) the benefits were anticipated to be relatively large. The researchers said they were conservative in their calculations, yet they found the total benefits of the four cases were about $7 billion, more than twice NASA's annual budget at the time.

The Denver Research Institute (DRI) studied the cost-benefit of specific NASA activities funded by the TU office,[126] including: (1) the monthly trade journal *Tech Briefs*, (2) Industrial Applications Centers, (3) the Computer Software Management and Information Center (COSMIC), and (4) applications teams. DRI measured program costs and program benefits, and concluded that the benefit-cost ratio for the total TU program was at least 6-to-1, with the individual ratios for the four TU program elements ranging from 3-to-1 up to 26-to-1. In 1979, DRI incorporated into a summary report: the results from a 1976 study of *Tech Briefs*, a 1977 study of the other program elements, and the 1976 Mathematica study results.[127]

NASA Space Program Econometrics: In the 1980s, NASA sponsored studies of the value added to the economy by the agency's overall space program, but they did not explicitly address the technology transfer role. Two studies by the Midwest Research Institute[128] showed paybacks on the NASA R&D investment ranging from 6-to-1 and 9-to-1. A Chase Econometrics Associates study showed paybacks of over 14-to-1.[129] Besides not addressing technology transfer, these studies also did not take into account any intangible, non-quantifiable factors.

NASA TU Program Elements – Chapman Survey: In more recent years, NASA technology transfer evaluation has changed to survey approaches. In the late 1980s, Chapman Research Group explored successful technology applications highlighted in NASA's annual publication called *Spinoff*.[130] The purpose was to identify benefits resulting from the applications highlighted in the publication each year, and to quantify those benefits where possible. Richard Chapman and his team examined 259 technologies through 600 telephone interviews (based upon some 3,000 calls) with 400 companies over eight months. The resulting 1989 report documented over $21.6 billion in economic benefits and countless intangible benefits.

As a companion piece to the *Spinoff* survey, Chapman followed up with a "characterization study" describing the lessons learned from doing that study,[131] particularly the difficulties in retrospectively collecting data for technology transfer events that occurred ten or twelve years earlier. Also, they were faced with inadequate records of company names, persons involved, addresses, telephone numbers, and even the technologies. Since that study, Chapman's consistent message for any technology transfer program has been to integrate measurement early-on into program activities so that it becomes a regular and systematic aspect of the operation. Another Chapman paper reiterated the message not to over-rely on quantitative measures.[132] He concluded, " . . .it is important to continually include some evaluation of quality *and* to be cautious about over dependence upon indications just because they are easily quantified—ie., avoid the attitude that . . .if you can't count it, it doesn't count."[133] He said the "chemistry" of the collaboration, commitment of top management, closeness of "fit" to the laboratory's technology and mission are key elements that numbers and dollars do not reflect. Chapman reiterated these two themes (about systematizing measurement and not over-relying on quantitative measures) in later papers, as well.[134] A 1994 paper compared the lessons learned from doing program evaluation studies with NASA, the Department of Agriculture, and the DOE National Renewable Energy Laboratory.[135]

The University of Tennessee Space Institute "characterized" successful companies featured in NASA's *Spinoff* publication from 1984 to 1991.[136] By surveying 287 companies with a 29 percent response rate, Brett Pichon and Bobbie Woodard developed a profile of characteristics for a "successful technology transfer company" (the company is involved in manufacturing, has fewer than 150 employees, etc.). The Tennessee Valley Authority and NASA's Marshall Space Flight Center (in Alabama) supported this study.

Individual NASA Field Centers Initiated Projects   NASA Langley – Bush Study: A study of NASA technology transfer by Lance Bush started out by examining all the field centers, but zeroed in on the NASA Langley

Research Center in Virginia.[137] Bush first compared overall NASA license royalties with those of U.S. universities and determined that NASA would have ranked 67th among them. He also performed a statistical analysis to identify the strength of correlations between NASA input, intermediate, and outcome measures. The quantitative input measures he obtained by interviewing NASA Langley researchers on their awareness, attitudes and perceptions of technology transfer; the outcome measures were the quantitative royalty figures. The somewhat varying correlations were explained by three case studies that showed the details of technology transfer can't be explained by only a simple royalty measure.

NASA Langley – Spinbacks Study: Richard Chapman studied the concept of "spinbacks"[138] for the NASA Langley Research Center.[139] In order to test the feasibility of undertaking a comprehensive study of the value of spinback, he examined nine cases. The exploration demonstrated that the spinback phenomenon is real, can be documented, and deserves attention.

NASA Marshall Surveys: Several news articles referenced an inaccessible "Fall 1994 Marshall report" with numbers related to the Marshall Space Flight Center. *Technology Transfer Business* and *FLC Newslink* both stated that a Marshall survey of 809 companies covering eighteen months generated 283 responses from firms receiving technical assistance and sixteen responses from firms with formal industry agreements with the Marshall center; surveys were mailed to 18 of 56 partners.[140] This survey indicated 665 jobs created or saved, 69 new products, $47.2 million increased sales, $10.2 million increased investment, and $11.5 million cost savings. The Marshall analysts extrapolated to the sample of 809 by applying Standard Industrial Classification (SIC) codes to numeric multipliers used by the Bureau of Economic Analysis in the census. This produced the following figures: 5,344 jobs created or saved, 182 new products, and $358.4 million in increased investment, an economic impact of 60-to-1. The *Technology Transfer Business* article noted that Marshall's extrapolations "raised eyebrows," and another news article highlighted the same metrics without mentioning the extrapolated numbers.[141] The Marshall center promulgated even other metrics on a state-by-state basis.[142] One source stated, "A nonresponse rate of 61% combined with a non-random sample brings serious concern to the extrapolation of the results to the sample population . . .In addition to the sampling issues of this study, there is some question as to the validity of the multiplying factors."[143]

NASA Southeast Alliance Surveys: Besides the Marshall Space Flight

Center in Alabama, the other Southeast region field centers are the Stennis Space Center in Mississippi and the Kennedy Space Center in Florida. Together, and with assistance from the regional technology transfer center in Florida, they are known as the Southeast Alliance for Technology Transfer. In the mid-1990s, NASA's southeastern region promulgated a variety of metrics related to the field center technology transfer programs in that region:

- A 1995 paper indicated the three southeastern centers sent questionnaires on technology transfer to 1,343 firms, with 508 (38 percent) responding.[144] The findings show that the centers' technology transfer activities in 1993 and 1994 created or saved over 7,400 jobs and provided over $185 million in direct economic benefits, with an estimated overall impact to the economy of over $654 million.
- The introduction to the Marshall Center's *1995 Research and Technology Report* stated that "approved surveys" showed, in a recent three-year time period, the three centers projected 10,500 jobs created/saved, 459 new products, and $988 million in economic impact.[145]
- The NASA publication *Tech Briefs* stated that a survey by the three centers indicated 13,200 jobs created or saved, 775 new products, and $1.2 billion in value to American businesses.[146]

Apparently, the Southeast Alliance study was an update of the Marshall study, and it was said to have the same nonresponse bias flaw.

Department of Energy Evaluation Got Caught in Politics　DOE Survey: In terms of in-house, system-wide evaluation of technology transfer, the Department of Energy (DOE) was a leader among the federal departments and agencies. The DOE Technology Utilization Office conducted a series of regional "partnership" meetings to gain feedback about the type of information industry would be willing to share with the government to help evaluate technology transfer, particularly CRADAs. A 1994 progress report[147] discussed how the department was developing a database to record customer satisfaction measures, as well as performance and effectiveness measures. The system would track indicators from partnership milestones to economic indicators (such as the number of jobs created, companies formed, new product sales, and costs avoided). It was designed to prevent unauthorized access of proprietary and confidential information.

Later in 1994, as the political fervor rose, members of Congress and the Congressional Budget Office raised concerns about whether the extensive

DOE investment in applied R&D programs[148] was cost-effective. In response, DOE's Technology Partnerships Office[149] contracted for a report on DOE R&D award-winners.[150] Data on 268 award-winners[151] indicated that 51 percent of the technologies were transferred to the private sector; for the 137 technologies that were transferred, 153 private sector partners were identified. The most common transfer mechanism was licensing (followed by exchange of data and software, and contracts and subcontracts). Also in 1995, DOE produced a series of "success stories" describing 61 technologies developed by its applied R&D programs and the economic benefits (eg., energy savings) of each technology.[152]

GAO Response: In response to this, GAO was somewhat critical of the *Success Stories* report's methodologies in testimony before Congress.[153] GAO reviewed fifteen of the 61 cases and found problems with eleven of them, ranging from mathematical errors to "unsupported links between the benefits cited and DOE's role."[154] GAO said the report described the successes of only a small percentage of DOE programs, and did not document the amount of money spent on the technologies. According to GAO, more rigorous cost-benefit analyses would have been appropriate. GAO's premise is that applied research should result in product sales, and program costs should be less than sales in order to be cost-effective.

This elicited a strong response from Congressman George Brown of the House Science Committee through a four-page letter to the Comptroller General.[155] Meanwhile, DOE continued developing its metrics system for the national laboratories.[156] However, Congress disbanded the DOE Technology Partnerships Office in 1996.

Individual DOE Laboratories Initiated Projects   NREL Survey: In spite of all this fervor at the DOE headquarters level, the individual DOE laboratories worked to develop metrics on their own. For example, the National Renewable Energy Laboratory (NREL) in Colorado contracted with Chapman Research Group to assess the benefits of technology transfer to successful NREL collaborators.[157] In 1994 and 1995, Richard Chapman and Dana Moran interviewed NREL partners involved in licenses, CRADAs, contracts, technical assistance, informal collaboration, reimbursable work-for-others, post-doctoral research, and NREL conferences. Of 156 partners interviewed, 66 provided estimates of sales (nearly $690 million) or savings (more than $20 million) resulting from their association with NREL and its technologies. They also measured benefits in intangible terms, resulting in "an array" of benefits to the partners.

Oak Ridge Manufacturing Assistance Evaluation: Certain DOE laboratories, such as Oak Ridge National Laboratory, have major technical assistance[158] initiatives in the manufacturing area. The Oak Ridge Center for Manufacturing Technology offers three services to its customers: rapid development and deployment of products and processes, problem-solving, and skills training. The center provides these services through several types of mechanisms, including CRADAs, technical assistance agreements, work-for-others, and user facility agreements. A 1995 booklet describes how the center has implemented a system for setting priorities and measuring results.[159] Accordingly, the center surveyed two hundred clients and received 140 responses. Extrapolating these responses to the eight hundred cases of assistance showed that the economic return (from a $3-4 million investment) could be measured in the "tens of millions of dollars," according to the center's Direct Assistance Manager.[160]

Evaluation of these laboratory-based manufacturing assistance programs have much in common with the federal-state Manufacturing Extension Partnership (MEP) program operated by NIST. The MEP centers have been evaluated by both the federal and state levels. At the state level, Michigan, New York and Georgia can be cited as interesting examples. Evaluators in Georgia developed a benefit-cost model, and found a combined net public and private benefit-cost ratio of 1.2-to-2.7.[161] In Michigan, the Industrial Technology Institute provides a Performance Benchmarking Service for the extension center's client firms so they can benchmark themselves against control groups of non-assisted firms (from which data has been collected).[162] The New York evaluation used state employment records to provide a control group.[163] In terms of focussing on technical assistance as a technology transfer mechanism, another group of providers of technical assistance include the regional technology transfer centers, but very little has been done in the way of evaluating their activities.[164]

DOD Measured Laboratory Transfer, Examined Programs   Office of Technology Transition Report: The DOD Office of Technology Transition, created in the early 1990s at the level of the Office of the Secretary of Defense, encompasses technology transfer in the military service laboratories. The 1993 National Defense Authorization Act requires the office to report annually to Congress on its survey of defense laboratories and implementation of its Federal Defense Laboratory Diversification Program. The offices' 1994 report[165] highlighted, for example, the following for fiscal year 1992: 246 CRADAs, 890 patent applications filed, nineteen licenses, and $274,000 in royalty income.

Battelle Benchmarking of Air Force: At the individual service level, the Air Force sponsored a broad-ranging study of CRADAs and commercialization in the early 1990s. The Battelle team conducting the study carried out a literature search, a survey, benchmarking interviews, and two workshops in order to gather information for a publication on best practices in partnering with the military.[166]

Congress and GAO Examined Technology Transfer, Multi-Agency House Committee Small Business Survey: In the late 1980s, the House Committee on Small Business asked its subcommittee staff to investigate the role of government laboratories in the competitiveness of U.S. firms. To accomplish this, the staff surveyed key agencies on technology transfer. They found that government technology transfer efforts were "under-staffed, under-directed, and only marginally focused."[167] This resulted in frustrated government scientists and discouraged businesses, and licensing revenues representing "a return on research investment of only .00005 percent."[168]

GAO Surveys: The General Accounting Office began evaluating technology transfer by monitoring agency and laboratory implementation of the earlier 1980 Stevenson-Wydler Technology Innovation Act.[169] In 1988, GAO began preparing for a major analysis of implementation of the 1986 Federal Technology Transfer Act,[170] and conducted a somewhat controversial evaluation through a sixty-page survey sent to 297 laboratories.[171] GAO produced a preliminary report on its pilot questionnaire in 1989,[172] and testified on more than one occasion regarding progress in the evaluation.[173] However, the final report and testimony were not that different from the preliminary report.[174] Evidently, a full report never was accepted for publication by the House Science Committee.[175] GAO's findings indicated "uneven" implementation of the act across agencies, according to five measures chosen as compliance indicators.

Only 44 percent of the laboratory directors were authorized at that point to negotiate CRADAs and only half had royalty-sharing programs. The report concluded that laboratory activities had not lived up to expectations, and that there was a need for massive awareness and outreach regarding technology transfer.

GAO followed this up with some analyses of miscellaneous aspects of technology transfer such as government laboratory patent licensing activities[176] and commercialization issues related to software copyrighting,[177] Also, the 1986 act required a study of royalty-sharing after five years.[178] GAO found that royalty-sharing had not increased the laboratory scientists' interest in patenting, and subsequent GAO testimony supported higher royalties.[179]

In 1993 and 1994, GAO focused on cooperative R&D agreements (CRADAs) beginning with comparisons among agencies.[180] In its comparison, GAO criticized DOE's partnership efforts, although DOE had been implementing CRADAs for a shorter period of time than the other two agencies examined.[181] After focussing on improving the use of CRADAs at DOE laboratories,[182] GAO focussed on the benefits of CRADAs generally.[183] For the CRADA benefits study, GAO found that agencies and companies benefitted from their collaborations, although they did not all achieve the same level of benefits. In any case, both the laboratories and companies accomplished their objectives, since the examined CRADAs resulted in the enhancement of laboratory R&D programs and the transfer of technologies into commercial products. Some of the CRADAs demonstrated a potential for long-term improvements to the nation's economy, health, and environment.

Most recently, Congress asked GAO to evaluate approaches for measuring R&D results, such as patents counts, peer reviews, bibliometrics, and return on investment.[184] Like OSTP, GAO found that output measures are specific to the mission and management of each federal agency, and that no single indicator exists to measure research results. GAO also said profit-related indicators used by the private sector could not be applied to government work.

Commerce Department Measured on a Government-Wide Basis   DOC Biennial Reports: The Department of Commerce is required by law to report to the President and Congress every two years on federal agency implementation of the technology transfer legislation. Per this requirement, DOC issued reports in 1989, 1993 and 1996.

While the 1989 report[185] was understandably brief, the 1993 report[186] focused on the cooperative research activities of both the 1986 Federal Technology Transfer Act and the 1989 National Competitiveness Technology Transfer Act; it also included NASA's cooperative research activities.[187] It provided process-based measures (eg., number of invention disclosures) as opposed to impact measures. For example, by 1992, there were 1,300 cooperative R&D agreements in place. Also, there was a two-fold increase in licenses from 1987 to 1991, and the number of royalty-bearing licenses increased from less than fifty percent (of total licenses) in the early 1980s to more than ninety percent.

The 1996 report was broadened to discuss not just technology transfer, but also technology funding programs such as the TRP, ATP, and SBIR programs.[188] It described the transition to the new paradigm of public-private collaboration, highlighted best practices by these programs, and offered recommendations for improving partnership effectiveness. The report contained a section discussing how success *should* be measured, yet provided only cursory updated statistics.

Interagency Committee Attempted Consensus-Building   In the early 1990s, the Interagency Committee on Federal Technology Transfer attempted a government-wide effort to establish agreed-upon metrics for measuring laboratory technology transfer. The Committee, chaired by the Assistant Secretary for Technology Policy at the Commerce Department, established several working groups in 1992, including a Working Group on Technology Transfer Measurement and Evaluation with members from fifteen agencies and organizations.[189] The group's goal was to develop a government-wide system for measuring technology transfer effectiveness and assessing economic impact. In 1994, the Interagency Committee issued a draft report from the working group containing an agreed-upon set of definitions and measurements and data collection framework.[190] The Department of Agriculture offered to test the framework using actual data,[191] however, the work was not finalized because the committee became inactive during much of the Clinton Administration.[192]

Federal Laboratory Consortium Contributed Reports   FLC Defense Conversion Report and Measures: A 1994 FLC report on defense conversion addressed the issue of how to measure success in technology transfer.[193] It broke new ground in suggesting the need to use measures of cultural change to measure progress and to motivate, noting the importance of being deliberate in choosing measures because "that which is rewarded is what motivates." The following year, the FLC testified before the House Science Committee with the this strong statement on performance metrics:[194]

> There is no issue more critical to federal technology transfer in 1995 than determining and using measures to assess its value to all participants. Disagreement appears to exist in two dimensions: (1) What types of measures truly represent the impact of technology transfer efforts? (Not just the process,

which is far easier to evaluate) and (2) Which measures are sufficiently sound that they can be used across a wide range of varying technology transfer efforts? (For example, a personnel exchange program within a government-owned, contractor-operated laboratory versus a CRADA between a government-owned, government-operated laboratory and a large industry, or providing technical assistance to a local small business versus licensing an existing patent to a large international firm). The volatility of technology also contributes to the difficulties of precisely measuring its transfer. The FLC agrees that the definition and adoption of performance metrics are of the highest priority and we are working with our member laboratories and agencies in this area.

Chapman Survey of FLC Mid-Continent Region: The FLC then commissioned a study by Chapman Research Group[195] which involved a survey of eleven laboratories in the FLC's eight-state Mid-Continent Region. The study identified both obstacles and best practices. Best practices involve: incorporating technology transfer into the laboratory's strategic planning; having an innovative and aggressive technology transfer officer; and having supportive top management at the laboratory. Obstacles include: lack of effective outreach and in-reach (to the laboratory scientists); and unsupportive agency headquarters personnel.

Outside Studies Compared or Combined Government, University Transfer
Bozeman NCRDP Studies: Barry Bozeman of Georgia Tech received NSF, DOE, and other funding since 1984 to develop a database for his National Comparative Research and Development Project (NCRDP). This ongoing project determined the performance and sources of influence in the U.S. R&D laboratory system, the extent of technology transfer activity, and the degrees of success. It also provided comparative data on the technical enterprises in other industrial nations. The NCRDP developed in several phases, depending upon funding. The NCRDP master database was primarily based upon questionnaires mailed to laboratory directors. Several effectiveness measures were employed, including subjective self-ratings and more objective measures. The early Bozeman work was based upon an input-output approach, where the outputs were the immediate or intermediate products of the technology transfer but, in later studies, more emphasis was placed upon ultimate impact. The findings were described in various articles highlighted below.

A 1988 Bozeman-edited symposium on evaluating technology transfer for the journal *Evaluation and Program Planning* presented Bozeman's early ideas on how technology transfer evaluation differed from evaluation of other policies.[196] In that series, he co-authored an article presenting a contingency framework for assessing technology transfer activities at U.S. national laboratories.[197] The model focused on: the technology transfer characteristics of the transferring agent, attributes of the transferred products, the transfer mechanisms, and attributes of the user. The article used Brookhaven National Laboratory as a case example.[198] Bozeman's colleague Michael Crow wrote another article in this journal series reporting on 32 laboratory case studies.[199]

Bozeman and Crow – Phase II: A 1988 article in *Public Administration Review*[200] by Bozeman and Crow (et al) built upon the earlier NCRDP work and presented results of 1987 data developed in Phase II of the project. This phase involved a survey and interviews with almost 1,000 government and university laboratories comparing technology transfer in both settings. The study showed that 52 percent of governmental laboratories and 38 percent of university laboratories indicated they considered technology transfer to industry to be a major mission of their laboratories.

A 1990 Bozeman/Crow *Policy Sciences* article again examined technology transfer according to government, university, and industry sectors, looking specifically at the amount of bureaucratization, cooperative research, and output for each sector.[201] Their sector-based classification explained the amount of "red tape," but politics and market-based influences better explained the amount of cooperative research and output. A 1991 article examined the question of red tape in more detail.[202] Data from surveys of 276 federal and state laboratory directors showed that laboratories involved in technology transfer did not have higher levels of red tape.

Bozeman and Coker – Phase III: For Phase III of the project, in 1990, the Bozeman team mailed questionnaires to more than 1,100 laboratory directors; 47 percent were returned. A 1992 article by Bozeman and Karen Coker[203] assessed laboratory technology transfer success, taking into account laboratory receptiveness to market influences. They considered three criteria: two based upon self-evaluations, and a third based upon the number of licenses issued. The results showed that multi-mission laboratories were more successful, especially if they had low levels of bureaucratization and either ties to industry or a commercial orientation in project selection.

A 1994 article by Bozeman in the *Policy Studies Journal*[204] reported on the government subsample of the Phase III data which involved 189 laboratories or about half of the larger government laboratories. A comparison of this data with Phase II 1987 data showed that government laboratory technology transfer activity increased by more than forty percent, representing what he called "considerably enhanced activity in a relatively brief amount of time."

A 1995 report to NSF by Bozeman, Coker, and Maria Papadakis,[205] presented findings from surveys of companies interacting with government laboratories during the previous five years. The data included 229 collaborative projects (e.g., CRADA, technical assistance, license, personnel exchange), 219 firms, and 27 laboratories. The study focused on industries' assessment of the benefits of working with federal laboratories, including monetary estimates of interactions and outputs. In addition to the cost-benefit of the interactions, the study examined: who initiated the interactions, how well the companies' objectives were achieved, barriers to working with the laboratories, and extent of commercialization of the output. The results indicated that 89 percent of the companies felt the collaborative projects were a good use of their company's resources. On average, the project benefits exceeded costs by 3-to-1. The average net benefit was more than $1 million, although nearly one-third reported net costs exceeded net benefits. The projects' job creation value was modest; the average number of jobs created was 1.5, and 90 percent of the projects created no net jobs. Overall, the projects exhibited a high commercialization rate; 22 percent of the interactions led to products on the market and 38 percent had products under development.

With his extensive background in technology transfer program evaluation, a pair of "before and after" articles by Bozeman serves to highlight the frustrations in this area. Fifteen years ago, in a confident piece in the "Public Management Forum" section of *Public Administration Review*,[206] Bozeman and Jane Massey presented some guidelines for investing in policy evaluation. They encouraged evaluating policies " ... where the direction of causality is more apparent . . .where direct effects are considered more significant than 'spillover' effects . . .where short-run benefits are claimed (avoiding premature evaluation) . . . and where the determinates of effectiveness can be controlled . . ." After fifteen years of technology transfer evaluation, Bozeman wrote another article[207] indicating that "many puzzles remain . . .There are more than enough methodological and practical challenges to keep us busy for some time." In this article, he asked questions such as, "What is a good batting average?" and "Can technology transfer be justified?"

Papadakis Evaluation: In 1995,[208] Papadakis carried her work with

Bozeman further, reviewing the NCRDP data and combining it with GAO and NSF data.[209] Her findings indicated that, because technologies are most likely to emerge from the mission-oriented laboratories, they are "the least likely to spin off and diffuse throughout the industrial base." Once the DOD, DOE, and NASA hardware needs are excluded, "most of the system's R&D output is fundamental knowledge, which flows through public domain literature and requires substantial additional processing to become commercial products." She concluded: (1) there is no reason to believe the current laboratory system can enhance competitiveness; (2) to enhance competitiveness, government laboratories must have explicit missions to do this; and (3) policy expectations (of commercial impacts) are inconsistent with policy requirements that laboratories conduct basic and applied research.

Roessner and Bean Surveys – Phase I: Beginning in the late 1980s, David Roessner at Georgia Tech and Al Bean at Lehigh University worked with the Industrial Research Institute (IRI), whose corporate members conduct about 85 percent of the industrial research in the U.S., to survey corporate opinions of working with government and university laboratories. The National Science Foundation (NSF) funded their initial 1988 survey. Roessner and Bean published the results in 1990 and 1991 articles.[210] They found that firms had a surprisingly high level of awareness of and interaction with federal laboratories, and many of them planned to increase their external R&D funding (such as for cooperative work). Roessner and Bean stated that they believed the firms including such external resources in their strategic planning would achieve stronger competitive positions than those that did not.

Roessner and Bean – Phase II: A 1992-1993 update survey by Roessner and Bean focused more specifically on industry interactions with the DOE national laboratories.[211] Both NSF and DOE funded this phase of their study which reported on the findings from 55 IRI member companies. They found that companies felt there was technology with commercial potential in the laboratories. However, since 1988, federal laboratories continued to "lag considerably behind" universities and other companies in being a source of external technology for industry. The survey also addressed questions related to problems and payoffs from companies interacting with laboratories. They found that companies interacted with government laboratories for long-term, less tangible payoffs rather than for commercialization opportunities. This caused Roessner and Bean to be concerned about the amount of trouble and expense needed in order for the government to make the connection with industry (eg., conducting seminars or incentives for scientists to make industry contacts). They also felt companies and laboratories should work to provide evidence of less tangible (but potentially higher-value) payoffs than profits.

Geisler Success Factors Survey: In terms of examining technology transfer process, in 1995, the National Science Foundation sponsored Eliezer Geisler's study of why federal laboratories succeed or fail at technology commercialization.[212] It involved 43 federal laboratories and their technology transfer offices, 51 industrial companies, and 428 scientists and engineers. The findings showed that in successful laboratories: (1) management supports cooperation with industry through incentives and the scientific personnel exhibit intrapreneurial attitudes; and (2) the cooperating companies support commercialization and their technical personnel perceive their laboratory counterparts as risk takers. Geisler also found that the incentives most likely to work are those creating a supportive environment for intrapreneurs (versus only financial incentives). Further, the reason industry and laboratories cooperate (e.g., access to technical resources) are different from the behavioral factors for successful commercialization.

Colorado Institute Survey of Best Practices: In 1993, the Department of Commerce and the Colorado Advanced Technology Institute sponsored the Colorado Institute for Technology Transfer and Implementation[213] to survey best practices in federal laboratories, universities, private companies, and technology brokers. The survey of sixty practitioners[214] produced vignettes describing 144 technology transfer best practices grouped into six core areas: strategy/policy, communication/organization, inventory, market assessment, resources, and reward/recognition. The final report highlighted findings related to: core practices, role and scope of intermediaries, universities and risk, the federal bureaucracy, public and private sector commitment, management of technology transfer, understanding the market versus understanding the technology, communication and data management, changing perceptions, and student resources.

University Models for Technology Transfer Evaluation   Two series of university technology transfer studies[215] that have been used as models of how government laboratories might go about developing metrics as a group. These studies were conducted by the Association of University Technology Managers (AUTM) and the Southern Technology Council. AUTM has examined licensing royalties and other measures annually since 1991. In 1996, AUTM compiled a five-year summary of its survey covering 1991 to

1995.[216] The findings showed continued growth in the numbers of inventions reported, patent applications filed, patents granted, licenses executed, and royalty income generated. Also, in 1994, AUTM provided qualitative data in a one-time "public benefits" survey forwarded to respondents of the licensing survey that year.[217] The report described products patented, licensed, and on the market as a result of university technology transfer, and summarized the number of "university-founded"[218] start-up companies since 1980. The Southern Technology Council (STC) has produced a series of benchmarking reports on university technology transfer in the southeastern region of the country since the early 1990s.[219]

The AUTM Economic Impact Special Interest Group has been compiling a bibliography of economic impact studies by individual universities, and identified 44 studies by 1997. These individual university methodologies are also applicable to government laboratory technology transfer evaluation. For example, the Massachusetts Institute of Technology (MIT) Licensing Office developed a model for measuring pre-production investment by companies licensing technologies from MIT, and then extrapolated to all university licenses based upon the AUTM data.[220] This study found the nationwide investment in university-based technology was estimated to be $2 to 5 billion a year. BankBoston surveyed entrepreneurial activity by MIT alumni, and found that MIT graduates created 4,000 companies, generating 1.1 million jobs worldwide and $232 billion in annual sales.[221] This type of study could also be applicable to government laboratories since more laboratories are beginning to institute entrepreneurial leave.

Another area where government laboratories could learn from university technology transfer is in regard to partnerships with consortia of companies. Evaluation of such partnerships would be comparable to efforts by the National Science Foundation to evaluate its Industry-University Cooperative Research Centers (IUCRC) program and Engineering Research Centers (ERC) program both of which involve industry-university partnerships.[222]

Michael Odza, editor of *Technology Access Report*, has questioned why government laboratories can't produce at least minimum technology transfer measures (eg., number of licenses and royalty amounts) across the board, as universities have.[223] By comparing university technology investments and resulting royalties with government laboratory investments, he roughly projected that government laboratories should be bringing in at least 700 new licenses each year.

A set of articles by Robert Carr compared federal laboratory and university technology transfer and analyzed best practices in government laboratory technology transfer. In the first article, Carr compared the AUTM data (on university licensing) with GAO data (on government laboratory licensing), showing that government laboratories lag certain universities in this area.[224] For example, MIT's royalty income from licensing activities in one year was over twice that of the entire DOE laboratory system the same year. He concluded that the explanation for the difference between the two sectors is due to the way each sector markets opportunities.

In the second of Carr's two-part series, he described best practices in technology transfer based upon a literature search and interviews with technology transfer professionals in both federal laboratories and universities.[225] The best practices addressed the following functions: organizing the technology transfer function, involving the science and technology staffs, capturing intellectual property, evaluating and patenting intellectual property, marketing technologies, preparing technologies for commercialization, transferring technology locally, using technology transfer intermediaries, and using technology search programs. The piece ended with a collection of conventional wisdom about technology transfer.

A subsequent Carr article noted that most studies have not measured the economic impact of technology transfer (as opposed to intermediate or process measures) because "measuring the economic value of technology transfer requires measuring the economic value of knowledge, an old problem."[226] He proposed development of case studies and surveys to determine the proper indicators to use, so that full-scale government-wide data collection could be based upon these refined variables.

Evaluation of Laboratory Incubators and Economic Development Projects
Government laboratories are beginning to adopt the concepts of economic development and incubators into their technology transfer and outreach programs.[227] For example, the federal facilities in Huntsville, Alabama (NASA Marshall field center and the Army Space and Strategic Command) are jointly implementing a technology transfer project through the local chamber of commerce which nearby university researchers recently evaluated.[228]

Not all incubators are necessarily high-tech oriented, although the ones associated with university campuses tend in this direction more than others.[229] Incubator evaluation work dates back to a classic 1987 study.[230] More recently, the National Business Incubation Association (NBIA) and the Southern Technology Council, along with other partners, identified best practices and surveyed 49 incubation programs and 126 incubator companies on results between 1990 and 1996.[231] The average company's sales increased by over 400 percent during its incubator stay, and 87 percent of incubator graduates remained in business. The study also analyzed the impacts of four incubators, and found they generated $402,000 in local tax revenues while receiving $81,000 in operating funds.

Several groups have produced handbooks on how to evaluate the impact of local economic development projects and incubators,[232] and NBIA has been working to implement standardized metrics within the incubator community. NBIA partner Louis Tornatzky has developed a proposal for standardizing definitions and measures across technology transfer communities so that benchmarking can be done.[233] If implemented, it would involve laboratories, universities, manufacturing extension, and other technology transfer organizations.

Technology Transfer Metrics – Summary

Irwin Feller, noted for his evaluation of state and university technology programs, recently offered his assessment of progress in the field of evaluating technology transfer.[234] He noted that considerable advances have been made in terms of agency commitments to evaluation and the standards of evaluation design, but he warned that latent issues will surface as the field evolves to a higher level of understanding. When that happens, he said, there will be calls to become even more rigorous and disciplined.

It has been argued that government laboratories would avoid some criticism by developing standard evaluation procedures so that each laboratory is reporting on a uniform set of measures using consistent approaches. University-based technology transfer (including incubators) and manufacturing extension are examples of technology transfer communities that have successfully developed standardized quantitative and qualitative indicators (in the form of systems for identifying, documenting and disseminating best practices and cases).[235] However, since many government laboratories and programs have already developed their own evaluation approaches, recommendations for standardization have caused debate. A major issue revolves around the purpose of the results. For example, are the findings to be used for justifying or improving programs? Either way, the laboratories point out that the record-keeping requirements for the company partners and the laboratories would be burdensome.

Another issue revolves around appropriate methodologies and there appear to be problems or misunderstandings over just about every method being used. As this review has shown, a common methodology in evaluating technology transfer involves an input-outcomes approach, where the technology transfer activity is the input. A problem with this approach is the difficulty in guaranteeing that outcomes are attributable to program inputs. Some studies used "multipliers" to estimate the economic impact generated by laboratory technology transfer. These types of analyses have been criticized, possibly because the results seem extravagant while the methods are not widely understood. Some have suggested that truly valid methods would necessitate the use of comparison and control groups. Yet there are not many examples of technology programs that have successfully done this. Two examples are the DOE-funded Energy-Related Inventions Program and the Michigan-based Performance Benchmarking Service.[236] Both efforts have been described as somewhat costly and time-consuming, yet worth the effort in terms of the credibility gained. Customer satisfaction ratings and peer review approaches require some systematizing and standardization to be useful tools. Benefit-cost analyses have been recommended as applicable, yet there are few examples of this being applied to technology transfer activities (other than manufacturing assistance), as opposed to R&D generally. Further, this approach is most applicable within a program area, and does not easily permit cross-comparisons.

It might help for one of the technology policy agencies, such as OSTP or the Commerce Department's Technology Administration, to support experimentation with some innovative ways to measure success. For example, the NIST MEP program has a pilot study using longitudinal Census Bureau data to compare the performance of client firms and non-assisted firms, while controlling for other factors that influence performance.[237]

Overall Summary and Conclusion – Study Fills Gap

This chapter attempted to show that technology transfer policies have evolved alongside technology funding policies and programs. The studies analyzing the government laboratory missions eventually focussed on the laboratories' role in technology transfer. Thus, the technology policy literature converged with the technology transfer policy literature.

As an epilogue to this chapter, it should be noted that some of the emotional rhetoric associated with the political divisiveness on technology policy (such as "industrial policy" and "picking winners") has subsided.238 For example, although a Corporate Subsidy Reform Commission was proposed in 1997 to identify existing "corporate welfare," it was announced early-on that R&D partnerships would be exempt from its purview. Even the phrase "critical technologies" has lost some of its earlier emotionally-laden connotations and both OSTP and CPC have standing committees on "critical technologies," although the status of federal programs in this area is not resolved.239 Furthermore, the Competitiveness Policy Council and the Council on Competitiveness sponsored a project which included a series of regionally-oriented meetings, designed in part to try to overcome the political rift.240 Within the DOE laboratory system, the latest laboratory review committee is continuing as the Laboratory Operations Panel, and has succeeded in making small incremental changes. Meanwhile, at least four Congressional bills are pending for reorganizing the DOE laboratories rather than abolishing the department. The R&D budget situation is calming down, as well; certain trade associations and professional societies are teaming together to back a bipartisan effort to recover the budget losses of the mid-1990s, and possibly even double the R&D budget over the next decade.

Regardless of the current lull, it has become apparent that the measurement and evaluation of science and technology programs is an area suffering from a general lack of experience. The measurement and evaluation of government technology transfer is similarly premature at this point. Agencies in both the Executive Branch and Congress (e.g., the General Accounting Office) have had problems producing meaningful results. For one thing, the relatively new technology development and transfer programs have not been in existence long enough to produce analyzable results. This is at a time when, due to budget pressures, all branches of government and the public are anxious to determine results.

Table 2.1 summarized some key technology program evaluations. Those relevant to government laboratories were presented in this chapter. The bibliography contains detailed references to additional related studies. This particular study is an assessment of the effect of technology transfer legislation and related policies. It evaluates outcomes and process based upon a qualitative analysis of a series of successful government cases using data collected from laboratory scientists and private partners. There have been many *individual* descriptive case studies of successful government technology transfer activity in a variety of literature sources. However, Table 2.1 made it apparent that there is a need for more *series* of case studies systematically developed using a consistent survey framework.241

The next two chapters detail the data collected in the interviews for this study, presented as a series of cases. Chapter 3 presents the pre-legislation cases, and Chapter 4 presents the post-legislation cases.

Notes

1   For purposes of this chapter, the trends are discussed separately although they are actually inter-related.
2   National Academy of Sciences, Panel on the Government Role in Civilian Technology, *The Government Role in Civilian Technology: Building a New Alliance*, Washington, D.C.: National Academy Press, ISBN 0-309-04630-0, 1992. Also, National Academy of Sciences, Committee on Science, Engineering, and Public Policy, *Science, Technology and the Federal Government: National Goals for a New Era*, Washington, D.C.: National Academy Press, 1993.
3   One of the last reports in the series was: Carnegie Commission on Science, Technology and Government, *A Science and Technology Agenda for the Nation: Recommendations for the President and Congress*, 1992.
4   Carnegie Commission on Science, Technology, and Government, *Science, Technology, and Government for a Changing World*, Concluding Report of the Carnegie Commission, ISBN 1-881054-11-X, April 1993.
5   See, for example, Congressional Budget Office, Using R&D Consortia for Commercial Innovation: SEMATECH, X-Ray Lithography, and High Resolution Systems, July 1990.
6   Office of Science and Technology Policy, *Report of the National Critical Technologies Panel*, March 1991. OSTP has updated its review more recently, as follows: Office of Science and Technology Policy, National Science and Technology Council, National Critical Technologies Review Group, *National Critical Technologies Report (1995)*, March 1996.
7   Department of Commerce, Technology Administration, Emerging Technologies: A Survey of Technical and Economic Opportunities, Spring 1990.
8   Department of Defense, Critical Technologies Plan for the Committees on Armed Services, United States Congress, March 15, 1990. An earlier version was Department of Defense, The Department of Defense Critical Technologies Plan, May 1989.
9   John Alic, Lewis Branscomb, Harvey Brooks, Ashton Carter and Gerald Epstein, *Beyond Spinoff: Military and Commercial Technologies in a Changing World*, Cambridge, Massachusetts: Harvard Business School Press, 1992.
10 See, for example, Aerospace Industries Association of America, Inc., *Key Technologies for the 1990s, An Overview*, Washington, D.C., November 1987,

updated version.

11 Council on Competitiveness, Gaining New Ground: Technology Priorities for America's Future, Washington, D.C., 1991.

12 Congressional Research Service, Critical Technologies: Legislative and Executive Branch Activities, 93-734 SPR, 1993.

13 Office of Technology Assessment, Redesigning Defense: Planning the Transition to the Future U.S. Defense Industrial Base, Washington, D.C.: U.S. Government Printing Office, 1991; and, Office of Technology Assessment, Building Future Security: Strategies for Restructing the Defense Technology and Industrial Base, Washington, D.C.: U.S. Government Printing Office, 1992.

14 Department of Commerce, Economic Development Administration, *From War to Peace: A History of Past Conversions*, 1993.

15 Deborah Shapley and Rustom Roy, *Lost at the Frontier: U.S. Science and Technology Policy Adrift*, Philadelphia, Pennsylvania: Institute for Scientific Information Press, 1985.

16 National Academy of Sciences, Committee on Science, Engineering and Public Policy, Panel on Technology and Employment, *Technology and Employment: Innovation and Growth in the U.S. Economy*, Richard M. Cyert and David C. Mowery, editors, Washington, D.C.: National Academy Press, 1987.

17 Nathan Rosenberg, Ralph Landau and David C. Mowery, editors, *Technology and the Wealth of Nations*, 1992. See also, Bruce R. Guile and Harvey Brooks, editors, *Technology and Global Industry: Companies and Nations in the World Economy*, National Academy of Engineering Series on Technology and Social Priorities, Washington, D.C.: National Academy Press, 1987.

18 Michael J. Boskin and Lawrence J. Lau, "Capital, Technology, and Economic Growth," *Technology and the Wealth of Nations*, Nathan Rosenberg, Ralph Landau and David C. Mowery, editors, Washington, D.C.: National Academy Press, 1992.

19 Richard R. Nelson, *High-Technology Policies: A Five-Nation Comparison*, New York: Columbia University, 1988.

20 See, for example, Francis W. Rushing and Carole Ganz Brown, editors, *National Policies for Developing High Technology Industries: International Comparisons*, Westview Special Studies in Science, Technology, and Public Policy, Boulder, Colorado: Westview Press, 1986.

21 General Accounting Office, *Competitiveness Issues: The Business Environment in the United States, Japan, and Germany*, Report to Congressional Requesters, GAO/GGD-93-124, August 1993.

22 National Governors Association and the Conference Board, *The Role of Science and Technology in Economic Competitiveness*, Final Report Prepared for the National Science Foundation, September 1987.

23 National Academy of Sciences, *The Positive Sum Strategy: Harnessing*

Technology for Economic Growth, Washington, D.C.: National Academy Press, 1986.

24 National Academy of Engineering, *Technology and Economics*, Washington, D.C.: National Academy Press, ISBN 0-309-04397-2, 1991.

25 See, for example, Wendy H. Schact and Glenn J. McLoughlin, *Technology and Trade: Indicators of U.S. Industrial Innovation,* Congressional Research Service Review, October 1986. See also, Glenn McLaughlin and Richard E. Rowberg, "Linkages Between Federal Research and Development Funding and Economic Growth," Congressional Research Service series, *Economic Policymaking in Congress: Trends and Prospects,* February 21, 1992.

26 Executive Office of the President, Office of Science and Technology Policy, *U.S. Technology Policy*, September 26, 1990.

27 Technology policy first appeared during the Carter Administration, then disappeared during the subsequent Reagan Administration.

28 This group should not be confused with the Competitiveness Policy Council, an independent advisory committee, or the private-sector Council on Competitiveness, both of which appear elsewhere in this chapter.

29 President William J. Clinton and Vice President Albert Gore, Jr., *Technology for America's Economic Growth, A New Direction to Build Economic Strength*, February 22, 1993.

30 White House, *Technology for Economic Growth: President's Progress Report*, November 1993.

31 Department of Commerce, *Commerce ACTS: Advanced Civilian Technology Strategy*, Draft for Public Comment, November 1993.

32 The Carnegie Commission and others had been calling for structural changes in the Executive Office for handling science and technology policy issues. See, for example, Carnegie Commission on Science, Technology and Government, *Technology and Economic Performance: Organizing the Executive Branch for a Stronger National Technology Base*, September 1991.

33 Office of Science and Technology Policy, National Science and Technology Council, Committee on Fundamental Science, *Science in the National Interest*, August 1994. The overall goal for the nation is to maintain leadership across the frontiers of scientific knowledge; sub-goals relate to connections between research and national goals, etc. This was the first in a two-part series; its companion piece was *Technology in the National Interest*, noted below.

34 See, for example, Neil MacDonald, "Tech Policy Veteran Puzzles Over Recent Federal Episodes," *McGraw-Hill Companies' Federal Technology Report* (November 20): 9-10, 1997.

35 The Commerce Department's Technology Administration assisted OSTP in producing reports such as: Office of Science and Technology Policy, National Science and Technology Council, Committee on Civilian Industrial

Technology, *Technology in the National Interest*, July 1996. This was followed by the OSTP biennial report to Congress: Executive Office of the President, Office of Science and Technology Policy, National Science and Technology Council, *Science and Technology Shaping the Twenty-First Century*, 1997.

36 Department of Commerce, Economics and Statistics Administration, Office of the Chief Economist, *Technology, Economic Growth and Employment*, 1994.

37 National Institute of Standards and Technology, *Technology and Economic Growth: Implications for Federal Policy*, Prepared by Gregory Tassey, Senior Economist, October 1995.

38 House of Representatives, Committee on Science, Space and Technology, *Technology Policy and its Effect on the National Economy: Report Prepared by the Technology Policy Task Force,* Washington, D.C.: U.S. Government Printing Office, 1988.

39 This council was created by Congress in the 1988 Omnibus Trade and Competitiveness Act. The President and Congress appoint twelve council members.

40 Competitiveness Policy Council, *Pursuing a New Technology Policy*, Report of the Critical Technologies Subcouncil, Erich Bloch, Chairman, May 1994.

41 Competitiveness Policy Council, *Saving More and Investing Better: A Strategy for Securing Prosperity,* Fourth Report to the President and Congress, September 1995. Previous CPC reports to the President and Congress included: *Promoting Long-Term Prosperity* (Third Report, May 1994; *A Competitiveness Strategy for America* (Second Report, March 1993); and, *Building a Competitive America* (First Report, 1992). An October 1993 "progress report," *Enhancing American Competitiveness,* discussed problem areas.

42 Previous CPC technology policy reports included: Competitiveness Policy Council, *Implementing Technology Policy for a Competitive America*, Report of the Critical Technologies Subcouncil, August 1993; and, Competitiveness Policy Council, *Technology Policy for a Competitive America*, Report of the Critical Technologies Subcouncil, March 1993.

43 Center for Strategic and International Studies, *Global Innovation/ National Competitiveness*, Washington, D.C., 1996.

44 The more recent 1996 version of the Council's policy assessment was largely a compilation of essays by well-known commentators: Council on Competitiveness, *Competitiveness Index 1996: A Ten-Year Strategic Assessment*, Washington, D.C., ISBN 1-889866-18-0, 1996.

45 Council on Competitiveness, *Technology Policy Implementation Assessment 1993*, Washington, D.C., 1993.

46 These focus areas grew out of the Council's 1991 assessment: Council on Competitiveness, *Gaining New Ground: Technology Priorities for America's Future*, Washington, D.C., 1991.

47 Vannevar Bush, *Science: The Endless Frontier*, first edition, 1945; 1990 edition published on the occasion of the Fortieth Anniversary of the establishment of the National Science Foundation which the publication helped to create.

48 Council on Competitiveness, *Endless Frontier, Limited Resources: U.S. R&D Policy for Competitiveness*, Washington, D.C., April 1996.

49 The report's appendices are in-depth assessments of R&D in six industry sectors that are central to U.S. competitiveness: aircraft, automotive, chemical, electronics, information technologies, and pharmaceuticals.

50 National Academy of Sciences, Committee on Criteria for Federal Support of Research and Development, Allocating Federal Funds for Science and Technology, Frank Press, Committee Chair, Washington, D.C.: National Academy Press, 1995.

51 Frank Press, "Needed: Coherent Budgeting for Science and Technology," Science 270 (December 1): 1448-1450, 1995.

52 The Congressional Research Service compiles regular reports on federal R&D funding and R&D funding for specific departments and agencies. See for example: Congressional Research Service, Research and Development Funding: Fiscal Year 1998, CRS Issue Brief prepared by Michael E. Davey, Science Policy Research Division, IB97023, Updated December 17, 1997; Congressional Research Service, The Department of Energy FY1998 Research and Development Budget and Issues, CRS Report for Congress prepared by Richard E. Rowberg, Science Policy Research Division, 97-233 SPR, Updated December 3, 1997; Congressional Research Service, The National Aeronautics and Space Administration: An Overview With FY1997 and FY1998 Budget Summaries, CRS Report for Congress prepared by David P. Radzanowski, Science Policy Research Division, 97-634 SPR, June 10, 1997; Congressional Research Service, Federal R&D Funding Trends in Five Agencies: NSF, NASA, NIST, DOE (Civilian) and NOAA, CRS Report for Congress prepared by Michael E. Davey, Science Policy Research Division, 97-126 SPR, January 17, 1997.

53 Congressional Research Service, Research and Development Funding in a Constrained Budget Environment: Alternative Support Sources and Streamlined Funding Mechanisms, 1996. See also, Congressional Research Service, Research and Development: Priority Setting and Consolidation in Science Budgeting, CRS Issue Brief prepared by Genevieve J. Knezo, Science Policy Research Division, IB94009, Updated January 15, 1998.

54 Generally, it is contrary to the role of government to identify and support winners. Illustrative editorials and commentaries on both sides of this topic include: "When the State Picks Winners," Editorial in The Economist (January 9): 13-14, 1993; Robert W. Rycroft and Don E. Cash, "Technology Policy Requires Picking Winners," Commentary in Economic Development Quarterly

6(3/August): 227-240, 1992.

55 Congressional Research Service, Science Policy Research Division, The Federal Role in Technology Development, CRS Report for Congress prepared by Wendy H. Schacht, 95-50 SPR, 1995, October 15, 1996 and updated January 12, 1998. See also Congressional Research Service, Industrial Competitiveness and Technological Advancement: Debate Over Government Policy, CRS Issue Brief prepared by Wendy H. Schacht, Science Policy Research Division, IB91132, Updated December 5, 1997; and, Congressional Research Service, R&D Partnerships: Government-Industry Collaboration, CRS Report for Congress prepared by Wendy H. Schacht, 95-499 SPR, Updated January 12, 1998.

56 White House, Council of Economic Advisors, Supporting Research and Development to Promote Economic Growth: The Federal Government's Role, October 1995.

57 Robert M. White, U.S. Technology Policy: The Federal Government's Role, Paper Commissioned by the Competitiveness Policy Council, September 1995.

58 Department of Energy, Secretary of Energy Advisory Board (SEAB), Report to the Secretary on the DOE National Laboratories, Prepared by the SEAB Task Force on the DOE National Laboratories, July 1992.

59 Department of Energy, Alternative Futures for the Department of Energy National Laboratories, Secretary of Energy Advisory Board Office, Task Force on Alternative Futures, February 1, 1995.

60 Such as adherence to budgets, adherence to project schedules, and research dead ends now avoidable.

61 For example: patents filed, inventions disclosed, estimates of cost savings from a given potential application or actual application of technology, lists of technical problems solved, quantity and quality of research papers published, laboratory/ university and laboratory/ industry interactions as well as other collaborative work anecdotes, including CRADA results. The report discouraged the use of two metrics, number of CRADA-related jobs (too "speculative"), and number of CRADAs (fails to measure output or different classes of CRADAs).

62 OTA was disbanded in 1995 under the auspices of the Republicans' "Contract with America."

63 Office of Technology Assessment, Defense Conversion: Redirecting R&D, Washington, D.C.: U.S. Government Printing Office, 1993. The first report, focused on companies, was: Office of Technology Assessment, After the Cold War: Living With Lower Defense Spending, Washington, D.C.: U.S. Government Printing Office, OTA-ITE-524, February 1992.

64 Congressional Research Service, DOE Laboratories: Capabilities and Missions, 93-752 SPR, 1993; more recently, see Congressional Research Service,

Restructuring DOE and Its Laboratories: Issues in the 105th Congress, CRS Issue Brief prepared by William C. Boesman, Science, Technology, and Medicine Division, IB97012, Updated January 9, 1998.

65 General Accounting Office, *DOE's National Laboratories: Adopting New Missions and Managing Effectively Pose Significant Challenges*, Testimony before the Subcommittee on Energy and Power, Committee on Energy and Commerce, House of Representatives, GAO/T-RCED-94-113, February 3, 1994.

66 General Accounting Office, *National Laboratories: Are Their R&D Activities Related to Commercial Product Development?* Report to Congressional Requesters, GAO/PEMD-95-2, November 1994.

67 General Accounting Office, *Department of Energy: National Laboratories Need Clearer Missions and Better Management*, Report to the Secretary of Energy, GAO/RCED-95-10, January 1995.

68 General Accounting Office, *Department of Energy: A Framework for Restructuring DOE and Its Missions*, Report to the Congress, GAO/RCED-95-197, August 1995.

69 General Accounting Office, *Energy Research: Opportunities Exist to Recover Federal Investment in Technology Development Projects*, Report to the Chairman, Subcommittee on Energy and Environment, Committee on Science, House of Representatives, GAO/RCED-96-141, June 1996.

70 As examples of studies in this area, which date back several decades, see: Battelle, *Interactions of Science and Technology in the Innovation Process: Some Case Studies*, Final Report, Prepared for the National Science Foundation, Columbus, Ohio: Battelle Columbus Laboratories, Contract NSF-C 667, March 19, 1973; Robert K. Yin et al, *A Review of Case Studies of Technological Innovations in State and Local Services*, Santa Monica, California: RAND Corporation, R-1870-NSF, February 1976; Federal Laboratory Consortium for Technology Transfer, Federal Laboratory-Industry Interaction Working Group, *Interagency Study of ORTA Organization and Operation and Lessons Learned Case Studies in Technology Transfer*, DOE/METC - 85/6019, May 1985.

71 Everett M. Rogers, *Diffusion of Innovations,* New York: Free Press, ISBN 0-02874074-2, First edition, 1962, Third edition, 1983, Fourth edition, 1995; see also, Everett M. Rogers, with the assistance of F. F. Shoemaker, *Communication of Innovations: A Cross-Cultural Approach*, New York: Free Press, 1971.

72 There is a wide variety of published material on the behavioral sciences and the examples are innumerable.

73 See, for example, the *Journal of Product Innovation Management*, bimonthly publication of the Product Development and Management Association (PDMA), Elsevier Science, Inc., publisher, ISSN 0737-6782. See also, Milton D.

Rosenau, Jr., editor, *The PDMA Handbook of New Product Development*, John Wiley & Sons, Inc.

74 Louis G. Tornatzky, J. D. Eveland et al, *The Process of Technological Innovation: Reviewing the Literature*, National Science Foundation, Division of Industrial Science and Technological Innovation, Productivity Improvement Research Section, May 1983.

75 Louis G. Tornatzky, Mitchell Fleischer et al, *The Processes of Technological Innovation*, Lexington, Massachusetts: Lexington Books, ISBN 0-669-20348-3, 1990.

76 The information-related "technology transfer" literature covers the gamut from archiving to database systems for disseminating information, related to the field of library and information science. For example, laboratory scientists write scientific and technical papers, but they are not responsible for publishing journals or maintaining databases. Researchers at Syracuse University's School of Information Studies recently studied how information on federal technology is transferred, surveying the distribution media and information-seeking behaviors. The findings are reported in: Rolf T. Wigand, Slawomir J. Marcinkowski and Igor Plonisch, "Transferring Technology on the Information Highway," *Technology Commercialization and Economic Growth, Technology Transfer Society Proceedings, 20th Annual Meeting, July 16-19, 1995, Washington, D.C.*: 267-276, 1995.

77 See, for example, C. Bruce Tarter, "National Laboratory Partnerships: What Works and What Doesn't," *AAAS Science and Technology Policy Yearbook 1998*, Albert H. Teich et al, editors, American Association for the Advancement of Science, ISBN 0-87168-611-2, p. 265-278, 1997; Wil Lepkowski, "R&D Policy: Cooperation is the Current Byword," *AAAS Science and Technology Policy Yearbook 1998*, Albert H. Teich et al, editors, American Association for the Advancement of Science, ISBN 0-87168-611-2, p. 223-236, 1997.

78 Alan S. Gutterman and Jacob N. Erlich, *Technology Development and Transfer: The Transactional and Legal Environment*, Westport, Connecticut: Quorum Books, ISBN 1-56720-021-4, 1997.

79 Fred E. Grisson, Jr. and Richard L. Chapman, *Mining the Nation's Brain Trust: How to Put Federally-Funded Research to Work for You*, Reading, Massachusetts: Addison-Wesley Publishing Company, Inc., ISBN 0-201-55015-6, 1992. See also Albert N. Link and Gregory Tassey, editors, *Cooperative Research and Development: The Industry-University-Government Relationship*, Norwell, Massachusetts: Kluwer Academic Publishers, 1989. Also, the FLC and National Technology Transfer Center have produced technology transfer handbooks to accompany their training courses in this area.

80 See, for example, Sally Rood and Diane Palmintera, *Tapping Federal Laboratories and Universities to Improve Local Economies: The Role of the*

Mayor and City Government, Washington, D.C.: U.S. Conference of Mayors, October 1988.

81 Annalee Saxenian, *Regional Advantage: Culture and Competition in Silicon Valley and Route 128*, Cambridge, Massachusetts: Harvard University Press, 1994; see also, Alistair Brett, David V. Gibson and Raymond W. Smilor, editors, *University Spinoff Companies: Economic Development, Faculty Entrepreneurs, and Technology Transfer*, Lanham, Maryland: Rowman & Littlefield Publishers, Inc., 1991.

82 See David Birch, *Job Creation in America: How Our Smallest Companies Put the Most People to Work*, New York: Free Press, 1987. On a related note, corporations are involved in technology transfer from the perspective of licensing in technologies or technology transfer within a corporation or a consortia of companies.

83 Department of Commerce, Technology Administration, Office of Technology Policy, *Listening to Industry: Business Views on Technology Policy*, Draft for Public Comment, June 1994. CRS summarized ten science policy studies, but this summary was not as specific about technology transfer. See Congressional Research Service, *Analysis of 10 Selected Science and Technology Policy Studies*, CRS Report to Congress, prepared by William C. Boesman, Science Policy Research Division, 97836 SPR, Updated October 24, 1997.

84 The Council's advisory committee was headed by its Distinguished Fellow, Erich Bloch, who also spearheaded subcommittee studies for the CPC.

85 Daniel F. Burton, *Industry as a Customer of the Federal Laboratories*, Washington, D.C.: Council on Competitiveness, 1992.

86 Bruce A. McKenney, *National Benefits from National Labs: Meeting Tomorrow's National Technology Needs*, Final Report of the CSIS National Benefits from National Laboratories Project, Washington, D.C.: Center for Strategic and International Studies, ISBN 0-89206-224-X, 1993.

87 Atlantic Council, *Transfer of Technology to Industry from U.S. Department of Energy Defense Programs Laboratories*, 1992.

88 Congressional Research Service, *Cooperative R&D: Federal Efforts to Promote Industrial Competitiveness*, CRS Issue Brief prepared by Wendy H. Schacht, Science Policy Research Division, IB89056, Updated December 5, 1997. See also, Congressional Research Service, *Cooperative Research and Development Agreements (CRADAs)*, CRS Report for Congress prepared by Wendy H. Schacht, Science, Technology, and Medicine Division, 95-150 SPR, Updated January 12, 1998.

89 Congressional Research Service, *Technology Transfer: Use of Federally Funded Research and Development*, CRS Issue Brief prepared by Wendy H. Schacht, Science Policy Research Division, IB85031, Updated December 5, 1997.

90 OTA undertook this study at the request of the House of Representatives Science Policy Task Force, so its report also appeared as committee print.

91 Office of Technology Assessment, *Research Funding as an Investment: Can We Measure the Returns? A Technical Memorandum,* OTA-TM-SET-36, Washington, D.C.: U.S. Congress, April 1986.

92 Ibid.

93 Office of Science and Technology Policy, National Science and Technology Council, Committee on Fundamental Science, Subcommittee on Research, *Assessing Fundamental Science,* July 1996.

94 Thomas P. Sheahen et al, "Evaluation of Technology Transfer by Peer Review," *Journal of Technology Transfer* 19 (3/4 December): 100-109, 1994.

95 Robert Solow conducted the original classic work. See, for example: Robert M. Solow, "Technical Change and the Aggregated Production Function," *Review of Economics and Statistics* 39: 312-320, 1957.

96 Edwin Mansfield, *Industrial Research and Technological Innovation: An Econometric Analysis,* Cowles Foundation for Research in Economics, Yale University, New York: W. W. Norton Books, 1968; Zvi Griliches, "Productivity, R&D, and Basic Research at the Firm Level in the 1970's," *American Economic Review* (March): 1986; Jeffrey Bernstein and M. Ishaq Nadiri, "Interindustry Spillovers, Rates of Return, and Production in High-Tech Industries," *American Economic Review Papers and Proceedings* 78: 429-434, 1988.

97 Edwin Mansfield, "Academic Research and Industrial Innovation," *Research Policy* 20 (February): 1-12, 1991.

98 See Zvi Griliches, "Research Expenditures, Edcation, and the Aggregate Agricultural Production Function," *American Economic Review* (December): 1964.

99 Edwin Mansfield, Anthony Romeo, M. Schwartz, D. Teece, S. Wagner and P. Brach, *Technology Transfer, Productivity and Economic Policy,* New York: W. W. Norton Books, 1982.

100 For example, Los Alamos National Laboratory in New Mexico comissioned a study of the laboratory's impact on the surrounding geographic region. New Mexico State University, the University of New Mexico and the DOE Regional Operations office conducted this study which found that, in 1995, the laboratory's funding of $1.2 billion accounted for nearly five percent of the state's economic activity, or $4.1 billion, and almost one-third of the economic activity in the three surrounding counties, or $3.4 billion of the region's estimated economic activity of $11.35 billion. See Los Alamos National Laboratory, *New Mexico Regional Impact Report,* May 1997.

101 Zvi Griliches, editor, *R&D, Patents, and Productivity,* Chicago, Illinois: University of Chicago Press, 1984.

102 The steps are: Determine gains to producers and consumers, determine the amount attributable to an innovation, subtract out losses, estimate labor market impacts and externalities, discount future benefits in order to determine net present value, and quantify the rate of return.

103 The National Science Foundation supported Mansfield's early work in developing this approach through seventeen case studies reported in two volumes. It was eventually summarized in: Edwin Mansfield et al, "Social and Private Rates of Return from Indusrial Innovation," *Quarterly Journal of Economics* (May): 221-240, 1977. Two other NSF-supported studies in this area are: Foster Associates, Inc., *A Survey of Net Rates of Return on Innovation*, Three Volumes, National Science Foundation, May 1978; and, Robert R. Nathan Associates, Inc., *Net Rates of Return on Innovation*, Three Volumes, National Science Foundation, October 1978.

104 For example, a study by CHI Research, Inc. examined patents in the 1993-94 and 1987-88 time frames, and found a tripling of the industry linkages to government R&D (as opposed to industrial R&D) from the early time frame to the later one; more than seventy percent of the key industry patent citations came from public science performed at universities, government laboratories and other public agencies. In general, this shows that technology is being transferred effectively from government laboratories or government-funded institutions. See Francis Narin, et al, "The Increasing Linkage Between U.S. Technology and Public Science," *AAAS Science and Technology Policy Yearbook 1998*, Albert H. Teich et al, editors, American Association for the Advancement of Science, ISBN 0-87168-611-2, p. 101-121, 1997. Bibliometric studies of R&D outcomes date back to the 1960s, when the Defense Department sponsored the Institute for Defense Analyses and RAND to retrospectively measure the increase in cost-effectiveness of defense systems assignable to DOD-funded research and technology. The study, called Project Hindsight (1969), found that technological advances were not based upon basic science. About the same time, the National Science Foundation sponsored the "Traces" study (*Technology in Retrospect and Critical Events in Science*) by the Illinois Institute of Technology Research Institute (1969) which countered the other study's claim. NSF still sponsors traces-type studies through SRI International.

105 Council of Economic Advisors, Supporting Research and Development to Promote Economic Growth: The Federal Government's Role, October 1995.

106 See, for example, N. Terleckyj, Effects of R&D on the Productivity Growth of Industries: An Exploratory Study, Washington, D.C.: National Planning Association, 1974.

107 M. Ishaq Nadiri, Innovations and Technological Spillovers, National Bureau of Economic Research (NBER) Working Paper Series, Cambridge, Massachusetts: NBER, Working Paper no. 4423, August 1993.

108 Edwin Mansfield, "How Economists See R&D," Harvard Business Review 59 (6/ November-December): 98-106, 1981.

109 Edwin Mansfield, "Social Returns from R&D: Findings, Methods and Limitations," Research/ Technology Management (November-December), 1991.

110 Richard L. Chapman, "Alternative Methods to Evaluate Technology Transfer," Technology Commercialization and Economic Growth: Technology Transfer Society 20th Annual Meeting Proceedings, July 16-19, 1995, Washington, D.C.: 1-9, 1995; see also, Richard H. White with An-Jen Tai, et al, The Economics of Commercial-Military Integration and Dual-Use Technology Investments, Alexandria, Virginia: Institute for Defense Analyses, IDA Paper P-2995, June 1995, reprinted 1997.

111 The commercializing firms may receive funds to perform R&D themselves rather than transfer it from a laboratory. See Irwin Feller and Gary Anderson, "A Benefit-Cost Approach to the Evaluation of State Technology Development Programs," Economic Development Quarterly 8 (2/May): 127-140, 1994.

112 Department of Commerce, National Institute of Standards and Technology, Setting Priorities and Measuring Results at the National Institute of Standards and Technology, January 1994.

113 National Institute of Standards and Technology, NIST Industrial Impacts: A Sampling of Successful Partnerships, NIST Special Publication 872, First printing September 1994, revised February 1996.

114 Tassey, Gregory, Rates of Return from Investments in Technology Infrastructure, National Institute of Standards and Technology, Program Office, 96-3 Planning Report, June 1996.

115 Gregory Tassey, Technology Infrastructure and Competitive Position, Norwell, Massachusetts: Kluwer Academic Publishers, 1992.

116 Ibid, 1996.

117 Gregory Tassey, The Economics of R&D Policy, Westport Connecticut: Quorum Books, ISBN 1-56720-093-1, 1997.

118 This series of NIST planning reports includes: "Economic Impacts of NIST-Supported Standards for the U.S. Optical Fiber Industry: 1981-Present" (1991); "Economic Impact on the U.S. Semiconductor Industry of NIST Research in Electromigration" (1991); "Economic Impact of NIST Research on Electromagnetic Interference" (1991); "An Evaluation of the Economic Impacts Associated with the NIST Power and Energy Calibration Services" (1995); "An Economic Assessment of the Spectral Irradiance Standard" (1995); and "Economic Evaluation of Radiopharmaceutical Research at NIST" (1997).

119 Albert N. Link, Evaluating Public Sector Research and Development, Westport, Connecticut: Praeger Books, ISBN 0-275-95368-8, 1996.

120 Often the case, he points out, because "one of the justifications for public sector

involvement is that the market has failed to provide sufficient quantities of such goods and services."

121 NIST supported Link to develop guidelines for the NIST program and project managers to familiarize them "with the motivation for and mechanics of an economic impact assessment." The guidelines provide step-by-step explanations for conducting and interpreting economic impact assessments on either completed or ongoing research projects. They discuss both the internal rate of return and the benefit-cost econometric models, the latter being more easily understood by both policymakers and public sector R&D managers, according to Link. They also explain the difference between internal rate of return and return on investment, and provide equations to compute net present values and other values. See, Albert N. Link, Economic Impact Assessments: Guidelines for Conducting and Interpreting Assessment Studies, National Institute of Standards and Technology, Program Office, Planning Report 96-1, May 1996.

122 NASA sponsors regional technology transfer centers, originally called "industrial applications centers," in each of the regions delineated by the Federal Laboratory Consortium.

123 Fiscal year 1977 and 1978 House of Representatives budget hearings and NASA Authorization Report.

124 Mathematica, Inc., Mathtech Division, Quantifying the Benefits to the National Economy from Secondary Applications of NASA Technology, Washington, D.C.: National Aeronautics and Space Administration, NASA Contract Report CR-2673/CR-2674, June 1975, revised March 1976; see also, Robert J. Anderson et al, A Cost-Benefit Analysis of Selected Technology Utilization Office Programs, Princeton, New Jersey: Mathtech, 1977.

125 The four areas were: cryogenic multi-layer insulation materials, integrated circuits, gas turbines in electric power generation, and computer programs for structural analysis (called NASTRAN).

126 Johnston, F. Douglas, with Martin Kokus, Jana Henthorn and Stephen Quist, *NASA Technology Utilization Program: A Cost-Benefit Evaluation*, Prepared for Office of Technology Utilization, National Aeronautics and Space Administration Denver, Colorado: Denver Research Institute, Contract NASW-3021, December 1979.

127 F. Douglas Johnston and Martin Kokus, *NASA Technology Utilization Program: A Summary of Cost-Benefit Studies*, Prepared for Office of Technology Utilization, National Aeronautics and Space Administration, Denver, Colorado: Denver Research Institute, Industrial Economic Division, NASA Contract NASW-3021, December 1977.

128 Midwest Research Institute, *Economic Impact and Technological Progress of NASA Research and Development Expenditures*, Three Volumes, Kansas City, Missouri: Midwest Research Institute, NASA Contract Report NASA-CR-

195946, September 1988. See also, Midwest Research Institute, *Economic Impact of Stimulated Technological Activity*, Three Volumes, Kansas City, Missouri: Midwest Research Institute, October 1971.

129 Michael K. Evans, *The Economic Impact of NASA R&D Spending*, Bala Cynwyd, Pennsylvania: Chase Econometrics Associates, Inc., April 1976.

130 Richard Chapman, Loretta C. Lohman and Marilyn J. Chapman, *An Exploration of Benefits from NASA Spinoff*, Littleton, Colorado: Chapman Research Group, Inc., Contract 88-01 with NERAC, Inc., June 1989.

131 Loretta C. Lohman and Richard L. Chapman, *"Lessons Learned" About the Collection of Spinoff Benefits Data*, Littleton, Colorado: Chapman Research Group, Inc., NERAC Contract #87-01, March 1989.

132 Richard Chapman, "Measuring Technology Transfer Success: Overcoming the 'If You Can't Count it, It Doesn't Count' Syndrome," *Technology Transfer Society 18th Annual Meeting Proceedings, June 26-29, 1993, Ann Arbor, Michigan*: 13 - 19, 1993.

133 Ibid.

134 Richard L. Chapman, "Alternative Methods to Evaluate Technology Transfer," *Technology Commercialization and Economic Growth: Technology Transfer Society 20th Annual Meeting Proceedings, July 16-19, 1995, Washington, D.C.*: 1-9, 1995.

135 Richard L. Chapman, "Case Studies in the Tracking and Measuring of Technology Transfer," *Technology Transfer Partnerships: Technology Transfer Society 19th Annual Meeting Proceedings, June 22 - 24, 1994, Huntsville, Alabama*, Kenneth E. Harwell, Kathy Wagner and Carl Ziemke, editors: 164 - 171, 1994.

136 University of Tennessee Space Institute and the Tennessee Valley Aerospace Region, *Technology Transfer Research Project: Identification and Analysis of the Factors Present in Successful Technology Transfer Cases*, Prepared by Brett Pichon and Bobbie Woodard, Sponsored by the Tennessee Valley Authority, June 17, 1993.

137 This study was originally produced as Bush's doctoral dissertation at Pennsylvania State University; Lance B. Bush, *An Analysis of Technology Transfer at NASA*, NASA Technical Memorandum 110270, Hampton, Virginia: Langley Research Center, July 1996.

138 The flow back to laboratories of technical advantage from technology transfer activities.

139 Richard L. Chapman, "An Exploration of the 'Spinback' Phenomenon," *Journal of Technology Transfer* 19/3-4 (December): 78-86, 1994.

140 Randy Barrett, "Will Metrics Really Measure Up?" *Technology Transfer Business* (Spring): 34-36, 1995. See also "Marshall Tech Transfer Generates Thousands of Jobs," *The FLC Newslink* (March): 2, 1995.

141 "NASA's Marshall Center Generates 5,300 Jobs," *Spotlight on Technology*, NASA Southeast Regional Technology Transfer Center & Southeast Regional Federal Laboratory Consortium (January/February): 5, 1995.

142 For example, since 1993, Marshall's impact on the state of Tennessee was: 1,547 jobs created or saved and 63 new products, with $171 million being the value of the jobs and products. See "NASA, Tennessee Agree to Renew Technology-Transfer Agreement," *McGraw-Hill's Federal Technology Report* (August 29): 5-6, 1996.

143 Bush, p. 52, 53.

144 Harry Craft, W. Sheehan and A. Johnson, "NASA's Southeastern Regional Initiative in Technology Transfer and Commercialization," *46th International Astronautical Congress, October 2-6, 1995, Oslo, Norway*, American Institute of Aeronautics and Astronautics, Inc., IAA-95-IAA.1.2.08, 1995.

145 Marshall Space Flight Center, *1995 Research & Technology Report*, Introduction: 2, 1995.

146 [untitled], *NASA Tech Briefs* (July): 23, 1996.

147 Department of Energy, *Our Commitment to Change: A Year of Innovation in Technology Partnerships*, September 1994.

148 About $1.65 billion in fiscal year 1995.

149 Officially, this office was the Office of the Deputy Under Secretary for Technology Partnerships and Economic Competitiveness, known as the Technology Partnerships Office, and previously called the Technology Utilization Office.

150 Department of Energy, *The Transfer and Commercial Impact of the U.S. Department of Energy's Award-Winning Technologies*, Prepared for Office of the Deputy Under Secretary for Technology Partnerships, U.S. Department of Energy, Prepared by Oak Ridge Institute for Science and Education, Training and Management Systems Division, February 1995.

151 For 36 years, *R&D Magazine* has conducted its annual R&D 100 Awards program. The awards are informally known as the "Nobel Prizes of Applied Research." DOE usually receives a large portion of these awards each year, in comparison to other federal departments and agencies.

152 Department of Energy, *Success Stories: The Energy Mission in the Market Place*, 1995.

153 General Accounting Office, *Energy R&D: Observations on DOE's Success Stories Report*, Testimony before the Subcommittee on Energy and Environment, Committee on Science, House of Representatives, GAO/T-RCED-96-133, April 17, 1996. See, also, General Accounting Office, *DOE's Success Stories Report*, GAO/RCED-120R, April 15, 1996.

154 Ibid.

155 House of Representatives, Committee on Science, Letter to the Honorable

Charles Bowsher, Comptroller General of the United States from Ranking Democratic Member George E. Brown, Jr., April, 17, 1996.

156 Moira M. Shea, "Technology Partnerships: Measuring Performance, The Integrated Technology Transfer System," *Technology Commercialization and Economic Growth: Technology Transfer Society 20th Annual Meeting Proceedings, July 16-19, 1995, Washington, D.C.*: 35-39, 1995.

157 Richard Chapman and Dana Moran, "Measuring the Results of Partnerships for Technology Transfer: Lessons Learned at the National Renewable Energy Laboratory," *Technology Transfer Models for Growth and Revitalization: Technology Transfer Society Proceedings, 21st Annual Meeting, July 21-23, 1996, Cleveland, Ohio*, William Grimberg, Sally Kickel and Lydia Skapura, editors: 145-154, 1996.

158 As noted, technical assistance is a technology transfer mechanism.

159 Department of Energy, *Setting Priorities and Measuring Results*, Oak Ridge Centers for Manufacturing Technology, 1995.

160 David Kramer, "Gauging Tech Transfers is Tough, DOE's Oak Ridge Operator Finds," *McGraw-Hill's Federal Technology Report* (June 23): 11-12, 1994.

161 Philip Shapira and Jan Youtie, *Assessing GMEA's Economic Impacts: Towards a Benefit-Cost Methodology*, GMEA Evaluation Working Paper E9502, Atlanta, Georgia: Georgia Tech Economic Development Institute, 1995.

162 This requires ongoing data collection, but it helps identify areas for program improvement; the non-assisted firms also receive benchmarking feedback in return for their data, which generates additional positive changes. See Kristin Dziczek, Daniel Luria and Edith Wiarda, "Assessing the Impact of a Manufacturing Extension Center," *Technology Transfer Metrics Summit Proceedings*, Sally A. Rood, editor, Chicago, Illinois, Technology Transfer Society: 186-198, June 1997.

163 Nexus Associates, Inc., *Evaluation of the New York Manufacturing Extension Partnership*, Final Report, Prepared for the New York State Science and Technology Foundation/ Empire State Development, Gen#95037, March 18, 1996.

164 A literature review of technology transfer to small firms said there is a "near total lack of information" on the subject of small firms and federal laboratory technology transfer, and few studies of the role of intermediaries in this process. (Mt. Auburn Associates, *Technology Transfer to Small Manufacturers: A Literature Review*, Final Report, Submitted to U.S. Small Business Administration, Submitted by Mt. Auburn Associates, Inc. with Regional Technology Strategies, Inc. Somerville, Massachusetts, August 1995.) The NASA-funded regional technology transfer centers (RTTCs), who facilitate technology transfer between government laboratories and industry, are examples of publicly-supported intermediaries. Nan Muir analyzed success

criteria for public intermediaries by surveying RTTCs, laboratories, and industry to determine the validity of quantitative metrics to measure the outcomes of their activities. (See Nan Muir, "Measuring Technology Transfer Success: A Study of Intermediary Agency Evaluation," *Technology Commercialization and Economic Growth: Technology Transfer Society 20th Annual Meeting Proceedings, July 16-19, 1995, Washington, D.C.:* 17-26, 1995.) She found a range of answers suggesting potential future work in this area. Other than this, there are few evaluations of public or private intermediaries for laboratories and universities. Private intermediaries working with universities (and occasionally with government laboratories) are more commonly known as "technology brokers." They market and license technologies, and sometimes take equity positions in new firms. This field is comprised of several large technology brokering firms of several hundred persons each, and hundreds of small brokerage outfits usually comprised of one or two partners who specialize in specific technology areas. There are few studies of private intermediaries.

165 Department of Defense, Director of Defense Research and Engineering, *Survey of Laboratories and Implementation of the Federal Defense Laboratory Diversification Program*, February 1994.

166 John Lesko and Michael Irish, *Technology Exchange: A Guide to Successful Cooperative R&D Partnerships*, Battelle and Economic Strategy Institute, 1995. The second edition appeared as follows: John Lesko, Phillip Nicolai and Michael Steve, *Technology Exchange in the Information Age: A Guide to Successful Cooperative R&D Partnerships*, Columbus, Ohio: Battelle Press, 1995.

167 House of Representatives, Committee on Small Business, Subcommittee on Regulation, Business Opportunities and Energy, *Technology Transfer Obstacles in Federal Laboratories: Key Agencies Respond to Subcommittee Survey*, Washington, D.C.: U.S. Government Printing Office, Committee Print 101-3, March 1990.

168 Ibid.

169 General Accounting Office, *Federal Agencies' Actions to Implement Section 11 of the Stevenson-Wydler Technology Innovation Act of 1980*, GAO/RCED-84-60, August 24, 1984.

170 General Accounting Office, *Technology Transfer: Constraints Perceived by Federal Laboratory and Agency Officials*, Briefing Report to the Chairman, Committee on Science, Space and Technology, House of Representatives, GAO/RCED-88-116BR, March 1988.

171 Some of the controversy over this questionnaire centered around its length and level of detail. Some were upset that it required onerous record-keeping requirements by technology transfer personnel, such as documenting every

phone call.

172 General Accounting Office, *Technology Transfer: Implementation Status of the Federal Technology Transfer Act of 1986,* Report to Congressional Requesters, GAO/RCED-89-154, May 1989.

173 General Accounting Office, *Implementation Status of the Federal Technology Transfer Act of 1986*, Statement of John M. Ols, Jr., Director, Resources, Community, and Economic Development Division, Before the Subcommittee on Science, Research, and Technology, Committee on Science, Space, and Technology, House of Representatives, GAO/T-RCED-89-47, June 1, 1989; General Accounting Office, *Implementation of the Technology Transfer Act: A Preliminary Assessment*, Statement of Carl E. Wisler, Director of Planning and Reporting, Program Evaluation and Methodology Division, Before the Subcommittee on Science, Research, and Technology, Committee on Science, Space, and Technology, House of Representatives, GAO/T-PEMD-90-4, May 3, 1990.

174 General Accounting Office, *Diffusing Innovations: Implementing the Technology Transfer Act of 1986*, Report to the Chairman, Committee on Science, Space, and Technology, House of Representatives, GAO/PEMD-91-23, May 1991. Also, General Accounting Office, *Diffusing Innovations: Implementing the Technology Transfer Act of 1986*, Statement of Kwai-Cheung Chan, Director of Program Evaluation in Physical Systems Area, Program Evaluation and Methodology Division, Before the Subcommittee on Technology and Competitiveness, Committee on Science, Space, and Technology, House of Representatives, GAO/T-PEMD-91-5, May 30, 1991.

175 The implication was that GAO was not sufficiently experienced in the technology transfer area, at that point, to be able to evaluate it. See Richard Chapman, "Alternative Methods to Evaluate Technology Transfer," in *Technology Commercialization and Economic Growth: Technology Transfer Society 20th Annual Meeting Proceedings, July 16-19, 1995,* Washington, DC: 1-9.

176 General Accounting Office, *Technology Transfer: Federal Agencies' Patent Licensing Activities*. Report to Congressional Requesters, GAO/RCED-91-80, April 1991.

177 General Accounting Office, *Copyright Law Constraints on the Transfer of Certain Federal Computer Software With Commercial Applications,* Statement of John M. Ols, Jr., Director in the Resources, Community, and Economic Development Division, Before the Committee on Commerce, Science and Transportation, United States Senate, GAO/T-RCED-91-91, September 13, 1991.

178 General Accounting Office, *Technology Transfer: Barriers Limit Royalty Sharing's Effectiveness,* Report to Congressional Committees, GAO/RCED-93-

6, December 1992.

179 General Accounting Office, *Technology Transfer: Improving Incentives for Technology Transfer at Federal Laboratories*, Testimony before the Subcommittee on Science, Technology and Space, Committee on Commerce, Science and Transportation, United States Senate, GAO/T-RCED-94-42, October 26, 1993.

180 General Accounting Office, *Technology Transfer: Implementation of CRADAs at NIST, Army, and DOE*, Testimony before the Subcommittee on Energy, Committee on Science and Technology, House of Representatives, GAO/T-RCED-93-53, June 10, 1993.

181 DOE was granted CRADA authority by the 1989 act rather than the 1986 act.

182 General Accounting Office, *Technology Transfer: Improving the Use of Cooperative R&D Agreements at DOE's Contractor-Operated Laboratories*, Report to Congressional Requestors, GAO/RCED-94-91, April 1994.

183 General Accounting Office, *Technology Transfers: Benefits of Cooperative R&D Agreements*, Report to the Vice Chairman, Joint Economic Committee, U.S. Congress, GAO/RCED-95-52, December 1994.

184 General Accounting Office, *Measuring Performance: Strengths and Limitations of Research Indicators*, Report to Congressional Requesters, GAO/RCED-97-91, March 1997.

185 Department of Commerce, *The Federal Technology Transfer Act of 1986: The First Two Years,* Report to the President and the Congress from the Secretary of Commerce, July 1989.

186 Department of Commerce, *Technology Transfer Under the Stevenson-Wydler Technology Innovation Act: The Second Biennial Report*, Report to the President and the Congress from the Secretary of Commerce, January 1993.

187 Carried out under authority of the 1958 Space Act.

188 Department of Commerce, Office of Technology Policy, *Effective Partnering: A Report to Congress on Federal Technology Partnerships*, Richard J. Brody, Project Director, April 1996.

189 These included: Ballistic Missile Defense Organization, Departments of the Air Force, Agriculture, Army, Commerce, Energy, Interior, Navy, Treasury and the Environmental Protection Agency, NASA, National Institutes of Health, National Science Foundation, Office of the Secretary of Defense, and National Technology Transfer Center.

190 Interagency Committee on Federal Technology Transfer, Working Group on Technology Transfer Measurement and Evaluation, *Collective Reporting and Common Measures: Draft for Comment*, Prepared by the Oak Ridge Institute for Science and Education (ORISE) Training and Management Systems Division for the U.S. Department of Energy's Technology Utilization Office, November 1994.

191 The U.S. Department of Agriculture had performed quite a bit of its own technology transfer evaluation through various means, including the use of independent evaluators such as Richard Chapman who conducted evaluation research with the Department of Agriculture in the late 1980s and early 1990s.

192 Late in 1997, the committee re-convened, but it is not clear whether it will continue.

193 Federal Laboratory Consortium, *Technology Transfer in a Time of Transition: A Guide to Defense Conversion*, 1994.

194 Tina McKinley, *FLC Chair, Lessons Learned in Technology Transfer: 20 Years of Federal Laboratory Consortium for Technology Transfer (FLC) Experience*, Prepared for the Committee on Science, Subcommittee on Technology and Subcommittee on Basic Research, U.S. House of Representatives, June 27, 1995.

195 Chapman Research Group, Inc., *Managing the Successful Transfer of Technology from Federal Facilities: A Survey of Selected Laboratories and Facilities in the Mid-Continent Region of the Federal Laboratory Consortium*, Federal Laboratory Consortium, 1997.

196 Barry Bozeman, "Editor's Introduction: Evaluating Technology Transfer and Diffusion," *Evaluation and Program Planning* 11: 63, 1988.

197 Barry Bozeman and Maureen Fellows, "Technology Transfer at the U.S. National Laboratories: A Framework for Evaluation," *Evaluation and Program Planning* 11: 65-75, 1988.

198 At the time, Bozeman was at Syracuse University, also located in New York.

199 Michael Crow, "Technology and Knowledge Transfer in Energy R&D Laboratories: An Analysis of Effectiveness," *Evaluation and Program Planning* 11:76, 1988.

200 Dianne Rahm, Barry Bozeman, and Michael Crow, "Domestic Technology Transfer and Competitiveness: An Empirical Assessment of Roles of University and Governmental R&D Laboratories," *Public Administration Review* (November/December): 969-978, 1988.

201 Barry Bozeman and Michael M. Crow, "The Environments of U.S. R&D Laboratories: Political and Market Influences," *Policy Sciences* 23: 25-56, 1990.

202 Barry Bozeman and Michael M. Crow, "Red Tape and Technology Transfer in the U.S. Government Laboratories," *Journal of Technology Transfer* 16 (2/Spring): 29-37, 1991.

203 Barry Bozeman and Karen Coker, "Assessing the Effectiveness of Technology Transfer from U.S. Government R&D Laboratories: The Impact of Market Orientation," *Technovation* 12 (4/ May): 239-256, 1992.

204 Barry Bozeman, "Evaluating Government Technology Transfer: Early Impacts of the 'Cooperative Technology Paradigm'," *Policy Studies Journal* 22 (2/ Summer): 322-337, 1994.

205 Barry Bozeman, Maria Papadakis and Karen Coker, *Industry Perspectives on Commercial Interactions with Federal Laboratories: Does the Cooperative Technology Paradigm Really Work?* Report to the National Science Foundation, Research on Science and Technology Program, Atlanta: Georgia Tech, Contract no. 9220125, January 1995.

206 Barry Bozeman and Jane Massey, "Investing in Policy Evaluation: Some Guidelines for Skeptical Public Managers," *Public Administration Review* (May/June): 264-270, 1982.

207 Barry Bozeman, "What We Don't Know About Evaluating Technology Transfer: Some Puzzles Seeking Solutions," *Technology Transfer Metrics Summit Proceedings*, Sally A. Rood, editor, Chicago, Illinois: Technology Transfer Society, 46-53, June 1997.

208 Maria Papadakis, "Federal Laboratory Missions, Products, and Competitiveness," *Journal of Technology Transfer* (April): 54-66, 1995.

209 GAO reports were described earlier; NSF, for many years, has produced annual compilations of national scientific and engineering indicators, including federal R&D statistics.

210 J. David Roessner and Alden S. Bean, "Federal Technology Transfer: Industry Interactions With Federal Laboratories," *Journal of Technology Transfer* (Fall): 5-14, 1990. See also, J. David Roessner and Alden S. Bean, "How Industry Interacts with Federal Laboratories," *Research-Technology Management* 34 (4/July-August): 22-25, 1991.

211 J. David Roessner and Anne Wise, *Patterns of Industry Interaction with Federal Laboratories: Final Report*, Georgia Institute of Technology, School of Public Policy, Martin Marietta Energy Systems, Inc., Oak Ridge National Laboratory, and U.S. Department of Energy Contract #19X-SK495C, May 1993. See also, J. D. Roessner and A. S. Bean, "Industry Interaction with Federal Labs Pays Off," *Research Technology Management* 36 (5): 38-40, 1993; J. David Roessner and Alden S. Bean, "Patterns of Industry Interaction with Federal Laboratories," *Journal of Technology Transfer* (December): 59 - 77, 1994.

212 Eliezer Geisler, *Why Federal Laboratories Succeed or Fail at Technology Commercialization*, Report to the National Science Foundation, 1995.

213 Lawrence K. Anderson and Brian D. Gurney, *Benchmarking Best Practices in Technology Transfer: Final Report*, Colorado Institute for Technology Transfer and Implementation, Colorado Springs, Colorado, Sponsored by Colorado Advanced Technology Institute and U.S. Department of Commerce, December 1993.

214 A literature search identified the best practitioners.

215 The 1980 Bayh-Dole Act encourages university technology transfer.

216 *AUTM Licensing Survey: FY 1991 - FY 1995, Five-Year Survey Summary*, Association of University Technology Managers (AUTM), Inc., Daniel E.

Massing, editor and Chair, AUTM Survey, Statistics and Metrics Committee, 1996. AUTM changed the survey categories for the 1996 survey so the 1997 report on the 1996 survey was not cumulative. See *AUTM Licensing Survey: FY 1996 Survey Summary,* Daniel E. Massing, editor, Association of University Technology Managers, Inc., 1997.

217 *AUTM Public Benefits Survey Summary of Results,* Prepared for Association of University Technology Managers (AUTM), Inc., Cranbury, New Jersey: Diane C. Hoffman, Inc., April 1994.

218 Companies for which initiation was dependent upon the licensing of university technology.

219 Louis G. Tornatzky, Paul G. Waugaman and Joel S. Bauman, *Benchmarking University-Industry Technology Transfer in the South: 1995-1996 Data,* Research Triangle Park, North Carolina: Southern Technology Council, Southern Growth Policies Board, July 1997; Louis G. Tornatzky, Paul G. Waugaman and Lucinda Casson, *Benchmarking Best Practices for University-Industry Technology Transfer: Working with Start-Up Companies,* A Report of the Southern Technology Council, Southern Growth Policies Board, Research Triangle Park, North Carolina: Southern Technology Council, October 20, 1995; Louis G. Tornatzky and Joel S. Bauman, *Outlaws or Heroes? Issues of Faculty Rewards, Organizational Culture, and University-Industry Technology Transfer*, A Benchmarking Report of the Southern Technology Council, Southern Growth Policies Board, July 1997.

220 Lori D. Pressman et al, "Pre-Production Investment and Jobs Induced by Massachusetts Institute of Technology Exclusive Patent Licenses: A Preliminary Model to Measure the Economic Impact of University Licensing," *Journal of the Association of University Technology Managers* 7: 49-81, 1995.

221 BankBoston, Economics Department, *MIT: The Impact of Innovation,* 1997.

222 NSF contracted Denis Gray to coordinate multi-year IUCRC evaluations and develop an evaluation handbook; see National Science Foundation, *Evaluator's Handbook: NSF Industry-University Cooperative Research Centers Program,* Raleigh, NC: I/UCRC Evaluation Project, 1997. NSF contracted with Irwin Feller and David Roessner to evaluate the ERCs; see Irwin Feller and David Roessner, "What Does Industry Expect from University Partnerships?" *Issues in Science and Technology* (Fall): 80-84, 1995. The National Academy of Sciences' Government-Industry-University Roundtable developed case studies of partnerships as background for a recent workshop on their measurement; see Industrial Research Institute, Government-University-Industry Research Roundtable, and Council on Competitiveness, *Industry-University Research Collaborations: Report of a Workshop, November 28-30, 1995, Duke University*, Washington, D.C.: National Academy Press, 1996. Also, Albert Rubenstein and Eliezer Geisler have analyzed industry-university cooperation

for many years; see, for example, Albert H. Rubenstein and Eliezer Geisler, "The Use of Indicators and Measures of the R&D Process in Evaluating Science and Technology Programs," *Government Innovation Policy: Design, Evaluation, Implementation,* J. David Roessner, editor, St. Martin's Press: 185-204, 1989.

223 Michael Odza, "What the AUTM Licensing Survey Statistics Mean for Federal Labs," *Technology Transfer Metrics Summit Proceedings*, Sally A. Rood, editor, Chicago, Illinois: Technology Transfer Society, 231-235, June 1997.

224 Robert K. Carr, "Doing Technology Transfer in Federal Laboratories" (Part 1), *Journal of Technology Transfer* 17 (2/3, Spring/Summer): 8-23, 1992.

225 Robert K. Carr, "Menu of Best Practices in Technology Transfer" (Part 2), *Journal of Technology Transfer* 17 (2/3, Spring/Summer): 24-33, 1992.

226 Robert K. Carr, "Measurement and Evaluation of Federal Technology Transfer," *Technology Commercialization and Economic Growth: Technology Transfer Society 20th Annual Meeting Proceedings, July 16-19, 1995, Washington, D.C.:* 221-230, 1995.

227 Ann Markesen and Michael Oden, "National Laboratories as Business Incubators and Region Builders," *Journal of Technology Transfer* 21 (1-2/Spring/Summer): 93-108, 1996.

228 Bernard J. Schroer, Phillip A. Farrington, Sherri L. Messimer and J. Ronald Thornton, "Measuring Technology Transfer Performance: A Case Study," *Journal of Technology Transfer* 20 (2/September): 39-47, 1995.

229 About 25 percent of the total 800 U.S. incubators are associated with technology centers, according to the National business Incubation Association.

230 Candace Campbell and David N. Allen, "The Small Business Incubator Industry: Micro-Level Economic Development," *Economic Development Quarterly* 1: 178-191, 1987.

231 Louis G. Tornatzky, Yolanda Batts, Nancy E. McCrea, Marsha L. Shook and Louisa M. Quittman, *The Art and Craft of Technology Business Incubation: Best Practices, Strategies, and Tools from 50 Programs*, Southern Technology Council, National Business Incubation Association, and Institute for Local Government Administration and Rural Development, ISBN 0-927364, 1995.

232 Peter Bearse, *The Evaluation of Business Incubation Projects: Comprehensive Manual*, National Business Incubation Association, for the U.S. Economic Development Administration, ISBN 1-887183-19-1, December 31, 1993. See also, Nexus Associates, *Guide to Economic Development Program Evaluation*, Belmont, Massachusetts: Nexus Associates, Inc., 1996.

233 See, Louis Tornatzky, "Pre-Concept Paper: Development of Recommended Standards for the Evaluation and Benchmarking of Technology Programs," *Technology Transfer Metrics Summit Proceedings*, Sally A. Rood, editor, Chicago, Illinois: Technology Transfer Society, p. 296-302, June 1997.

234 Irwin Feller, "Technology Transfer: How Do We Know What Works, Summary Comments," *Technology Transfer Metrics Summit Proceedings*, Sally A. Rood, editor, Chicago, Illinois: Technology Transfer Society, 224-230, June 1997.

235 This process, its underlying logic, and the criticism are all similar to what the NIST Manufacturing Extension Partnership program has experienced in the past half-decade.

236 Marilyn Brown et al, "Evaluating Technology Innovation Programs: The Use of Comparison Groups to Identify Impacts," *Research Policy* 24 (4): 669-684, 1995.

237 Preliminary results indicate that program participation is related to productivity growth but not sales growth. See Ronald S. Jarmin, *Measuring the Impact of Manufacturing Extension*, Washington, D.C.: Center for Economic Studies, U.S. Bureau of the Census, August 1996, revised January 1997.

238 Lewis M. Branscomb, "From Technology Politics to Technology Policy," *Issues in Science and Technology* (Spring): 41-48, 1997.

239 The CPC "Subcouncil on Critical Technologies" and its annual reports was noted earlier.

240 The final report will be released Spring 1998. An interim report is: Harvard University, Center for Science and International Affairs, *Investing in Innovation: Toward a Consensus Strategy for Federal Technology Policy*, Project on Technology Policy Assessment Steering Committee: Lewis Branscomb, Richard Florida, David Hart, James Keller and Darin Boville, Sponsored by the Competitiveness Policy Council, April 24, 1997.

241 In 1989, through a FLC Southeast Region/ Martin Marietta Energy Systems, Inc. grant to Virginia Tech, the FLC sponsored development of a series of case studies about successful technology transfer. Alistair Brett, who was previously jointly affiliated with Virginia Tech (as Director of Technology Management and Transfer) and the state of Virginia's Center for Innovative Technology, compiled cases on twelve spin-off companies from eight federal laboratories. However, this series was intended for ultimate use in FLC-sponsored training on technology transfer rather than for measurement or evaluation purposes. See Alistair M. Brett, "Federal Laboratory Spin-Off Companies: Development of Case Studies for Training in Effective Domestic Technology Transfer," Virginia Polytechnic Institute and State University, August 9, 1989, unpublished.

3 Early Award Winners

This chapter describes the pre-legislation cases from 1985 and 1986. It introduces the cases using the "Level II Analysis" described in Chapter 1's section on research design. Using information from the 1994 FLC Winner's Document, all of the FLC 1985-86 awards are organized by federal department or agency and topic area. The topics are based on the interview topics defined in Chapter 1's section, "Core Elements of the Government Technology Transfer Process." The following topics are included in the introductory section: technology applications, role of the laboratory researchers and other personnel, intellectual property, technology transfer mechanisms, user groups, user benefits/economic impact/outcomes, government gains, and elapsed time. In order to avoid excessive duplication, topics are addressed primarily with illustrative examples rather than a comprehensive survey of all the cases.

Following the Level II analysis of the 1985-86 group, six selected cases are examined in greater detail.[1] Examples for the introduction to the Level II analysis are largely drawn from the cases not selected. After presentation of the six cases, the final section of the chapter groups the key data from the six cases according to the topics.

Introduction – Level II Analysis, Pre-legislation Awards

Departments/Agencies and Technology Applications

The eight 1985-86 awards were distributed between laboratories of departments and agencies as follows:

- Agriculture (one technology) – a technology to fight insect and weed contamination of water and vegetation.
- Air Force (one technology) – a purchasing system.
- Commerce (two technologies) – a modem-accessible electronic bulletin board used by the National Institute of Standards and Technology (NIST) to advertise technologies and a program to ensure quality in radiation therapy calibrations.

- Energy (two technologies) – a technology that allows tree root control by public works departments and a technology to detect air pollution and fire in buildings.
- NASA (one technology) – advanced materials for use in electronics, aerospace and other industries.
- Navy (one technology) – equipment to classify waterway sediment.

Roles of the Laboratory Researchers and Other Personnel

The eight 1985-86 awards involved 15 researchers, 14 male and one female. The role of these laboratory researchers varied. Some of the researchers developed computers programs. For example, a group of scientists at a military laboratory developed a purchasing product assessment system. Developing this system involved working with federal agencies issuing RFPs, industry users of the purchasing system, and the standards community. Also, commercialization of the system involved working with private software vendors. Another example of program technologies was a group of scientists who created a dial-in electronic information exchange. Creating this exchange involved assembling hardware and software, organizing information, developing a marketing campaign, and designing a support system. Two research teams developed new instruments: one measured river sediment, and the other measured building pollution while also detecting fires. The remaining 1985-86 technology developers worked with peers in outside institutions to evaluate and test the technologies. Several teams went further and sought potential manufacturers for their technologies.

Intellectual Property

From the Level II analysis, only two of the 1985-86 awards appeared to involve patent applications. But interviewing the researchers revealed that other technologies were patented and one of the apparently patented technologies was not patented.

Technology Transfer Mechanisms

The technology transfer mechanisms included licenses, CRADAs, marketing in trade publications, and other communication. As with the patent data, the data on the licenses and CRADAs was incomplete. Few of the awards as described highlighted CRADAs, so it is assumed that CRADAs were not signed as of 1993. The team that developed the purchasing system transferred the technology through hands-on training seminars and follow-up assistance to thousands of government and industry personnel. The team also marketed the system by writing trade journal articles to publicize its availability for commercialization. The electronic bulletin board on national security and computer issues was a transfer mechanism in itself and helped to move laboratory technologies and products into the market effectively. On this dial-in system:

> . . .The menu directs callers to both NIST products and information about security, conferences, new security and computer standards, product evaluation, new and established software, and hardware systems. Participants are able to exchange information on technical topics with each other, and discuss services, vendors, and opportunities for technology applications. Recent topic additions include data management activities and applications, validation services for test devices, and conformance to security standards.[2]

User Groups

There were many examples of outside users and partners among the 1985-86 awards. The electronic bulletin board provides an example. The 1994 FLC "Winner's Document" noted that more than 10,000 users had access to the system 24 hours a day, seven days a week via modem. Some portions of the bulletin board were available through the Internet at that time. "Laboratories, agencies, and private organizations outside NIST volunteered contributions to the bulletin board, and NIST has expanded the range of information available," according to the document.

User Benefits/Economic Impacts/Outcomes

The 1985-86 technologies had great potential benefits, according to the Winner's Document. For example, the write-up about the electronic bulletin board said, "You can save U.S. businesses millions of dollars a year with information about security systems, computers, new technologies, and sophisticated software." As another example, the purchasing system integrated components and services from a hundred vendors as demonstrated at a 1989 software exposition. Major Fortune 500 companies who adapted and marketed the system's projects commercially documented savings of 33 percent.

Government Gains

Some of the awards indicated both current and future gains for the government. Two of the awards had specific future applications that would benefit the government. The building leak detector technology developed at a DOE laboratory was to be used by NASA to certify the leak integrity of modules built for the proposed space station. The Navy-developed penetrometer technology supported the Navy's emerging initiative to compile oceanographic information on coastal areas.

The purchasing system project had already generated state-of-the-art national standards and integrated current ones. The system met DOD, ANSI (American National Standards Institute), and Open Systems Interconnections standards and was current with more than thirty other standards common to industry.

Elapsed Time

For three of the eight 1985-86 awards, the award was made because of the speed with which the technology was transferred. For example, the Winner's Document contained statements such as, "The researchers told the users what they needed to know in the fastest . . .way," and "The program was moved into general use quickly and painlessly . . ."

On the other hand, two of the awards involved long periods of time. For example, one 1986 award was first conceived in the 1970s. In another award, the researcher started working on the technology in the early 1980s, was nominated for the FLC award and listed in the case list in 1985, but was not given an award until 1992.

Selected Pre-legislation Cases

Table 3.1 displays basic data for the six awards selected for further research beyond the 1993 data highlighted above. The six cases are presented in the next section. They are:

- penetrometer for seabed classification/measurement,
- advanced thermoplastic polymer material,
- substance tracer technology,

Table 3.1 Selected Pre-legislation Case Characteristics

#	Year	Technology	Agency/Lab	Researcher(s)	Partner(s)
1	'85	Penetrometer for Seabed Classification and Measurement	Naval Oceanographic Office (Stennis Space Center, Mississippi)	Mr. Carey Ingram	Dr. Joseph Suhayda (Louisiana State University), Sippican Corp.
2	'86	Advanced Thermoplastic Polymer Material	NASA – Langley Research Ctr. (Hampton, Virginia)	Dr. Terry St. Clair	Hoescht-Celanese Corporation, M&T Chemical, Mr. Milton Evans (High Tech Services, Inc.)
3	'86	Substance Tracer Technology	DOE – Brookhaven National Laboratory (Long Island)	Dr. Russell Dietz	Natl. Assoc. of Home Builders and AIM, Inc., Perfect Sense, Inc., Electric Power Research Inst., Consolidated Edison, Tracer Labs, Vacuum Instr., John Booker
4	'86	Slow-Release, Alginate-Based Herbicide/ Pesticide	Agric. Research Serv. – Southern Regional Research Center (New Orleans, La.)	Mr. William Connick	Grace-Sierra, Dr. James Walter (Thermo Trilogy Corp.), Dr. Ramon Georgis (Biosys, Inc.), Mycogen Corp., EcoSciences Corp.
5	'86	Controlled-Release, Chemically-Imbedded Herbicide/ Pesticide Material	DOE – Pacific Northwest National Laboratory (Richland, Washington)	Dr. Peter Van Voris, Dr. Dominic Cataldo, Frederick Burton	Mr. Harry Barnes (Reemay, Inc.), Mr. Rodney Ruskin (Geoflow, Inc.), Mantaline Corp.
6	'85	Radiation Therapy Quality Assurance	DOC – Natl. Institute of Standards and Technology (G'burg, Md.)	Dr. Robert Loevinger	Dr. Geoffrey Ibbott (American Association of Physicists in Medicine)

- slow-release, alginate-based herbicide/pesticide,
- controlled-release, chemically-imbedded herbicide/pesticide, and
- quality assurance for radiation therapy dosage measurement.

Case 1 (1985) – Penetrometer for Seabed Classification/Measurement

Role of Laboratory Researchers and Other Personnel   In the early 1980s, Pearl River County, Mississippi, asked what was then called the Marine Geological Laboratory to help them survey the mineral resources in the lower East Pearl River. They were interested in determining whether the mineral deposits at the bottom of the river could help them offset the cost of dredging.

The field survey was performed by a team comprised of personnel from the Navy laboratory, the state of Mississippi's Department of Economic Development, and the Pearl River County Development Association. Mr. Carey Ingram, an oceanographer at the laboratory, served as the lead scientist for the team.

The survey inspired Mr. Ingram to develop a new penetrometer to measure the river bottom. Mr. Ingram actually invented two different penetrometers for the survey. He adapted one from existing equipment. For the other penetrometer, he served as consulting scientist during prototype development by Louisiana State University (LSU) in Baton Rouge, Louisiana. During the prototype development phase, he co-authored a laboratory report on the technology with LSU engineers.[3]

The Technology and Applications   Mr. Ingram adapted one of the penetrometers from a soil-testing device used by the Army Corps of Engineers, a "deep-sea towed-side scan sonar." The sonar was diver-operated so its depth capabilities depended upon the diver's depth limit. Mr. Ingram based the more sophisticated penetrometer on an "XBT" (Expendable Bathy Thermograph) manufactured by the Sippican Corporation, based in Massachusetts.

The penetrometer collects data from the bottom of a stream or river through a probe. It then analyzes the data, identifying the sediment below the mud line, and develops a depth profile of the river bed. The XBT-based penetrometer included a ballistically-shaped probe and a recorder for measuring water temperature. The probe fell through the water, hit bottom, and then collected information through the recorder. Mr. Ingram modified the XBT to include an accelerometer in place of a temperature measurement tool (or thermistor).

The penetrometer comprises: a wire for the probe wound on a spool (secured with a release pin), a launcher that releases the spool of wire from a canister holding the spool, a transducer recorder, a calibration unit with telephone jack terminals (to change the polarity of the signal going to the computer from analog to digital), and a Zenith Z-100 microcomputer/printer set-up. A miniature accelerometer inside the probe measures deceleration once the probe hits bottom.

The penetrometer probe is released from a ship into the channel. As the probe falls and hits bottom, the Zenith micro-computer classifies the sediment particles (e.g., soft mud, sand, course gravel) based upon calculations related to impact velocity, depth, strength, mass, shape and diameter. This data is combined with boat direction data to determine geological cross-sections in real-time.

The Laboratory Mr. Ingram is now part of the Special Support Division of the Naval Oceanographic Office. The Naval Oceanographic Office and a "detachment" of the Naval Research Laboratory (NRL) are co-located on the John C. Stennis Space Center site near Bay Saint Louis, Mississippi, on the Gulf of Mexico coast. The NRL is known as the Navy's "corporate laboratory" and is headquartered in the Washington, D.C. area. The Stennis Space Center is one of nine NASA field centers. The Stennis Center houses not only the NASA field center and Navy offices, but also 21 other federal agencies from the departments of Commerce, Defense, Interior, and others. The research activities conducted by these agencies range from exploration of space and the oceans to promoting environmental quality.

Each agency at Stennis has developed special technical capabilities and facilities. The Naval Oceanographic Office deploys twelve ships and three aircraft to conduct ocean surveys and other data collection for safe and accurate ocean navigation. The office's operational center provides the Navy's Regional Oceanography Centers with real-time ocean front and eddy information. Among other programs, the Naval office has a pilot program called "Adopt-a-Ship" which introduces young people to oceanographic survey practices. Along with information for ships and aircraft, the Naval office capabilities include a wide array of instrumentation and communications technologies for naval applications like remote sensing. This includes an oceanographic prediction system using a Cray supercomputer to support Navy initiatives. The Navy offices in Mississippi also house various libraries such as: an atmospheric master library containing all the standard models and databases used by the Navy, a library of over 20,000 technical reports on oceanography, and a library on geomagnetic data and analysis for the U.S. Department of Defense.

University Involvement   The Naval Oceanographic Office signed a contract with LSU's Civil Engineering Department for initial prototype design, construction, field testing, and evaluation of the penetrometer. As a result of a meeting with the university developer, Dr. J. N. Suhayda, Sippican Corporation was interested in modifying its commercially available XBT product for additional sales. Dr. Suhayda and the Sippican engineers discussed how the new penetrometer might be commercialized. Dr. Suhayda and Mr. Ingram worked with the county users to discuss potential future survey needs. They also considered other scientific requirements, as well as budgetary limitations. Subsequently, the university proved that the technology worked.

Funding, Financing   The university and corporate contract work on the penetrometer was performed for the Naval Oceanographic Office Special Projects Branch using Navy funds.

Intellectual Property   Two patents were intended to be based on this penetrometer technology: a Diver-Operated Sea Floor Penetrometer and an Expendable Bottom Penetrometer that is operated from a ship at speeds of up to 15 knots. However, Mr. Ingram said the Navy did not want to pursue his patent application.[4] The agency might not have wanted to spend the money required to apply for a patent. Also, if the patent application process does not look like it will yield a return on the investment, the organization may choose not to pursue this route.

The university also did not file for any rights, although intellectual property was discussed. It was not a conscious or deliberate decision; Dr. Suhayda says they just never got around to this.[5] Once the funding ended, the university felt there was no organization to authorize it taking a proprietary interest in the technology since there was a classified aspect to the technology.

It was not possible to contact the Sippican Corporation for this case; however, neither Mr. Ingram or Dr. Suhayda indicated that the company had sought to patent the technology.

Technology Transfer Mechanisms　　The survey work was performed as technical assistance to Pearl River County. After LSU produced the initial prototype, the Sippican Corporation produced six "holotypes" (or prototype clones) under contract to the Navy.

User Groups　　State and local governments often have responsibility for river channel maintenance. This maintenance requires river bottom surveys which are often conducted by drilling holes into the core, an expensive procedure for identifying existing minerals. The penetrometer offers an alternative.

Penetrometer technology *could* also contribute to Department of Defense requirements for collecting oceanographic data on coastal areas. Another user group is the oil industry, which needs this type of information for offshore oil exploration.

Barriers to Commercialization　　Mr. Ingram said that at the time of this case, a patent counted about as much as a publication toward professional advancement. At the time, Navy scientists received $50 to 100 per patent. This amount of money was not much incentive to do technology transfer work, but recently CRADAs have helped to make the process easier, he said.[6]

Other Factors　　The Pearl River is a few miles to the west of the Stennis site and separates Mississippi from Louisiana. The surrounding counties and municipalities are very involved with the Stennis Center in a variety of ways; this case is only one example. For example, local elementary and high school students track the endangered sea turtle through a pilot project sponsored by the NASA Teacher Resource Center. The project, called "Close Encounters of the Endangered Kind," uses information provided by the National Marine Fisheries Service at Stennis.

User Benefits/Economic Impact/Outcomes　　The penetrometer was used to map river bottom cross-sections and to identify existing minerals for Pearl River County. However, the penetrometer technology was never fully developed and transferred to the operational survey community or any other user group. It is possible that the six Sippican Corporation holotypes fulfilled the Navy's need for non-civilian (possibly classified) use at the time.

In the early 1990s, new Navy requirements sparked renewed interest in this area of technology. As a result, the Naval Oceanographic Office joined with the Navy's Civil Engineering Laboratory in Dayton, Ohio, to improve computer calculations for another type of penetrometer, the Acoustic Doppler Penetrometer, developed by Sonatech, Inc. These improvements made this penetrometer technology more cost-effective than the Expendable Bottom Penetrometer, which ended up on the shelf.

International Activity   None.

Government Gains   It is not certain how the Navy used the six instruments delivered by Sippican Corporation under a contract basis.

Elapsed Time   The request for technical assistance came in to Stennis from Pearl River County in 1982. The patent application for the Expendable Bottom Penetrometer was filed with the Navy Patent Counsel in July 1983, which was still considering it and the diver-operated version in early 1985. Mr. Ingram and Dr. Suhayda co-authored the laboratory report in December 1984. The university developed the prototype in 1984, and it was field tested during the summer of 1984. The Sippican Corporation produced the six prototype clones in the latter half of the 1980s. The Expendable Bottom Penetrometer technology has never been fully-developed or mass-produced. As a side note, Mr. Ingram was nominated for the FLC award in 1985. The case is listed as a 1985 case in the FLC Winners' document, but for some reason the award was not made until 1992.

Case 2 (1986) – Advanced Thermoplastic Polymer Material

Role of Laboratory Researchers and Other Personnel   Dr. Terry L. St. Clair is a chemist at the NASA Langley Research Center who creates new materials used in aerospace systems. Dr. St. Clair invented a new type of plastic material, a thermoplastic, and developed a process for producing it. Once Dr. St. Clair was sure the material had reliable properties and could be mass-produced, he began discussing it with manufacturers in order to transfer both the material and the production process. He scheduled a number of workshops, seminars, and lectures for a wide variety of non-aerospace organizations. In addition, he played a role in technology transfer conferences co-sponsored by NASA and the Aerospace Industries Association of America for both Fortune 500 and small companies. During the final development stages, he worked with potential producers and users to refine both the material and the manufacturing process for their purposes.

Technology and Applications   Composite materials are often used in place of metals to decrease weight and add strength. Composites materials are made of fibers (usually carbon fibers) impregnated with a resin or adhesive in a matrix-type format. Most resins limit the composites to applications involving temperatures not exceeding 350 degrees Fahrenheit. The advanced composite materials used by NASA must be able to withstand temperatures as high as 450 to 500 degrees Fahrenheit for up to 10,000 hours. To find a resin that can be used in higher-temperature environments involves long-term experimentation exploring the nature of a new resin as well as methods to process it. The material must also be volative-free during processing and have a high modulus. If all of these objectives are met, NASA makes its new formulations available to commercial sources and suppliers of advanced materials to convert into materials that can be mass-produced on a cost-effective basis.

Dr. St. Clair met all of these objectives by inventing a new polymer, generically called "polyimide sulfone." It combines the properties of two classes of polymers, polyimides and polysulfones, thus "polyimide sulfone" (or PISO2). One part of the compound, the polysulfones, is easy to process but very soluble. However, when combined with polyimides, the resulting polyimide sulfone resists solvents. The combination, therefore, enables polyimide sulfone be used in applications where solvents such as aviation fluids, adhesives, and other corrosive types of fluids are present. Polyimide sulfone can be used as a matrix-type resin to bind together the fibers reinforcing composite materials. In addition it can be dissolved in solvents and molded like a foam to form products. The resulting composite materials and other products are lightweight and can be used in many types of industrial applications. They are thermally stable when exposed to high temperatures (700 degrees Fahrenheit) for long periods of time. All of this makes PISO2 superior to current plastics used in engineering because the quality and longevity of the ultimate products are improved. It is relatively easy for manufacturers to process because it requires only moderate temperatures and pressures to produce, and therefore it can be produced at a relatively low cost. In addition, during processing, it does not release volatile chemicals so it has low-toxicity.

The Laboratory   The NASA Langley Research Center is one of nine NASA field centers that perform research to advance aeronautics and space flight.

Langley, which is located in Hampton, Virginia, employs over 5,000 people and has a budget of around $500 million. The center has unique facilities such as a space simulation complex, a pyrotechnic test facility, a scientific visualization system, a bolt tension monitor, over forty wind tunnels that test speeds as high as Mach 20, and a model-building wind tunnel.

University Involvement HTS' first SBIR contract was a joint SBIR with Dr. Bruce Norman, a chemical engineering professor at Rensselaer Polytechnic Institute (RPI), who helped to do the development work. RPI is in Troy, New York, where HTS was formerly located.

Funding, Financing High Technology Services, Inc. (HTS), a small, minority-owned materials testing and consulting firm in Troy, New York, received Phase I and II NASA Small Business Innovation Research (SBIR) contracts to adapt PISO2 to specific applications like specialty coatings for electronic devices and to improve its processing. The company's intention was to eventually mass-produce it for sale to other companies. HTS originally produced the material as a solution called polyamic acid, but it became apparent that they could achieve more processing flexibility if it were available in powder form. So the company set to work to develop a commercial process for producing it in powder form and to identify related applications.

Also, the New York State Energy Research and Development Authority (NYSERDA), a state public-benefit corporation, awarded HTS cost-shared contracts for materials development work. NYSERDA tries not to compete with venture capital groups and banks, so it funds smaller companies that have technologies they are trying to get ready for the market. NYSERDA invests some $17 million each year in R&D projects, at no more than $250,000 per project. Recipient companies repay NYSERDA based on 1.5 percent of sales, until the original investment is returned. NYSERDA contracts can cover equipment purchases, third-party testing, or hiring of new employees, but are not used to pay a company principal's salary.

Intellectual Property A graduate student from the Massachusetts Institute of Technology helped Dr. St. Clair with scale-up work on the technology at NASA Langley. They filed the invention disclosure together and are listed as co-inventors on the patent. Two patents for the material and the production process were issued in August 1983 and December 1984.

Dr. St. Clair explained that, where there is a graduate student, professor, post-doctoral fellow (such as through the National Research Council), or contractor (such as from Lockheed Martin) working on the technology, that individual can elect to assign his or her rights to the government so that the agency owns the patent in its entirety. Alternatively, the individual can elect to retain his or her portion of the rights (half in this case). The MIT student chose the latter, and MIT filed the patent on behalf of the student.[7] The individual from the university is not required to allow their university, as agent, to receive any portion.

Technology Transfer Mechanisms   Soon after the process was patented, NASA approved manufacturing licenses for Celanese Corporation and M&T Chemical, Inc. These companies produce high-performance plastics, resins, and composite materials. Several years later, HTS also obtained rights through its SBIR contracts and the Bayh-Dole Act. HTS also has licenses for several other NASA patents, but the company's core product line is based on the PISO2 technology. The company was founded in 1983 by its president, Mr. Milton L. Evans, who had worked at General Electric Company for two decades in a variety of positions, including scientific, marketing, and management.

Dr. St. Clair[8] noted that when the federal government grants a technology license, there are many rules surrounding the royalty-sharing arrangements. If there is more than one federal inventor, they share the inventor's portion of the incoming royalties (20 to 25 percent at NASA). NASA gives the inventor(s) all of the up-front money for a license agreement. For example, if $100,000 is provided by a licensee up front, then NASA would give each of four joint inventors $25,000 the first year of the license. If the agreement calls for six percent of gross sales in running royalties, each year NASA would take 75 percent of that total, and then split the remaining 25 percent among the four inventors. By law, an inventor can receive no more than $150,000 per year over and above his or her salary.

Although royalty-sharing arrangements provide incentives for laboratory scientists to work harder at technology transfer, Dr. St. Clair[9] pointed out that various implementation issues need to be examined in coming years. For example, multiple inventors are now often involved in technology transfer activities which may call for mechanisms to resolve potential disputes over royalty-sharing. In this regard, NASA has an "Inventions and Contributions Awards Board" which could address this need at the agency level. He added that it would be important for such a mechanism to appear objective in terms of favoring all the various technology areas.

User Groups   The resin industry is a $5 billion-a-year industry. During the 1980s, new types of resins provided significant benefits to the aerospace, transportation, electronics, chemical, and consumer goods industries. It is anticipated that these benefits will continue to accrue through the 1990s as new markets and uses for thermoplastic materials are stimulated.

Polymer resins are an ingredient in composite materials which are used in aircraft structures. Laminating resins are also used in printed circuit boards, because matrix-type resins offer protection from degradation induced both electrically and chemically, as well as through radiation.

Barriers to Commercialization   The barriers HTS has encountered have been mostly financial. It is hard for a small company to bring a complex technology to market. Testing and development work is expensive, as are the raw materials. Mr. Evans[10] notes, however, that NASA has been very cooperative (he has met with NASA Administrator Dan Goldin) in providing the amount they required to get off the ground. NASA provided "a few thousand dollars," according to Mr. Evans,[11] to buy material samples to evaluate its possible use in space applications. The agency is interested in getting them linked with others to license the material and get it commercialized. NASA has also provided HTS with some publicity: an article on PISO2 in NASA's *Spinoff 1992* highlighted HTS as much as the material.[12]

User Benefits/Economic Impact/Outcomes   Celanese License: NASA licensed the technology to Celanese Corporation who intended to manufacture the product because the company did not have any materials in this temperature range like its competitors such as General Electric. After the license was signed, the company assigned researchers to the project. However, shortly thereafter Hoechst AG, the German parent company, bought out Celanese, and the company's CEO retired. Hoechst Celanese Corporation decided not to pursue this and dropped the research. Dr. St. Clair[13] felt that Hoechst Celanese could have been competitive with General Electric, producing the material at about $1 to $10 per pound, depending upon the ultimate uses. NASA technology transfer had incomplete records and Celanese had no record of the license being terminated in writing, however, it is assumed that Celanese abandoned the license.

M&T Chemical License: NASA licensed PISO2 to M&T Chemical, Inc. M&T Chemical, a medium-sized company, had previously bought a small start-up company founded by a former General Electric scientist who had a number of patents on materials in the polyimide family and a marketing partner. Presumably, M&T wanted to round out their portfolio in polyimide materials to make the entire portfolio more valuable commercially. M&T Chemical started prototype work on the technology and began marketing PISO2 as "M&T-4605" in 1985. Subsequently, Elf-Acquitaine took over M&T Chemical after the M&T employee with the related patents had left the company. Elf-Acquitaine sold the PISO2 technology rights to National Starch, and M&T is presently out of this business altogether. National Starch is not actively pursuing commercialization in spite of keeping up with the licensing fees. Since NASA does not currently have any record of other licenses on the PISO2 technology, it is presumed that the license has either expired, or reverted back to NASA, or is inactive.

High Technology Services SBIR: HTS obtained rights to adapt PISO2 for specific applications through Phase I and II NASA SBIR contracts. Originally located in Troy, New York, HTS now does business as High Technology Systems, Inc. outside of Albany, New York. The company produces the material "Techimer 4001" in both powder and solution forms and has introduced it to the marketplace. The material's principal application is as a matrix resin for composite materials. But it has a niche market according to HTS' founder Mr. Evans,[14] because it can be molded into thermoplastic products and if it doesn't work out, it can be melted, re-processed, and re-used.

HTS is currently marketing Techimer 4001 as a high-performance thermoplastic that can be used to make aircraft structural adhesives or coatings that protect interior electronic components from high temperatures and radiation. According to Langley personnel, HTS manufactures and sells the material in small quantities that do not allow them to get the volume up or the price down. For larger volume sales, HTS contracts with a manufacturer to produce Techimer for them on a toll basis. HTS then sells the material for two to five times the cost of producing it, so the cost ends up being too high. Mr. Evans[15] confirmed that although Techimer is still a viable product for them, the response has not been that positive. There have not been major sales partially due to competition by DuPont.

HTS is working to develop an even better process and new customers. For example, they are exploring the material's use in a flame-resistant foam for both aerospace and marine applications. Apparently, it does not burn, drip, or smoke when exposed to a flame. In addition, the material is under evaluation for use by Praxir (formerly Union Carbide) and other fabricators, processors, and end users, which might increase sales. Mr. Evans[16] hopes that, eventually, molded aircraft and automobile parts will be made from the material.

International Activity   HTS has supplied samples to European countries, including Germany, France, and the United Kingdom, but nothing has developed. Firms in China and Japan are also interested in the material.

Government Gains   According to Dr. St. Clair,[17] it's possible that NASA may revisit the PISO2 material as a candidate for renewed industrial marketing efforts resulting in more widespread commercial availability for ultimate government applications (along with the possibility of increasing related R&D funding).

Dr. St. Clair pointed out that there is a positive factor regarding royalties being disbursed to the relevant laboratory (rather than being returned to the U.S. Treasury, generally). The laboratory (or NASA field center, as in this case) might decide, for example, to provide a portion to inventors working in non-patentable technology or technologies related to the agency's mission such as space exploration that may not be commercializable or income-oriented.

Economic Development, Technical Assistance   Mr. Evans recently attended a NASA-sponsored mentor/protégé workshop in Springfield, Virginia, where HTS was introduced to Lockheed Martin. The two firms have since been in contact to examine how Lockheed Martin might use PISO2 in a defense or aerospace application. The New York State Energy Research and Development Authority also invited HTS to NYSERDA-sponsored programs.

Elapsed Time   As noted, the material and the process were jointly patented by NASA and MIT in 1983 and 1984. In 1985, soon after the process was patented, NASA approved the first manufacturing license. In 1990, HTS was awarded the NASA Phase I SBIR contract. HTS's Phase II SBIR contract was finished in 1992.

Case 3 (1986) – Substance Tracer Technology

Role of the Researcher and Other Laboratory Personnel   Dr. Russell N.

Dietz at Brookhaven National Laboratory (BNL) won an award from the FLC for his perfluorocarbon tracer technology. He now heads the Tracer Technology Center in the laboratory's Department of Applied Science, Environmental Chemistry Division. The Tracer Technology Center has a staff of about five laboratory scientists. Like other DOE laboratories, Brookhaven has its share of visiting researchers. In the case of the tracer technology, other personnel who were involved in various aspects of its testing or development included: a visiting scientist from Israel, a student collaborator, and a physics teacher from a local high school.

Since the early 1980s, Dr. Dietz and collaborators have promoted the technology in a number of publications. They published articles in scientific and technical journals including air pollution and microbiology journals. Their papers appeared in the proceedings of trade organizations such as the American Society for Testing and Materials. They published papers for BNL and government agencies (i.e., the National Oceanographic and Atmospheric Administration). They wrote and presented at seminars sponsored by the Electric Power Research Institute. They also published test and demonstration reports. An innovative approach they have incorporated is the use of a videotape on the technology's applications.

In order to determine other applications for utilities, in 1994, Dr. Dietz conducted a demonstration sponsored by Long Island Lighting Company (LILCO) on the use of the tracer technology to provide on-line measurement of air inleakage into its power stations, as well as underground pipelines and underwater cables. Another LILCO demonstration certified the leak tightness and developed an acceptable protocol for testing its oil-fired gas turbine systems. Similar demonstration projects were also performed with Union Electric in St. Louis and Boston Edison.

Technology and Applications The technology works via an injector or other type of apparatus releasing a very small amount of a perfluorocarbon "tracer" into the atmosphere or underground cable pipes. The tracer acts as a simulated pollutant and "tags" the surrounding substance, such as the air and/or underground fluid. One or more tiny sampling devices are used to follow the tracer and track its course over time. The level of sensitivity of the sampling devices is so high that only very small amounts of tracer are used, so the technology is environmentally safe. There are five per fluorocarbon tracers that are routinely used. The data collected by the sampling devices is analyzed at the laboratory and used to create models of ventilation flow, heating leaks, air impurities, or other systems it is tracing.

These devices are left in place for anywhere from a week to a month. They can be placed on street lights for outdoor monitoring purposes. Outside, the sampling devices can measure the movement of air pollution of all types, even the fallout from a nuclear disaster. In a house or building, they would be placed every 500 feet or so. In the case of underground cables, a "sniffer" is used to test the air over the pavement since the tracer will permeate the soil and enter the atmosphere through cracks in the pavement.

The two major application areas for this technology are leakages in the air and underground. There are a variety of specific usages in each area. The following sections will divide up the discussions of the technologies by these two areas plus a miscellaneous category.

Ventilation Analysis: The original application for this technology was measuring building ventilation and air leaks. Many R&D-type institutions, both public and private, need to know this information about buildings and homes in order to normalize indoor air quality. For example, designing and implementing an isolation room or operating room for a hospital requires ventilation challenges because the spread of contaminants must be guarded against while maintaining a comfortable environment. The technology could be used to certify the performance of heating, ventilating, and air conditioning (HVAC) systems.

Underground Leak Detection: Another application area researched by Dr. Dietz early-on involved the electric power industry. The tracer technology has been used, for example, to pinpoint leaks in underground power transmission cables so that fewer excavations are needed per leak site. It has also been used to detect leakage of air into condensers for power station systems. In addition, the technology has applications for natural gas and fuel oil utilities and industries, because it has the potential to identify leaks in underground tanks and pipes, or to study petroleum reservoirs.

Measuring the tracer concentration for petroleum reservoirs around the world works like building ventilation. Fluids are injected into the injection wells to push the petroleum up to the production wells. When the injection fluid is tagged with tracers, it helps them to understand timing to reach production wells and other engineering characteristics. In the past, producers used radioactive tracers; in 1986, they started using perfluorocarbons for tracers.

In addition to these original applications, there are a number of applications for this technology with "potentially large markets . . .The potential for new applications and markets is significant," according to Dr.

Dietz.[18]

The technology can be used to detect fire. Tests have been performed demonstrating the capability of the tracer technology to detect and locate thermal overheating of electrical components in a system. The tracer is mixed with insulating paints and other materials to form insulating coatings around electrical wiring and components. When the system is monitored, if the temperature increases, this overheating causes the tracer vapor to be emitted. Sampling devices placed throughout the system not only detect the overheating, but also localize the problem to a specific section of wiring.

Other application areas include explosives detection, environmental monitoring, disaster emergency management, and instrumentation sales. The brochure for the laboratory's Tracer Technology Center explicitly states that the center is seeking private partners interested in proprietary agreements related to the food packaging industry (for screening the seal integrity of food packaging), semiconductor component leaks, and equipment leak certification. Dr. Dietz emphasized, "The opportunity exists for establishing commercial services based on the technology."[19] Also, opportunities exist related to sales of the instruments used for doing the analyses. In particular, the paper noted that there was a need for several instruments to be commercially prototyped and manufactured according to specific end-user requirements.

The Laboratory   Brookhaven National Laboratory is a government-owned, contractor-operated DOE laboratory. Until 1997, BNL was operated by Associated Universities, Inc., a non-profit research management organization sponsored by nine universities: Columbia, Cornell, Harvard, Johns Hopkins, MIT, Princeton, Pennsylvania, Rochester and Yale. BNL's original mission focused on the peaceful aspects of nuclear science such as nuclear medicine. It has since broadened to include other aspects of high-energy physics and energy conversion and storage. BNL now includes oceanography, atmospheric and environmental sciences, and other multidisciplinary frontiers of science.

Since the end of World War II, BNL has been located at the former site of Camp Upton in the center of Long Island in Upton, New York. BNL has an annual operating budget of almost $300 million and a staff of about 3,500. The laboratory has large, complex research facilities (called the "big machines") such as reactors, accelerators, and superconductors. They will allow scientists to observe phenomena that have not occurred since the Big Bang. The laboratory has worked to make its facilities accessible to university and industry scientists through cost-shared cooperative research programs based upon proposals for use. It has earned a reputation beyond the scope of most individual institutions for making its unique facilities available. For example, fourteen major corporations including Dupont, Kodak, Exxon, IBM, and Mobil, have an ongoing presence at the synchrotron radiator, which is the world's largest synchrotron. Basic research may be performed at any of the laboratory's designated user facilities, subject to availability of that facility. Even proprietary research may be performed at the laboratory, as long as BNL and the outside user enter into a formal Proprietary User's Agreement. In such a case, the user would pay full cost recovery to BNL for machine time and any related technical services provided by the laboratory. Also, the user has the option to take title to any inventions resulting from work at the facility and to consider all data generated at the facility as proprietary.

University Involvement   Universities often do air-quality studies for builders and installation manufacturers in response to EPA requirements. BNL is sometimes requested to assist in those studies.

Funding, Financing   The Tracer Technology Center at BNL makes roughly $75,000 to 100,000 per year for its services, doing business with about fifteen to twenty customers each year. The cost of each service is based upon a BNL pricing structure depending partly upon internal cost requirements, just as is the case with the other BNL-provided technical services and facilities. There are probably only three or four other laboratories in the world that have similar capabilities. For example, BNL built an analytical capability for the Japanese government, and the Netherlands copied the technology. So, there is a demand for the service, albeit not enough to justify offering it commercially.

Each project is performed under a separate contract issued at the time of the activity, with the user issuing a purchase order for the service. However, the early projects done with the electric utilities were considered R&D projects to demonstrate the feasibility of the concept. The Consolidated Edison of New York (Con Ed) tests were conducted under a DOE contract with the Electric Power Research Institute (EPRI): they were paid for by Con Ed, EPRI, and Empire State Electric Energy Research Corporation (ESEERCO). Con Ed is a private utility that has 25 percent of the underground cables in the United States. EPRI is an R&D consortium of member companies. ESEERCO is a non-profit R&D corporation in New

York, a conglomerate of electric utilities in New York. The procedures used in each case vary, depending upon the application area. For example, for the building ventilation, BNL provides a kit containing the measurement device to the customer who places it on-site. After the device has collected data on-site for an appropriate amount of time, the kit is returned to BNL. The analysis is forwarded to the customer when completed. For the utility work through EPRI, Underground Systems, Inc. tags the feeders; then, the laboratory personnel drive around with the laboratory equipment and locate the leak.

Intellectual Property   Dr. Dietz said laboratory researchers tended to publish rather than patent in past years. Along these lines, the early applications for this technology were not patented. In recent years, however, Associated Universities, Inc. (AUI), the laboratory's contracting operator, developed a more active program of establishing AUI ownership, through patenting, of BNL technologies.

The tracer technique for locating underground line leaks is a proprietary, knowledge-based system that belongs to EPRI. A patent was never filed in this area or any other area except for a patent for an early warning pre-fire detection system (using tracers) issued to AUI in 1993. This is an extremely sensitive system to detect potential fires before they ignite. Dr. Dietz and Gunnar I. Senum are listed as the inventors.

Technology Transfer Mechanisms   Ventilation Analysis: Dr. Dietz built a complete tracer analytical system for the NAHB Research Foundation so that organization could provide the ventilation/leak detection service commercially for building analysis. In doing this, he further developed the technology so that it would be easier to use and more economical. Initially, the National Association of Home Builders' (NAHB) Research Foundation in Rockville, Maryland, was interested in making this technology available as a service to the NAHB constituents. NAHB works with home manufacturers, building material suppliers, commercial building owners, owners' associations, and utility companies. The intention was to provide a nationwide monitoring service to home owners that assessed the air leakage of homes and building materials. NAHB would provide customers with a statement certifying the level of need for weatherization to reduce energy consumption in old homes or certify the energy efficiency of new homes. The tracer technology was transferred to the NAHB Research Foundation through an exclusive licensing agreement between BNL and a private company, AIM, Inc., in Washington, D.C. BNL helped to set up this company so that they could perform the service for NAHB. AIM's analysis laboratory was set up at NAHB; it was necessary to have the equipment in order to be able to perform the tests.

For a related application, Dr. Dietz developed and tested various ways to evaluate the performance of hospital isolation room ventilation systems. He worked in partnership with Stony Brook University Medical Center and a company called Perfect Sense, Inc., of the city of Islandia on Long Island. The company is in the ventilation business, and their main line of work is making sensors and system controls that respond to pollution.

DOE headquarters put in place a CRADA providing funds for BNL to participate with the company. The company sponsored its own time working on the CRADA, which is standard procedure. Bob Vandella, the company's president, was hoping the instruments could be implemented on-line to warn the hospital of ventilation problems in, for example, isolation rooms, operating rooms, emergency rooms, intensive and critical care units, and HIV and tuberculosis wards.

Underground Leak Detection: Dr. Dietz conducted several tests of the electric utility application of the technology in conjunction with Con Ed, EPRI, and ESEERCO. Tests were conducted in 1988, 1990, and 1992 on both simulated and real leaks of dielectric fluid within subsurface pipes. A companion project was carried out with Cablec Utility Cable Company to determine whether the tracer technology works on their pipes.

The demonstrations of the tracer technology in pinpointing the Con Ed leaks were successful because they showed that the technology was more precise than conventional methods. The demonstration project with Cablec Utility Cable Company also successfully verified that the tracer technology worked on their pipes. As a result, EPRI is promoting this service to companies that are members of the EPRI consortia. Underground Systems, Inc. has a license with EPRI to provide the underground cable testing commercially. The company developed a special injection system used to introduce the tracer into a pipe. This is a case of commercialization of both the instrumentation and the service.

Although the underground line leak work was quite successful, BNL found it was not as easy to quantify and accurately pinpoint a leak in the other utility areas, so BNL is not promoting related application areas.

The Tracer Research Corporation, an established commercial laboratory in Arizona for detecting leaks from underground storage tanks, had its own instruments and procedures that did not involve this tracer technology.

Communication between Dr. Dietz and the company led to Tracer performing some tests with BNL to see if they wanted to begin using the tracer technology. In one of these research projects, a tracer was used to tag underground storage tanks in seven cities in Massachusetts (Hanover, Concord, Lexington, Peabody, Milton, Revere, and Freetown) and the results were compared to conventional warning signals for determining the integrity of underground tanks. The tracer was more accurate (100 percent) in detecting leaks.

Other Applications: A new spinoff of this technology is being explored which would "tag" explosives related to clandestine bombs in airports. For this application, BNL loaned one of their devices to Vacuum Instrument Corporation on Long Island, New York. The company is working on developing its own commercial prototype sniffer. It is called COPS, for continuously-operating perfluorocarbon sniffer. This system can be used to pinpoint tagged fluids in less than ten seconds. Vacuum Instruments is a company that provides leak detection services for commercially manufactured components such as air conditioner condensers and heat exchangers, automobile engines, etc. They were interested in seeing if the BNL technology would work better and/or more cost-effectively than their existing technique.

Pre-fire detection is not being actively pursued at the present time. Dr. Dietz and his colleagues have not found an angle for promoting a viable commercial product, although the concept works well. The laboratory's marketing materials state that licenses on this patent are available on an exclusive or nonexclusive basis, and that the "competitive advantage" is such that a system is "commercially practical." Obviously, AUI has also developed an active licensing program for making its technologies readily available for commercialization. In fact, in a BNL Office of Technology Transfer brochure, the tracer technology was listed as the third example, although 62 technologies are listed as available from a list that is updated as of mid-1996.

User Groups   BNL provides the tracer technology service for customers in the United States, Canada, and occasionally overseas. In addition to all of the projects and users highlighted in this case (utilities, hospitals, petroleum companies, etc.), laboratory service users include small R&D laboratories, utilities, air-quality agencies performing studies, and universities. For example, BNL recently performed the building ventilation service on 1,500 homes for a customer in the Netherlands. The petroleum reservoir work has been done for U.S.-based companies, only.

Barriers to Commercialization   The tracer technology-type services, in the various applications noted, continue to be offered commercially in a small way through Stieff R&D, Perfect Sense, and John Booker. Generally, each of the companies that has entered this area has found that the market does not offer enough of a return on investment for the technology to serve as the main service or product line for a stand-alone commercial entity. Each service provider has realized only a few thousand dollars worth of revenues in this area, which represents a side venture for each of the companies.

Other Factors   BNL obtains its perfluorocarbon tracers from British Nuclear Fuels, Ltd. at about $150 per kilogram. American companies make perfluorocarbon compounds using various electrochemical processes that don't work as well with the BNL tracer methodology. As far as the equipment BNL must purchase to do this work, the samplers and related components (like the adsorbent material) are manufactured by Gilian Instrument Corporation, Computer Control Corporation, and Bios International all in New Jersey; and, Rohm and Haas Company and Supelco Incorporated in Pennsylvania.

User Benefits/Economic Impact/Outcomes   Ventilation Analysis: The NAHB service was advertised for four years. However, it never materialized as a business opportunity because it did not bring in enough of a return on the investment to support a stand-alone commercial laboratory using just the tracer technology for analyzing ventilation and air leaks. Once it became apparent that the business would not succeed commercially, all three parties (BNL, NAHB, and AIM) discussed the issue to resolve what to do. Although one of the options involved having NAHB continue to promote the service with an additional cost for the value-added, in the end BNL bought back the equipment from the company, and the service is no longer available through NAHB.

The part of AIM that was involved now exists under the name Stieff Research and Development Company, Inc. headed by Lawrence Stieff, vice president of AIM. AIM/Stieff R&D have customers in Sweden, among other locations, so they still make use of BNL's service and Dr. Dietz's scientific and technical support. When BNL bought the equipment back from AIM, Inc., they combined it with the Tracer Technology Center's other analytical systems.

The CRADA work with Perfect Sense, Inc. made it apparent that offering hospitals on-line ventilation testing was not feasible, and that it would be more reasonable to provide the service on an as-needed basis. For one thing, there are no laws or regulations requiring hospitals to do this on an on-going basis, so there is no demand for the service. The company now offers various levels of off-line, independent tests to determine whether air leaks are occurring under continuous operating conditions in specialized hospital rooms. The company's brochure notes that the tracer technology provides the "most cost-effective and accurate testing technology in comparison with alternative methodologies." It also notes that the company's solutions work toward compliance with guidelines issued by the U.S. Centers for Disease Control.

Underground Leak Detection: The estimated cost of each demonstration BNL performed with utilities was $20,000 plus the cost of necessary excavations. A leak usually requires five or more excavations, costing $10,000 to 30,000 each, but with the tracer technology only two excavations are required. Over time, BNL did about twenty "leak hunts" for the utility industry. About sixteen of the twenty electric utility leak hunts have been for Con Ed. Since EPRI is mostly a privately funded consortia, it is not known what kind of revenues are being realized, on the whole, from provision of this service to its members.

With the Tracer Research Corporation joint tests with BNL, although the tracer—and not their conventional method—was one hundred percent correct in identifying leaks, the company decided not to use the tracer technology.

Other Applications: Vacuum Instruments Corporation planned to market a tracer instrument and related systems to continue validating and certifying the leak integrity of components of the bomb sniffer. The same technique could be used to certify pressure vessels such as fire extinguishers and other materials used in buildings. However, the instrument uses a radioactive foil that requires licensing by the Nuclear Regulatory Commission to be used for trade, and the company has encountered some difficulty in obtaining that radiation licensing. The company says it still plans to pursue this area, but it is not a high priority.

An example of commercialization of instrmentation, John Booker and Company in Texas collaborated with BNL to build instruments for an Italian customer. A dual trap analyzer, one of the monitoring instruments needed for underground cable testing, was built to customer specification and is available from John Booker. John Booker also provides analysis systems "following the Brookhaven concepts."

International Activity   Dr. Dietz conducted an experiment for the Commission of European Communities that simulated pollutant clouds analogous to the Chernobyl nuclear disaster and the Bopal, India, chemical disaster. Called the European Tracer Experiment (ETEX), two tests were conducted in October and November of 1994. A tracer was released to test and improve the computerized models that predict the atmospheric dispersion of these types of pollutants. This type of activity has significance for global emergency planning and management: in the event of such an emergency, it could result in earlier warnings thereby saving lives. In other international activity, a demonstration project was conducted in England with British Railway.

Government Gains   BNL provides the tracer technology service to other government users. For example, in 1991, the Marshall Space Flight Center in Huntsville, Alabama, demonstrated the tracer technology to determine the leak tightness of future NASA space station modules. At the time, another approach being considered by NASA to determine module tightness was the pressure decay approach. The demonstration showed, however, that the tracer technology would require less time and would be more accurate. So NASA planned to leak-certify the space station modules using the tracer technology.

Another government user was the U.S. Environmental Protection Agency. In 1992, the technology was used for atmospheric tracing to study whether the haze in the Grand Canyon came from the Los Angeles basin or from the Mohave power generating station. The stack was monitored for fifty days, a multimillion dollar effort. The EPA program monitor, Marc Pitchford at the EPA office in Las Vegas, was quoted as saying there was no way EPA could have done this without the BNL technology.[20]

Elapsed Time   As of 1996, the tracer technology had been under development at BNL for over twenty years with many milestones. In 1986 when Dr. Dietz received his award from the FLC, the laboratory had entered into its license agreement with NAHB and AIM, Inc. The service was offered through NAHB (for building ventilation) until about 1990. The very successful underground electric utility tests were conducted from the late 1980s to the early 1990s. The tests to determine other electric utility applications didn't take place until 1994. As noted, the underground leak detection is now still offered as a service to EPRI member companies. The petroleum reservoir work started as the regulations were changing in the mid-1980s. The inter-agency projects with NASA and EPA took place in 1991 and 1992, and the major European project took place in 1994. The explosion detection R&D continues. The instrumentation improvement work and instrumentation sales and services are ongoing.

Case 4 (1986) – Slow-Release, Alginate-Based Herbicide/Pesticide

Role of Laboratory Researchers and Other Personnel   The researcher in this case, Mr. William J. Connick, Jr., is a research chemist in the Commodity Utilization Research Unit of the U.S. Department of Agriculture (USDA) Southern Regional Research Center (SRRC) in New Orleans. His unit specializes in biocontrol formulations to control agricultural pests. While researchers like Mr. Connick deal with chemistry, he has teamed with a variety of scientists at other facilities dealing with the biological sciences. This case shows how and why researchers from the two fields came together to develop a new area of research.

   Mr. Connick began his work focusing on chemical applications, but he quickly expanded from chemical pesticide/ herbicide applications to living, biological applications when Dr. Paul C. ("Chuck") Quimby approached him after a conference presentation. Dr. Quimby was a researcher at the USDA Southern Weed Science Laboratory in Stoneville, Mississippi. The Stoneville laboratory focused on safe methods of weed control to reduce losses in crops and increase their production, particularly southern crops. Mr. Connick commented to Dr. Quimby that he didn't know that fungi killed weeds. Thus began a long and fruitful collaboration with the Stoneville researchers including Dr. Quimby, Dr. Harrell Lynn Walker, and Dr. C. Douglas Boyette, a graduate student at the University of Arkansas at the time who is now a senior scientist at the Stoneville laboratory.

   Mr. Connick also contacted other researchers based on their articles. He saw an article in USDA's *Agricultural Research* journal by a scientist from the Biological Control of Plant Diseases group at the USDA center in Beltsville, Maryland. Mr. Connick subsequently teamed up with the Beltsville scientists, including Dr. Debra R. Fravel, a plant pathologist at the Soilborne Diseases Laboratory at the Beltsville complex, to formulate an alginate-based fungus to control plant diseases. Through the course of this work, Dr. Jack A. Lewis, Mr. Connick's counterpart at the USDA Beltsville complex, screened hundreds of fungi for different formulations. Similarly,

Mr. Connick also worked with the USDA/ARS Aquatic Weed Research Laboratory at Ft. Lauderdale, Florida, which researches the control of aquatic weeds through biological methods, among others.

Mr. Connick authored or co-authored articles on each of the technology's applications areas. The article that "started it all" was a 1982 article in the *Journal of Applied Polymer Science* by Mr. Connick dealing with the slow or controlled release of chemical herbicides.[21] The first article on the biological control of weeds using alginate granules was a landmark paper in his series of papers. It was co-authored with Dr. Walker at Stoneville and published in a 1983 edition of *Weed Science.*[22] Another landmark paper was the first paper on the biocontrol of soil-borne plant disease using alginate granules by Mr. Connick and Dr. Fravel (et al) at the Beltsville center, which appeared in a 1985 edition of *Phytopathology.*[23]

The fundamental importance of some of this early work is illustrated by the extent to which Mr. Connick and colleagues are cited in subsequent work. An overview paper by Mr. Connick, entitled *Pesticide Formulations: Innovations and Developments,* appeared as a chapter in a 1988 publication based upon an American Chemical Society symposium.[24] In this paper, each of the highlighted applications began with the 1983 Walker/Connick paper on biological control. Although Mr. Connick did not mention it,[25] the paper's literature citation noted that Mr. Connick wrote an early paper on alginate-based herbicides for aquatic weed control as far back as 1979 for the International Symposium on the Controlled-Release of Bioactive Materials.

Because of professional connections made through these papers and conferences, Mr. Connick has worked with scientists in many government agencies, universities, and industry over the years. Mr. Connick was in touch with scientists engaged in insect control research. Professor R. C. Axtell, an expert in mosquito control at the North Carolina State University Department of Entomology in Raleigh, read Mr. Connick's papers and contacted him. This resulted in the preparation of floating alginate granules that disintegrate in water to release a fungus that kills mosquito larvae. The testing of this formulation is described in various papers by Prof. Axtell, most notably an article in the 1987 *Journal of the American Mosquito Control Association.*[26] Similarly, researchers at the University of Idaho Department of Plant, Soil and Entomological Sciences cite Mr. Connick's work numerous times in a 1990 article they wrote for the *Journal of Economic Entomology* about their testing of the alginate technology to control aphids in cereals such as brans.[27]

The professional collaborations eventually lead to company collaborations. The first occurred when the Beltsville group enhanced an existing collaboration with Grace-Sierra Crop Protection Company, based in California, to further develop plant disease application. Collabortions with companies are detailed in the section on Technology Transfer Mechanisms.

Technology and Applications   The technology is a process for incorporating chemical pesticides and herbicides into alginate that has been formed into little granules or beads. Alginate is a natural polymer derived from seaweed. It is used in dental gels for impressions, and in the food industry as a thickening agent for foods like puddings and pie fillings. For this technology, the alginate is used as a matrix or medium, and living things such as weed-killing fungi are incorporated into the alginate granules. When dry, the granules stabilize the fungi. Then, when wet, the fungi sprout on the granule surface and release spores in a sustained way over days or weeks.

Alginate formulations are effective against not only weeds, but also plant and soil diseases, and even insects. The effect is the slow release of a weed/pest killer that is superior to traditional chemical spraying because spraying can have negative side effects. Because the delivery of the biocide is direct, the dosages are lower than those used in spraying. Therefore, an indirect effect of this more targeted weed and pest control is reduced groundwater contamination and surface water pollution.

There are a wide range of uses for this technology. It can be used for groundwater protection, and the control of weeds, insects, and plant diseases. More specifically, the technology has been used to fight crop diseases (greenhouse-type root rot), mosquito larvae, aphids that attack cereals, watermilfoil (a submerged aquatic weed), and aflatoxin (a carcinogenic fungus that contaminates certain kinds of nuts and grain). Most recently, the technology is being used for long-term bioremediation of toxic chemicals in the soil.

The aflatoxin applications also appear to have been initiated by Mr. Connick and other researchers at SRRC. Alflatoxin is a carcinogen produced by certain fungi that infect cottonseed, peanuts, tree nuts, and corn kernels. For this application, the alginate formula is used bio-competitively, whereby non-toxic strains of a toxin-producing fungus (*Aspergillus flavus*) are applied in large quantities so that the "good" fungi colonize and prevent the harmful fungi from gaining a toe-hold by "out-competing" them. An article on this was written by the New Orleans researchers in a 1995 edition of *Biocontrol Science and Technology*.[28]

The Laboratory   The SRRC in New Orleans is one of four major research centers for the USDA Agricultural Research Service (ARS) which is headquartered at the Beltsville, Maryland, Agricultural Research Center. The ARS is USDA's in-house research arm. For over fifty years, the New Orleans center has specialized in research on agricultural commodities produced in the southern part of the country, such as cotton, peanuts, and rice. This includes textiles, yarns, fibers, and chemical finishes for textiles. Examples of developments that have come from the SRRC include flame retardant treatments for cotton fabrics, frozen orange juice concentrate, "smart fabrics" that change with weather conditions, and stronger and more vibrant naturally-colored cottons. In addition to the four major ARS centers, there are a hundred ARS laboratories nation-wide.

University Involvement   In the university research arena, a vast amount of work has spun off from this technology worldwide since Mr. Connick's early work in this area. For example, University of Arkansas researchers have published papers on their studies on use of the alginate technology for various applications like weed control (in a 1988 edition of *Plant Disease*[29]) and fighting destructive soybean nematodes (in a 1995 supplement to the *Journal of Nematology*[30]). Similarly, cooperative research between the ARS New Orleans center and Tulane University resulted in a paper co-authored with a professor and researchers in the university's Department of Cell and Molecular Biology[31] and ultimately the filing of a patent application for bioremediation applications of the technology. Mr. Connick said the technology had wonderful success as a model research system for many types of projects because it was easy to implement; it was reliable, because the micro-organisms and concepts could be tested in the laboratory; and it was easy to tell whether an organism worked or not.

Although all of this evolved from the research that began at SRRC in New Orleans, not much of the related university research is being done in formal collaboration with the ARS facilities in New Orleans, or any of the USDA laboratories. For one thing, much of the university interest in the technology has been for research purposes rather than for licensing and commercialization.

Funding, Financing   Mr. Connick did not mention any special USDA funding. His research and technology transfer activities were performed under SRRC's research budget.

In the product development stages, Biosys, Inc. supported a number of individual university researchers both in the United States and overseas to perform field research developing efficacy data. This was usually done on a consulting basis without necessarily involving the universities, institutionally. Overall, the products were tested by a number of growers, distributors, and USDA Extension Service entomologists.

The only other support worthy of note was that various companies (eg., the Thiele Kaolin Company of Georgia) provided materials such as the clay used in the USDA/ARS alginate studies. A clay called kaolin is commonly used as a filler in the alginate formulations.

Intellectual Property   USDA's current portfolio of patents numbers over 800 for the four major ARS centers along with its system of a hundred laboratories. This system is served by seven patent advisers and five technology transfer coordinators located throughout the country. More than 200 licenses have been issued by ARS' central Office of Technology Transfer (which happens to be located at ARS headquarters in Beltsville). USDA licensees have included companies, individuals, researcher foundations and/or universities, and technology management groups. When a USDA patent is licensed, the inventor can share in royalties "to a small extent."[32]

The alginate technology was patented in the 1980s and 1990s. The first patent on the use of alginates with chemical herbicides was issued to USDA in 1983 ("Controlled release of bioactive materials using alginate gel beads"). Mr. Connick is listed as the inventor. Another related patent was issued in 1986 on incorporating living organisms into alginate for biological control.

The first two patents on the use of the alginate technology for the biological control of soil-borne plant diseases were issued to USDA in 1987 ("Preparation of pellets containing fungi and nutrient for control of soilborne plant pathogens"). The inventors on both patents are Mr. Connick and Dr. Lewis and Dr. George C. Papavizas of the USDA/ARS center at Beltsville.

The first two patents on the use of the alginate technology for biological control of weeds were issued to USDA in 1988 ("Method for the preparation of mycoherbicide-containing pellets"). Mr. Connick and Dr. Walker and Dr. Quimby of the ARS Stoneville laboratory are registered as the inventors. Both of the two sets of 1987 and 1988 patents have the same title and text, but different filing dates and patent numbers. These are what is known as patent "dividends," the later patent involving different applications. In this case, for example, one patent would be for killing weeds and the other for killing insects.

In addition, a joint patent application was filed early in 1996 by USDA/ARS and Tulane University in New Orleans with Mr. Connick listed as a co-inventor. The patent, which is still pending, is in the area of bioremediation of toxic chemicals in soil. Growing from alginate granules, the fungi metabolize the toxic chemicals over time to render the chemicals harmless.

Mr. Connick had some comments about USDA's eagerness toward CRADAs and the effect of this on inventor's intellectual property rights and returns.

Technology Transfer Mechanisms   The early technology transfer mechanisms in this case were a number of research papers and presentations at national meetings like those of the American Chemical Society and the Controlled-Release Society. As the USDA publications state, ARS has an aggressive technology transfer program and actively pursues relationships with companies. Out of all the government's R&D establishment, USDA was the first government entity to sign a CRADA under the 1986 Federal Technology Transfer Act in July 1987. As of mid-1990, USDA had entered into more than 500 CRADAs with over 100 companies, associations, universities, and other agencies. The technology transfer mechanisms used with each company partner are as follows:

Grace-Sierra Crop Protection Company: Because of U.S. Environmental Protection Agency (EPA) regulations in the early 1980s, W.R. Grace and Company felt that American growers would not be able to compete internationally, and so the company sold its water-soluble horticultural products and related business to a greenhouse company. Soon after this, W.R. Grace merged with Sierra Chemical to form Grace-Sierra Horticultural Products, which produced and sold fertilizers and weed-killing herbicides.

In the 1980s, cooperative research funding provided the only effective way for Grace-Sierra to collaborate economically with the government in a way that protected patents. A Grace-Sierra contact said the problem was that until recently USDA researchers "lived off publications" because that is how they were professionally graded; as a result, back in the 1980s, it was very difficult for collaborating company partners to obtain patents for technologies. Since CRADAs were not possible in the early to mid-1980s, Grace-Sierra successfully proposed cooperative research funding from

USDA/ARS to work on biological herbicides as alternatives to chemicals.

At some point, Grace-Sierra Horticultural Products became Grace-Sierra Crop Protection Company. The cooperative work between Grace-Sierra and USDA/ARS-Beltsville evolved into several CRADAs in the late 1980s and early 1990s, and Grace-Sierra was granted a license on the preparation of alginate beads with a biocontrol fungus agents for controlling plant disease. The CRADAs, however, were not all with Grace-Sierra, as the technology passed from company to company in the midst of numerous company ownership changes.

Biosys, Inc.: Another company, Biosys, Inc., a biological pest-control company headquartered in Palo Alto, California, based a product on the alginate technology. The product did not directly infringe the patent, so it did not warrant going through the licensing process, but the early USDA research in this area laid the groundwork to enable the product. Biosys obtained information about the technology through Mr. Connick's papers. Ramon Georgis of Biosys called Mr. Connick to ask if he could visit him at the New Orleans center; they were interested in the fact that the process was being applied to living organisms, as opposed to inorganic chemical substances.

Mycogen Corporation: An exclusive license was granted to Mycogen Corporation of San Diego, California, on all the related patent applications involving biocontrol of weeds. Dr. Walker was instrumental in transferring the alginate technology that he and Mr. Connick invented to Mycogen. In fact, Dr. Walker left government service and went to work for Mycogen in Ruston, Louisiana, for several years.

EcoScience Corporation: Scientists at the University of Florida's Center for Aquatic Plants were looking for better ways to control aquatic watermilfoil. As with regular weeds, the traditional control methods, which offer only temporary relief, involve chemical herbicide treatments and/or expensive mechanical cutting. The researchers were seeking biological control alternatives that would not harm the other plant life and living things in the environment. They tested an experimental product that was developed by EcoScience Corporation of Massachusetts. EcoScience had used procedures described in the 1983 paper by Dr. Walker and Mr. Connick to develop the product. EcoScience has a 20,000 square foot research/administrative headquarters facility in Worcester, Massachusetts, and a manufacturing facility in Amherst, Massachusetts. Mr. Connick's contacts were with the company researchers who developed the product, Dr. James Stack, based in Orlando, and Dr. David Miller in New Jersey, as opposed to the university researchers who were testing it.[33]

Mr. Connick continues to test, evaluate, and refine the technology for new uses. The group at SRRC in New Orleans continues to work on the original chemical applications of this technology, in particular, experimenting with changing the release rates. Their research has become focused on reducing pesticides and herbicides leaching into groundwater in order to reduce groundwater pollution. Two SRRC scientists published a paper on this in 1995 in the *Journal of Environmental Science Health*.[34]

User Groups   Since there are a number of uses for the technology, there are also a number of potential user groups. Users for this technology would include farmers: professional greenhouses; nurseries; landscape firms; homeowners; and anyone involved in plant and crop diseases, weed control, or insect infestation. It is particularly useful to users for which chemical spraying is not an option. With the growing level of interest being placed on protecting the environment and worker safety, products derived from plants and biological organisms offer a viable alternative to increasingly unacceptable chemical herbicides and pesticides. Biological controls are specific only to their host plant or pest; in addition, they have minimal toxicity and are highly effective.

Barriers to Commercialization   In the commercial arena, there are cost issues associated with scaled-up manufacturing of the technology in all of the application areas because the alginate is expensive. So, a user may end up choosing a less expensive method of control. Dr. Georgis[35] reports that scale-up to the 80,000-liter level was a problem on the Biosys products, but the problems were eventually solved.

Another problem is that the product shelf life often doesn't exceed several months, depending upon the product and whether it is in liquid or granule form. As a result of this limited shelf life, introduction of the product overseas has been limited. Nevertheless, Biosys has applied for patents on the formulation in Japan, Australia, Canada, and the European Patent Office.

Other Factors   Mr. Connick emphasized several times that he gets a great deal of personal satisfaction from the impact being made by his technology, not just in this country but worldwide.[36] He said he is amazed at the way the original technology has evolved and branched off into new areas, even without his knowing it. With incremental technology transfer, such as this, it seems easy to lose track of what is happening with the various aspects of the technology and its applications. This is illustrated by a perception on Mr. Connick's part that there were not any alginate-related formulations on the market at the present time even though Biosys and others were marketing related products. Also, it is possible for various scientists to inaccurately gain or lose credit for "cooperative" work by others.

User Benefits/Economic Impact/Outcomes   Since Mr. Connick's original laboratory discovery, a chain of events has led to commercial activity, all of which is helping to maintain the competitiveness of American agriculture in the global market.

Grace-Sierra Crop Protection Company: Grace-Sierra came out with an off-the-shelf product called GlioGardTM Biological/ Microbial Fungicide, the first product to biologically control plant disease through a fungus. GlioGard was introduced regionally in the early 1990s through test markets. It was marketed as an alginate granule preparation that was effective in controlling "damping-off" diseases in greenhouse plants. The active ingredient was *Gliocladium virens*, a naturally occurring non-toxic soil fungus that was antagonistic to harmful plant fungi, thereby controlling what was known as root rot or "damping-off" disease. It was for use in greenhouse ("non-terrestrial") growing conditions or for nursery production of plants eventually to be transplanted as field crops. The product had a reasonable shelf life and was easy to handle; it was incorporated into the soil by hand or using a tiller at the rate of a 25-pound bucket to 25 cubic yards of soil. When the product was added to the soil, moisture was absorbed by the GlioGard granules; this re-hydrated and activated the beneficial micro-organisms the granules contained. In the test marketing, GlioGard received a 90 percent acceptance rating, which was very favorable feedback. GlioGard was on the market about two years, and Grace-Sierra sold "thousands and thousands" of pounds.[37] Although it took three years to gain EPA approval, the Grace-Sierra product became the first bio-fungicide product registered by the EPA.

Just as the product received EPA approval, W.R. Grace sold Grace-Sierra to O.M. Scotts Company. During the course of the research between USDA/ARS and Grace-Sierra, there were questions about each fungus they screened and each alginate/fungus formulation they tested. The issues were related to economics (e.g., development costs) and how to scale up from pilot plant to mass production. In fact, upon scaling up the manufacturing of GlioGard, the O.M. Scotts Company ran into problems preparing the alginate beads. At higher levels of production, they became more labor-intensive to produce, and more quality control was required. Plus, the alginate/clay mixture was expensive. So the overall costs were higher. Scotts embarked upon a six-month crash course and changed the formulation. In the end, they re-engineered a portion of Mr. Connick's original technology to make it work (but making it different enough, according to Mr. Connick,[38] that it didn't infringe upon the patent). In any case, according to the company research, Dr. James ("Jim") Walter,[39] the alginate bead preparation contained the fundamental active fungal ingredient that could be traced back to the original work at both the Beltsville and New Orleans USDA/ARS sites.

The change in formulation subsequently required some minor adjustments to EPA's regulatory requirements. After all that, O.M. Scotts Company let the product rights revert back to W.R. Grace and Company. At that point, the product name was changed to SoilGard™ 12G Microbial Fungicide. It was manufactured through the Grace Company's Biopesticides Division. The company decided to distribute it on its own rather than through distributors.

In the summer of 1996, W.R. Grace "changed hands" and sold off several of its products, including SoilGard, as well as the intellectual property rights and related personnel, to Thermo-Electron's Thermo Trilogy Corporation. Apparently, Grace-Sierra recently notified the laboratory that it would not be renewing its license for the next year, although the annual renewal fee had been paid every year until then. Presumably, Thermo Trilogy will need to re-negotiate the technology license with the laboratory. Thermo Trilogy is currently a relatively small private company that may go public next year.

Dr. Walter is now director of research and development at Thermo Trilogy. He has worked for each of the succession of companies involved with the technology through USDA/ARS-Beltsville beginning with the Grace Company. Mr. Walter[40] says the base technology has survived despite changes in ownerships, distributors, and formulations, and the resulting product is still on the market and is successful.

Biosys, Inc.: Biosys took the alginate formulation technology and first adapted it for use on window screens, not as granules or beads. The screens were used to trap beneficial nematodes, which are naturally occurring worm-like organisms that seek out and kill insects that damage plant roots. The beneficial nematodes are distinguished from destructive parasitic threadworm-like nematodes that destroy about $100 billion in crops, such as beets, worldwide each year. Mr. Connick learned about the Biosys product from a 1995 American Chemical Society publication containing a chapter on the nematode formulation by Dr. Georgis, D.B. Dunlop and P.S. Grewal of Biosys.[41] The nematodes were combined with alginate technology to create a product called BioSafe[R]. BioSafe controlled soil-based insects in their early post-egg stages—as larvae and grubs—as opposed to the adults above ground that feed on leaves.

Being a natural product, BioSafe was exempt from EPA registration. Biosys signed a marketing and distribution agreement with Ortho's Consumer Products Division of Chevron Chemical Company located in San Ramon, California. BioSafe was marketed toward "natural gardeners" for use in vegetable gardens, around fruit trees, with ornamental plants, and in flower beds. It killed flies, beetles, and a type of gnat, but did not harm beneficial insects such as ladybugs or earthworms. It had a five-month shelf life and was on the market until 1995 for a total of about six years. It was available commercially in two sizes: BioSafe[R] 20, treating 480 square feet, and BioSafe[R] 100, treating 2,400 square feet.

Dr. Georgis[42] said BioSafe was a popular product because it was both safe and effective. It was also discovered that it killed housepet fleas, as well. However, when Ortho sold its retail line to Monsanto, they subsequently canceled seventy percent of their products, including BioSafe.

Meanwhile, Biosys signed a two-year CRADA with another ARS laboratory, the USDA Subtropical Agricultural Research Laboratory at Weslaco, Texas, after the laboratory discovered a new species of insect-killing nematodes in 1990. This ARS laboratory, which is six miles from Mexico by the Rio Grande, conducts crop and fruit insect research, and biological pest control research. The CRADA was revised when another species was discovered, and Biosys is now a co-licensee of the technology.

Dr. Georgis of Biosys insisted that this technology was related to the original work by Mr. Connick.[43] Mr. Connick was uncertain how closely related their work was to his original work, and he was careful to point out that he was not involved in the Biosys work.[44] Biosys' CRADA involved at least nine company researchers—including formulation scientists and field development scientists—in addition to the ARS scientist, Dr. Jimmy R. Raulston and his group.[45] According to Dr. Georgis, an excellent relationship was put into place (and still exists) between Biosys and the ARS researchers. The Biosys portion of the nematode-related research was completely funded by the company. The research evolved into a new line of Biosys products, three of which are now on the market in both granule and liquid formulations: Vector^R MC for controlling mole crickets in turf grass or Bermuda grass; Lesco™ Vector^R MC, for insect control; and BioVector^R 355 for controlling citrus weevils. The two Vector products are marketed by Lesco, Inc. of Rocky River, Ohio, while the BioVector product is marketed and distributed by Biosys.

Although there have been company location moves and name changes, all three products have been very successful. Based upon introduction of the first two products, Biosys' market share increased from 27 to 84 percent in the citrus market (compared to the company's previous product). The third product is doing well in only its first year of introduction. In 1995, 20,000 acres of Florida citrus groves were treated with BioVector 355, and it is expected that this figure will more than double in 1996.[46] Early trials of Vector MC were conducted at one of the world's largest golf clubs in Savannah, Georgia, which previously had spent up to $100,000 per year on cricket control. These BioSafe products are less expensive than chemical pesticides and do not need to be applied as often. For example, the granule version of BioVector is sold in 25-pound drums, each of which will treat twenty to thirty acres of citrus groves (or somewhat less coverage if used in nurseries) at about $20 per acre. The liquid version is sold in a similarly-priced 2.5-gallon jug that must be used within two days. Furthermore, these products can be mixed in with fertilizers during "fertigation," or mixed with registered pesticides. However, seasonal applications must be carefully timed during the year, and it is important that watering occur after each treatment.

USDA/ARS has filed a patent application on discovery of the new nematode. Biosys has developed proprietary knowledge on its formulation and production, and a related patent application is in process. Biosys buys the nematodes from Archer-Daniels Midland, which grows a new batch every thirty days since they are time-sensitive.

EcoScience Corporation: EcoScience, which raised over $80 million in equity capital and public stock offerings in 1992, experimented with a product called Aqua-Fyte™, a bio-herbicide to control watermilfoil in environmentally sensitive lakes and waterways. A company annual report stated that extensive laboratory and field testing was a key part of EcoScience's product development activities. The company received an Experimental Use Permit from the EPA for large-scale trials of the product. The experiments involved laboratory comparisons of treated waterweeds and untreated controls. They also provided the product to the University of Florida researchers for use in their tests of watermilfoil control products.

Mr. Connick[47] believes the product was used for experimental purposes only, and that the company doesn't produce it any more.

Mycogen Corporation: Mycogen Corporation maintained its license for a number of years; however, they exercised the right to terminate it about 1993. Apparently, the company decided not to continue in that direction. No products made it to the marketplace.

International Activity   The International Atomic Energy Agency Laboratories, an international consortium based in Austria, that includes the International Atomic Energy Agency, is researching applications for this technology. The Agrochemicals Unit of this laboratory did a search of laboratory research, and their work cited Mr. Connick. They wrote an early paper on the use of alginates for chemical pesticides and have done several papers since then on the use of alginates with herbicides (e.g., 1994 edition of *Pesticide Sciences*[48]), all citing Mr. Connick's research.

Other overseas researchers have also published studies based upon Mr. Connick's work. For example, a 1992 article by scientists at the Institute for Chemical Research in Belgium cites Mr. Connick.[49] A 1993 article by Chinese researchers at Beijing Agricultural University also cites him.[50]

Government Gains   Mr. Connick's pending joint patent with Tulane for bioremediation relates to applications that could be used to clean up toxic wastes at military sites, explosive sites, chemical dump sites, and old refineries. It could also be used in Superfund environmental clean-up activities.

Elapsed Time   In 1978, Mr. Connick developed the chemical pesticide applications at his Center in New Orleans and the first patent was granted in 1983. In 1981, he began working on the weed-killing applications with the Stoneville laboratory. He published his first article in 1983, and was issued a patent in 1988. About 1984, he teamed up with the center in Beltsville to begin work on plant diseases and ultimately teamed with Grace-Sierra Company. The first publication with the Beltsville group was in 1985, and a related patent was granted in 1987. The time frame for the commercial activity was as follows:

- The first Grace-Sierra product, which was under development for a number of years, hit the market in 1991.
- The Biosys nematode products from its CRADA with ARS-Weslaco took three to four years from basic research to market introduction. The

scale-up of formulation and production to the 80,000-liter level took six months and four or five people. Upon development of the finished product, the marketing effort lasted eight months.

- The EcoScience/University of Florida experiments took place in 1992 and were described in a 1993 article in *Biological Control* by two university researchers.

Case 5 (1986) – Controlled-Release, Chemically-Imbedded Herbicide/ Pesticide Material

Role of Laboratory Researchers and Other Personnel   The researchers for this case were Dr. Peter Van Voris, a biologist and senior program manager of the Earth and Environmental Sciences Center at the Pacific Northwest National Laboratory (PNNL); Dr. Dominic A. Cataldo, a plant physiologist; and the now-retired Dr. Frederick G. Burton, a biophysicist. These DOE researchers, with widely varying backgrounds at PNNL, came together to conceive of a new technology for controlling underground root growth where plant roots were unwanted. The researchers' serendipitous convergence into this technology area was the result of a wisecrack at a bridge game related to the idea of putting plants on birth control! At the time, Dr. Burton was doing research at DOE funded by the World Health Organization on a slow-release contraceptive device. Dr. Van Voris' team was made up of a variety of backgrounds, including soil scientists, computers modelers, etc.

The Technology and Applications   The technology is a controlled-release root barrier that incorporates a chemical herbicide into a plastic or rubber material that slowly releases root inhibitor over time. When applied this way, the herbicide is not harmful to plant or animal life. Outside of its "zone," vegetation can flourish naturally, so it doesn't harm the environment. Also, the material is safe for children to be around, and it avoids the need to spray chemicals.

Because the technology was originally applied to radioactive sites to control root growth, it needed to release a steady dose of the herbicide over a hundred years. It also needed to control roots without killing the plants, and it needed to be non-soluble so it wouldn't be washed away. The trick was finding the right herbicide along with a material to which it could be attached such as a polymer or other synthetic rubber that would hold it in place over an extended period of time.

The PNNL group of researchers experimented with a number of herbicides that would control roots, including phosphoric acid, but they all involved cost and risk factors and high maintenance. Eventually, they found a commercially available herbicide called Treflan^R, produced by DowElanco, containing the active ingredient trifluralin. Trifluralin is not systemic, meaning it is not absorbed by the upper leaves of plants or trees, nor will it harm nearby landscaping, birds, mammals, or insects. Also, the concentrations required in order to inhibit root growth are low. It is biodegradable, meaning it decomposes in soil. If it were to be applied directly to soil rather than molded into plastic or a geotextile fabric, it would rapidly decompose. In addition, it is not very water soluble compared to other herbicides. It doesn't leach in water, and it is non-toxic. So, overall, it was an economical and ecologically sound choice for this technology. Treflan, which was already registered with the EPA, gained EPA approval in April 1990 specifically for use in landscaping and food crop applications.

Three major commercial applications resulted from this barrier: (1) protection for underground plastic watering pipes; (2) a fabric used under sidewalks roads and other structures; and (3) herbicide sewer gaskets. There are a number of other applications still being developed such as controlling insects and rodents.

For the commercial products being produced, the trifluralin active ingredient is shaped into small pellets and bonded onto another material, then slowly released in controlled uniform doses over time. In order to target the technology to particular applications, the researchers varied the herbicide's concentration and tested it against the roots of a group of control plants. They also experimented with several different types of candidate carrying materials that appeared to protect the herbicide from degradation. And, they varied the thickness of the various carrier materials. The technology could, thus, assume varying shapes and sizes. An "accelerated-diffusion" apparatus was used to test the long-term effectiveness of the pellets combined with other materials, since the carrier material needed to be relatively strong, yet inexpensive to process.

For two of the commercial products, the Treflan was fused into plastic pipes and fittings molded in the factory. For one of the products, the sewer gasket, the PNNL researchers actually designed, constructed, and tested the prototype themselves. In order to develop a prototype for the irrigation pipe application, they worked with Mr. Rodney Ruskin of Agrifim International.

For another commercial product, pellets or dots of the herbicide were permanently bonded several inches apart onto a geotextile fabric.

Geotextiles are the textile-like materials placed over soil to help prevent erosion or serve a variety of engineering and/or landscaping purposes. They are becoming increasingly popular. Dr. Van Voris discovered a geotextile fabric in his local hardware store called Typar[R], manufactured by DuPont. They began testing the herbicide in conjunction with this material after trying several others. They also experimented with and discovered different ways to join the herbicide with the material.[51] PNNL was fortunate to have been working with DuPont, a multi-billion dollar chemical company, before its line of Typar products came on the market.

Tests of the root barrier technology's ability to protect the asphalt shields from the plant roots were conducted at the two DOE Colorado sites. Despite a Nobel laureate and other experts brought in by PNNL management saying the technology wouldn't work,[52] the PNNL research team persisted, and the tests were successful.

The Laboratory   PNNL, formerly Pacific Northwest Laboratory, is a government-owned contractor-operated DOE laboratory in Richland, Washington, operated by Battelle Memorial Institute, the world's largest and oldest contract research organization. Battelle has operated PNNL since its establishment in 1965. It has an annual operating budget of almost $450 million and a staff of 4,000. PNNL is a multi-program laboratory. In this era of non-proliferation, the laboratory's modern mission involves solving broad-based environmental waste problems with one very major exception. PNNL is involved with radioactive waste disposal because it is adjacent to the DOE Hanford Site of fifty-year-old underground storage tanks containing defense nuclear wastes. The Hanford Site has an annual operating budget of over $1.2 trillion and staff of over 11,000. Because of this proximity, PNNL tends to focus on environmental restoration. One of the user facilities available at PNNL is the Hanford National Environmental Research Park. Another focal area for PNNL is global climate change. PNNL has earned a reputation for anticipating future national needs and strategizing and planning programs to meet those needs.

The technology was originally developed for the DOE system to protect its radioactive waste disposal sites from root penetration. At DOE's Hanford Site next door to PNNL, patrols ride around the site's 570 square miles to check the tumbleweed for radiation using Geiger counters. Similarly, in addition to many others, there are two DOE radioactive waste disposal sites in Colorado. These sites contain, not bomb remains, but uranium remains, called "tailings," from uranium mining, extraction, and milling conducted in the Grand Junction area. The tailings emit radon, a gas which emanates from the ground. As a result, DOE put an asphalt barrier called "Petromet" on top of the Colorado waste sites. Eventually, ultraviolet light from the sun damaged the asphalt, which cracked, and the radon gas came up through the cracks. So DOE put soil on top of the asphalt to protect it from the sun's ultraviolet light, and they stabilized this soil with plants. However, the plants eventually broke up the asphalt shield below them.

University Involvement   There was no university role in this case.

Funding, Financing   The total cost for bringing the technology to its first stage of development at PNNL was less than $250,000 over a three-year period. Rockwell International, which managed the Hanford Site at the time, contributed $20,000 to the development effort. An additional $600,000 from the Office of Nuclear Energy at DOE headquarters (due to applications at the nuclear waste disposal sites), allowed the work to continue and supported the Colorado tests. However, PNNL's managing operator, Battelle Memorial Institute, said the project didn't have any market potential and stopped supporting it. According to the team, their blessing was needed in order to get DOE money.

Only five percent of Dr. Van Voris' group's time is devoted to long-term controlled-release technologies. There are no sources of funding in the DOE's national laboratory system for this technology area. Therefore, the small amount of funding they receive includes private funds from several companies for long-term product development and public funds from the military for pest control.

The other 95 percent of the PNNL researchers' time is spent working on DOE mission areas and other agencies' problems. This work is not related to the herbicide/pesticide technology. While the weapons laboratories are considered the "big dogs" within the DOE system, PNNL is funded out of the Environmental Restoration and Waste Management (DOE/EM) offices at DOE, and the laboratory must compete for funds with other DOE laboratories doing work in this area. As with its other laboratories, DOE/EM decides the overall budget for PNNL, and hears pitches from the PNNL divisions in order to figure out how to allocate that budget. Dr. Van Voris[53] described this competition as "ruthless" and "cut-throat," both within PNNL and the overall DOE system. He advocated that the best way to survive in this type of environment was to practice strategic planning to handle both the laboratory level and headquarters level.

Intellectual Property   Back in 1979, the PNNL researchers filed an invention disclosure on this technology to control roots. Since then, there have been seven invention disclosures on various related applications.

The first patent was filed in 1983. However, the U.S. Patent and Trademark Office said that unless testing was actually conducted for a hundred years, the claims in the patent application were unacceptable, so what ended up being patented was the process. By 1989, seven additional patents were pending.

Technology Transfer Mechanisms   After the successful tests of the technology, dozens of firms were contacted as part of a concerted technology transfer outreach effort to find licensors. Also, during the 1980s, the researchers wrote a number of papers and made presentations (e.g., the *Journal of Controlled Release*,[54] *Nuclear Technology*,[55] *Water Engineering and Management*,[56] various Pacific Northwest Laboratory papers[57] and marketing materials), and contributed to books.[58] Later, after the technology was licensed, the irrigation application in particular received a great deal of attention in the printed media. For example, various combinations of PNNL laboratory researchers, Mr. Ruskin, and Agrifim company scientists presented and published in proceedings for the International Micro-Irrigation Congress,[59] American Society of Agricultural Engineers' National Irrigation symposiums,[60] International Erosion Control Association conference, American Society for Enology and Viticulture,[61] and others. In addition, there were articles by outside researchers involved in the technology that appeared in publications of the American Water Works Association, Hawaii Water Pollution Control Association, and Pan Pacific Green Industry conference. Trade journal publicity, which is more industry-oriented (while peer-reviewed scientific and technical journals are oriented toward university researchers), appeared in *Agriculture Engineering*,[62] *Irrigation News*, *Grape Growers*, *Nut Growers*, and others.

The technology transfer mechanism used with each partnering company is described below.

Agrifim Irrigation International: Agrifim Irrigation International, Inc., N.V. in California's San Francisco Bay area, obtained an exclusive worldwide license from Battelle to apply the technology to keep roots from invading underground watering and agricultural irrigation pipes. Agrifim sub-licensed its Battelle license exclusively to Geoflow™ Subsurface Irrigation, an Agrifim division.

Geoflow has been manufacturing and installing below-ground drip irrigation products for two decades. These systems are buried six to eight inches below ground, with one to two feet between the drip lines. They have many advantages over traditional above-ground watering systems. For one thing, watering with a drip system usually requires only a few minutes a day every second or third day. This type of irrigation has been used traditionally in agriculture, particularly vineyards. In fact, Geoflow has had a long-term relationship with the U.S. Department of Agriculture's (USDA)/Agricultural Research Service (ARS) Water Management Research Laboratory in Fresno, California. In addition to agriculture, the Geoflow watering products are now being applied to landscaping, sports/commercial turf applications such as football fields, and others.

However, one of the main drawbacks to underground systems is the intrusion of roots over time. Geoflow's founder and CEO Rodney Ruskin (originally with Agrifim) worked with the PNNL researchers using their technology to create a product called Rootguard[R], which protects underground systems from clogging caused by roots. The Rootguard technology in the plastic hoses prevents root tips from growing close to the water emitters in the hoses. Geoflow specializes in products with Rootguard protection and, in fact, is the only company that guarantees root control. Geoflow products containing Rootguard are engineered to last twenty years and warranted for ten years. If roots intrude within the first five years, Geoflow replaces the product for free; within the second five-year period, they are replaced at a proportional discount.

Reemay, Inc.: For the geotextile-based products, the fabric tested by the laboratory researchers was DuPont's Typar[R]. In late 1986, DuPont sold its division producing non-woven products and its Typar product line to the privately-owned InterTech Group of North Charleston, South Carolina. A subsidiary of InterTech Group located in Old Hickory, Tennessee, a chemical manufacturer called Reemay, Inc., bought the Typar product line and this technology. Reemay signed an exclusive license with PNNL to manufacture a geotextile fabric containing the herbicide pellets. Mr. Harry Barnes, a Reemay manager,[63] said the technology transfer process from the laboratory went well. The product was designed to protect two to three inches of soil from root penetration for 7 to 125 years. Incorporating the herbicide into a geotextile harbored the herbicide so that it did not quickly seep into groundwater, lakes, or streams.

Mantaline Corporation: To manufacture herbicide sewer gaskets, the Mantaline Corporation of Ohio obtained an exclusive license with PNNL.

The company developed and tested a polymer gasket impregnated with the herbicide, which prevented roots in sewer pipes for 25 to 50 years. Mantaline's product was called Root Shield™.

A fourth license was issued for the application called GrowGuard that involved putting a treated cord into pavement cracks. This application was tested at O'Hare International Airport in Chicago.

User Groups   Users of this technology include farmers, municipalities, facilities maintenance companies, and even homeowners. The technology has the potential to save these users millions of dollars each year in maintenance costs or in reduced water consumption. For example, routing out sewer lines is a $300 million annual cost in the United States. Local government public works departments are using the technology to protect roads, highway joints, sidewalks, and landscaping from roots and weeds. Curbs and gutters ruined by roots cost approximately $15 to $20 a foot to repair, and sidewalks costs $5 per square foot to repair. When these infrastructure costs are added to the cost of landscaping-related labor, it becomes apparent that the cost of installing a root control barrier represents a savings in the long run. Other areas where the technology could be applied includes both private and public driveway construction, swimming pools, and tennis courts.

The technology applications are currently being expanded to include insect and rodent control. In fact, the insect applications are currently being researched and tested, although not yet ready for market. For example, roach control is a $500 million market in the United States, and a $1.1 billion market globally. Termite control is a $130 million market in the United States (chemicals only) and a $.5 billion market globally (including labor).

Long-term application areas being targeted include biodegrading telephone poles and railroad ties, and decaying buried power/gas lines. There are twenty million telephone poles and five million railroad ties in the United States. The expected life cycle of each pole is 15 to 20 years. A million of them need to be replaced each year. The technology could be used as a fungicide to keep the poles and railroad ties from rotting. Presently, the labor involved in testing, applying Kreosote treatments, and replacing those ties and poles is a $5 billion global industry. Similarly, underground power lines and wires involve a potentially huge commercial market because tree growth and rodent attacks on buried power and gas lines is a major problem.

Barriers to Commercialization   According to Dr. Van Voris[64] there are some inherent problems that make industry adverse to commercializing controlled-release technologies. First, chemicals produced by DuPont, Dow Chemical, and others are sold by the carloads. With slower-acting controlled-release products, smaller quantities are needed, so the products are sold by the truckload rather than by the carload, and it is difficult to break the mindset toward larger quantities.

Another problem is that industry is interested in resales, so companies are reluctant to sell products that are effective as long as two years. As a compromise, the companies are now developing a material that lasts six months; the amount of the pesticide incorporated into the product is still only one-third of the amount that would be needed in another form of product for a mere three-month time frame.

In addition, there has not been a consumer demand for certain of the products because there were no existing/competing products. This makes it difficult to create visibility for new products entering the market. An example is the use of the root-controlling fabric under city sidewalks. Although buckled sidewalks pose a potential liability for local governments, there is still not a visible market for the product. And it doesn't help that appointed local officials maintain the attitude that they would be spending money for a product that would outlast them in their jobs.

As a result of all these problems, the technology has not resulted in an "instant revolution" as predicted, but rather evolution. Dr. Van Voris[65] said the so-called "green revolution" is just now becoming mainstream, and product spinoffs from this technology are just beginning to emerge. He adds that it is important to look at not just commercial value, but also other intangible benefits such as the fact that the technology has opened the door for other breakthroughs or that it may be reducing the risk of cancer and toxic exposure by children. From these perspectives, the technology's real value won't become apparent for fifteen to twenty years.

From the company perspective, Mr. Barnes of Reemay echoed that "these products are a different animal to market."[66] Reemay, for example, was a chemical and textile firm; but, with introduction of the Biobarrier product, the company is now in the pesticide market at a time when customers are cutting down on the use of chemicals that require expertise about wind and weather conditions. Customers must be convinced that the technology is not going to be harmful. Meanwhile, management saw instant success and didn't see this hidden problem. In addition, these products do not involve traditional marketing issues because the company is having to create the markets for the products. With Biobarrier, the product must be specified into the design of, say, a golf cart path while it is still on the drawing board. The golf course designer needs to be forward-thinking enough to know that a particular tree might ruin the path or the golf cart as it ages. As Mr. Barnes said, "You've got to sell the technology before you can sell the product."[67] To do this, Reemay exhibits at conferences of the Golf Course Superintendents, American Public Works Association, and the Water Department Directors. Reemay is also experiencing marketing problems with the weed control product that it is marketing to landscape designers rather than commercial landscaping firms, because landscaping firms are also in the business of weed control and are considered competition.

Other Factors   Dr. Van Voris' frustration with the DOE system and its lack of support for his team to continue its work on further technology development was apparent from indirectly related comments.[68]

User Benefits/Economic Impact/Outcomes   The FLC Winners' publication said that the product has resulted in millions of dollars of revenue to the product manufacturers. Dr. Van Voris notes that, from the perspective of private industry, $20 to 40 million in sales would be more desirable than the amounts currently being made. Because each of the companies here are privately-held, it is difficult to obtain and report quantifiable data on sales or workforce aspects.

Agrifim Irrigation: In the United States, products of Agrifim's division Geoflow have been installed mostly in California, Texas, and Hawaii. The four main areas for using underground watering systems with Rootguard[R] are for irrigating agricultural crops, landscape watering, turfgrass treatments, and now irrigation via recycled wastewater. The company lists a total of fifty agricultural sites where Rootguard has been installed, with users ranging from Silverstar Farms in Brush Prairie, Washington, to New York Vineyard in the State of New York. The sites involve crops as varied as alfalfa, asparagus, cantaloupes, citrus, cotton, hops, all types of nuts, prunes, raspberries, ginger, sugar cane, and vineyards of all types.

In the area of landscape watering systems, since the late 1980s, Geoflow has been servicing new markets such as public parks, golf courses, roadside median strips, military bases, shopping centers and storefronts, university campuses, senior citizen centers, rooftop gardens, and so on. The sampling of twenty-eight landscape site installations in the company literature ranges from K-Mart and Taco Bell to the Price Club.

Geoflow has also installed underground systems using Rootguard at turfgrass sites such as football fields and baseball diamonds. The fifteen examples of users provided by the company range from Travis Air Force Base in California to the Water Valley School in Texas.

Until Rootguard came on the market, it was difficult to do underground watering using recycled wastewater (also called "gray" water) from homes and other sources because roots were particularly attracted to this nutrient-rich water. It is now the water policy of several states (eg., Arizona, California, Florida) to convert a certain percentage of wastewater into reclaimed water for watering lawns, agricultural irrigation, and other non-potable applications. Geoflow's line of products in this area are called Wasteflow^{TM}. Wasteflow systems have been installed at the Fort Myers, Florida, airport; at two resort developments in Hawaii; at a Brigham Young University site; and in Melbourne, Australia. Although the use of Wasteflow systems reduces the need to build water treatment plants, it appears that these systems are roughly twice as expensive as the Rootguard systems. For example, 1,000 feet of half-inch hose containing Rootguard costs a little over $200, whereas the same item in the Wasteflow line costs about twice as much.

In other new areas, The Toro Company now has a sub-license for the technology from Geoflow Inc. to allow Toro to make and market underground watering systems for back yards and similar recreation areas. Another group has been testing a new product for controlling termites and fire ants at the USDA/ARS center in New Orleans. A foreign patent has been obtained for this application and the company will soon be launching and promoting the product worldwide. In the future, Agrifim plans to focus on aquatic weed applications of the technology. Other applications that haven't been exploited include swimming pools and landfills. Meanwhile, the laboratory is seeking new partners to pursue the sewer gasket application.

Reemay, Inc.:   Reemay's line of fabric-like products is called "Biobarrier," a name the company trademarked in 1988. The company's major product is the Biobarrier^R Root Control System, which slowly degrades in the ground. In fact, Reemay won an award in Europe for the product's unique approach to slowly releasing the chemical herbicide. The diverse interest in Reemay's fabric-like product now includes architects, state and local public works departments, construction companies, and sports companies.

However, Reemay's market was not always that diverse. Mr. Barnes[69] said the company spent a great deal of money and effort trying to sell to the nuclear waste market and ultimately did some work with DOE's Hanford Site and its Savannah River Plant. However, Biobarrier "wasn't going anywhere" as a product in this area in spite of the fact that this was the original application area and in spite of the great need for solutions in this area. It appears that this is a case of differing expectations. Reemay thought it would get more help or support from Battelle. The technology was licensed from Battelle with the thought that, as the licensor, Battelle would help them "get their foot in the door" in the area of nuclear applications and other parts of the DOE system. According to Mr. Barnes, the individual operating sites within DOE's Defense Programs are becoming competitive profit centers that contract out many of their activities, similar to the private sector downsizing trend of the 1980s. So when a company such as Reemay licenses from one laboratory within the system, that does not necessarily help in marketing to other sites or laboratories.

In the past two years, the Biobarrier product has branched to applications in municipalities where tree roots are uplifting sidewalk and road pavements. The Reemay fabric is designed to last fifteen years; whereas the expensive and labor-intensive process of pruning roots must be done every one to five years. Public works departments have tried everything from plastic sheeting to lumber to concrete barriers to control root problems. Biobarrier is easier to install, and the result is a decrease in accidents and related liability. Mr. Barnes said Biobarrier "found its home" in this market segment, and has started to move. In fact, Reemay considers that it is at the front end of the curve and the future looks bright for Biobarrier products. There are several factors contributing to this trend. First, more people are taking on leisure activities that require smooth paths (eg., roller-blading, skate-boarding, jogging, bicycling, and walking). Also, because of new state "green laws," more parks and scenic paths are being built and more new trees are being planted. In the past, Mr. Barnes says, if a tree damaged a sidewalk, the tree would be cut down. So, the product is "at the right place, at the right time," word is spreading, and there are now many customers. For example, horticultural experts at Virginia Tech and Radford University advised the city of Bristol, Virginia, to replace trees with five-foot trunks growing in three-foot spaces, and to use the root control product. Other examples include the city of Sanibel, Florida (thirty miles of bicycle paths); the city of Carmichael, California (nine miles of sidewalks in seven parks); the Portland, Oregon, airport (two miles of median strips); and, Sarasota County, Florida (focus on preventing tree damage by mowers, as well as mower damage by trees).

In late 1994, Reemay introduced a second product in the Biobarrier line called Biobarrier[R] II Preemergence Weed Control System. This is another fabric-like product that can be custom-contoured around existing landscaping designs by cutting it with a knife. Then the weed killing sheets can be covered with mulch, bark chips, gravel, sand, or soil. Biobarrier II controls weeds for at least a decade. Cutting long-term maintenance costs by using this product requires careful planning during design and construction phases for commercial and industrial landscaping. Users can choose between repeated chemical applications or high landscaping labor costs. Another advantage of Biobarrier II is that it releases the exact amount of herbicide that is biologically required, so landscapers don't have to calculate complicated ratios for diluting chemicals in water.

As a private company, Reemay does not readily divulge its exact product revenues, although Mr. Barnes said that the volume has not been as great as expected. The Reemay license is still in effect, and the company is paying royalties to Battelle. He added that Biobarrier is an expensive product to manufacture. This expense has to be passed on to the company's customers, who must be convinced of the product's value.

Reemay is also trying to create a new market for the protection of underground fiber-optic cable which gets crushed by root growth.

Mantaline: For two years, Mantaline (with Dr. Van Voris' help) sought an exemption from the Federal Insecticide, Fungicide, and Rodenticide Act. In 1987, Mantaline received regulatory approval from the U.S. Environmental Protection Agency to manufacture gaskets incorporating the herbicide. In 1988, Mantaline's President Robert ("Bob") Merian was quoted in a DOE publication as saying, "The possibilities for Root Shield are quite exciting . . .Product reception at city test sites is exceedingly strong . . .we are proceeding at full speed."[70] However, the sewer gasket failed as a commercial product and as a solution to infrastructure problems. It seems that public works officials were not eager to spend money for this product whose benefits would outlast the time frame of their own job. The Mantaline license is now dead.

International Activity   Geoflow proudly advertises that its Rootguard products are manufactured in the United States and, as noted from the examples, sold worldwide, having been installed in several locations in Spain, Canada, Australia, and New Zealand. However, its "made-in-the-USA" label applies only to basic products. Special Rootguard products containing both an herbicide and a bactericide to control water pollution are being produced in Israel, Columbia, India, and South Africa under license.

Reemay is only marketing in the United States. However, the company is doing full-scale product testing of two products to control termites and cockroaches in Australia. In addition, the cockroach product is being marketed in Japan.

Government Gains   Applications for the technology include fighting insect repellants, a problem being addressed by a variety of federal agencies. For example, the U.S. Department of Defense transfers funds to the U.S. Department of Agriculture and to the Army Corps of Engineers to reduce the amount of pesticides used at military sites. It is now mandated that military bases deploy pesticide strategies. According to Dr. Van Voris, the only way to do this long term is to support infrastructure maintenance by using this technology. Another application being considered is soldiers' uniforms that repel mosquitos and other insects over a long period of time.

Elapsed Time   The technology was conceived in the 1970s. The research geared up in 1978 and the first invention disclosure was filed in 1979. The tests in Colorado were conducted during the early 1980s. The journal articles, paper presentations, and other publications were written by the PNNL researchers primarily in the 1982 to 1986 time frame. At that time, DOE did not have a clear patent/licensing procedure in place for its contractor-operated laboratories, and it took some time to work the technology through the DOE system. The first patent application was filed in 1983. Eventually, the patent rights were released to Battelle, which started negotiating the first license in 1983. The license was eventually granted in 1985 to Mantaline. Mantaline's sewer gasket received regulatory approval two years later. The licenses with Reemay and Agrifim were both signed in 1986. Joint journal articles between the PNNL researchers and Mr. Ruskin of Agrifim were written primarily in the 1988 to 1993 time frame. Reemay's Biobarrier root control product has been commercialized for eight years and on the market for four to five years.

Case 6 (1985) – Radiation Therapy Quality Assurance

Role of Laboratory Researchers and Other Personnel   Dr. Robert Loevinger joined what was then called the National Bureau of Standards (NBS)[71] in 1968 as a radiation specialist in the dosimetry group. He worked in the Ionizing Radiation Division, which is located in the Radiation Physics

Building at NBS' Physics Laboratory Gaithersburg, Maryland. He retired twenty years later in 1988, working there on a part-time basis since then. Before joining NBS, Dr. Loevinger was a medical physicist. Today, virtually all radiation therapies in the United States are traceable for their accuracy to the measurement and calibration programs that were led by Dr. Loevinger at NBS.

In the radiation community, Dr. Loevinger is considered "Mr. Dosimetry." He has published over a hundred technical papers. His early-1980s article in *Medical Physics* outlined the basis for a new x-ray measurement protocol for calculating cancer patient dosages of radiation therapy.[72] Dr. Loevinger has received numerous other honors. In 1980, for example, he received a Department of Commerce Silver Medal for his work at NBS. In 1982, he was chosen to present an annual memorial lecture by the Health Physics Society. Dr. Loevinger, now retired, is also recognized as an international expert through his participation on influential committees in his field. He was a member of a radiation standards committee for the American National Standards Institute and was on the Medical Internal Radiation Dose Committee for the Society of Nuclear Medicine. Further, he continued working for some time with the Science Council of the American Association of Physicists in Medicine (AAPM), the subject of this case. He was also a permanent consultant for the AAPM Radiation Therapy Committee and a member of two of the committee's tasks forces.

Dr. Loevinger's work to establish a national quality assurance system for radiation therapy measurements was accomplished in cooperation with the Radiological Society of North America (RSNA) and the American Association of Physicists in Medicine (AAPM). The Radiological Society includes doctors specializing in radiology and medical physicists who work in conjunction with those radiologists. AAPM was founded in the 1950s and headquartered in College Park, Maryland. AAPM is the only national society in the United States whose major concern is the physics aspect of radiation therapy. Its membership is composed of radiation scientists and technologists from radiation centers in over 1,000 hospitals and clinics in the United States. They serve 1,200 radiation therapy facilities, which treat a total of 600,000 cancer patients each year. RSNA, the older of the two organizations, was the first medical group to require more standardization because of the growing number of persons needing radiation therapy for cancer (as opposed to diagnostic radiation used to x-ray broken bones, for example).

The AAPM meets twice a year: once in conjunction with RSNA and once on its own, because medical physicists are a "fiercely independent" group, according to Dr. Loevinger.[73] At one of these AAPM/RSNA conferences, Dr. Loevinger pointed out that national methods were needed in this area. His comment was received with interest, and a task group[74] under the Radiation Therapy Committee was appointed. The Radiation Therapy Committee, in turn, was under the AAPM Science Council. At first, the committee had only six to eight members, but it grew to include representatives of all the large cancer organizations. Dr. Loevinger was on the committee, but did not chair it because it would look too much like "the government telling them what to do." It took about two years for the committee to get started, but once it did, it was very effective. Because AAPM's task groups were temporary, the task group was changed to a subcommittee.

Technology and Applications   As in other highly technical areas, radiation measurements are traceable to the international measurement standards. Among the base units of measurement in the international measurement system are: length, time, mass, weight, temperature, and electrical charge. Other units, such as energy, are derived units. Defined physical quantities can be obtained from these basic units.

In the field of radiation therapy, the key quantities are (1) exposure and (2) absorbed radiation dose.[75] A dose is the quantity of radiation absorbed by a mass of tissue. The absorbed dose depends upon the strength and distance of the x-ray beam and the duration of exposure.[76] The process of measuring the dosage of x-rays or radiation is known as dosimetry, and the portion of the x-ray machine doing the measuring is known as an ionization chamber. The chamber measures the electrical charge in the air. The complete dosimetric system is comprised of the ionization chamber, an electrometer and its capacitor, and a voltmeter.[77]

Hospitals can build their own x-ray machines or buy them.[78] In either case, each x-ray instrument requires detailed documentation that shows on a step-by-step basis how the final dosage measurements are derived from the national and international standards. Thus, calibrations are the fundamental step in the determination of tumor doses for cancer patients. This x-ray dosimetry traceability system is the technology that was transferred.

Before the national x-ray dosimetry system was established, in principle, radiation measurements in the United States were done according to the international standards for dosimetry. In practice, however, these measurements were done in a vague, uncontrolled, and amateurish way in many institutions around the country. NIST's standards existed and NIST calibrated the x-ray machines for some national laboratories, certain Veterans Administrations hospitals, and various other institutions. But, there was no systematic process in place for hospitals, x-ray laboratories, and clinics so that patients could be assured that the dosage of radiation they were receiving would be accurate and consistent with the national standards.

However, instrument calibration is only the first step in the system, as established. There is also a protocol for calculating the amount of a patient's radiation dose (the subject of Dr. Loevinger's award-winning paper). The physics of establishing and documenting traceability relates to calculating the dosages. With traceability, the proper dose each patient should get from a radiation worker operating an x-ray machine or a megavoltage linear accelerator can be traced back to NIST. For example, for cancer patients, overall radiation therapy should be accurate to five percent.[79] For the overall radiation therapy to be accurate to five percent, the x-ray's beam and the machine have to be calibrated to one to three percent accuracy. A five percent level of error in each would result in a ten percent error in the overall treatment.

The Laboratory   The National Institute of Standards and Technology (NIST) was founded in 1901 with the explicit mission to work with industry to maintain national standards of measurement so that measurements are done in consistent units and terms. The physics laboratory that is the subject of this case is one of several major laboratories at NIST. Other laboratories deal with building and fire research, electronics, computing, manufacturing, materials, and chemistry.

As national laboratories for industry standards, many of the NIST facilities available for cooperative and proprietary use focus on measurements of all types. This includes, for example, facilities for measuring acoustics, high-voltages, and the flow of water and other fluids.

The standard-setting system in the United States is predominantly voluntary. NIST assists associations and groups establishing standards and funds research related to standards. NIST staff members make presentations and consult with industry groups and other organizations on the importance of traceability to national standards. About one third of NIST's employees, or 1,175 people, are members of standards committees.

To assist this process, NIST provides more than five hundred calibration services to ensure that users of precision instruments around the country achieve measurements of the highest possible quality. These services link a customer's precision equipment to national standards. More specifically, for calibrations, NIST personnel check, adjust, or characterize an instrument or device. When these instrument calibrations are done, customers are assured that measurements are consistent with national standards. These NIST instrument calibration services are available for a fee to outside organizations such as hospitals, universities, industry, other calibration laboratories, nuclear energy establishments, the U.S. Department of Defense, and other government laboratories. NIST's calibration services have grown over the years. For example, in 1984, the laboratory provided 468 calibrations; in 1995, NIST performed 9,200 calibrations.

NIST calibration services encompass seven major areas. These are the measurement of: dimensions, time, frequency, mechanics, thermodynamics, electromagnetics, and optical and ionizing radiation. The subject of this case is ionizing radiation. Credibility of ionizing radiation measurements has been a critical issue for the U.S. radiation medical diagnostics and therapy communities. Ensuring the accuracy of radiation doses in this country depends heavily upon NIST calibrations, reference materials, and laboratory accreditation services. NIST scientists work to disseminate the standards and technology required for reliable measurement of ionizing radiation to the medical community. In addition, they monitor and evaluate radiation measurement needs and participate in radiation research, metrology development, and quality control activities.

The NIST scientist in this case initiated the development of a method for improving the accuracy of field measurements used in radiation therapy in this country through a national system of secondary standards laboratories. As a result, in this area, NIST has had a strong influence on the design and implementation of a quality assurance program accredited under AAPM. In support of this program, NIST continues to provide technical expertise for traceability to national standards through on-site assessments.

University Involvement   As noted, some of the hospitals involved in the national quality assurance system are university hospitals.

Funding, Financing   Each of the quality assurance systems described in this case are voluntary and self-supporting, that is, the Regional Calibration Laboratories (RCLs) are not subsidized. Yet there are no official incentives such as regulations for tertiary-level users to utilize the system, with the exception of Nuclear Regulatory Commission regulations for specific units. Unofficially, legal implications and AAPM protocols are the incentives.

According to an AAPM-conducted, NIST-funded study, there are incentives to become an RCL, including prestige, convenience for the operating institution, and its role as an outreach function for the operating institution.[80] On the other hand, it also appears that there are incentives not to become an RCL. Not all of the RCLs had reached a financial break-even point and financial self-sufficiency as of the time of the study conducted in the late 1970s. The situation differs from one institution to the next in that certain of the for-profit institutions housing RCLs are doing well and performing efficiently because they must be financially self-sufficient. Other institutions have agreed to house an RCL mostly to meet their own in-house needs, and in those cases the RCL laboratory is subsidized by the institution itself. None of the RCLs is subsidized by the government or other outside source. Each RCL laboratory provides its own equipment for the purpose of performing calibration services.

Dr. Loevinger's work with AAPM and its committees was covered by his government salary. The AAPM committee sponsoring the national program was initially hoping to receive some monetary support from the American Cancer Society until the Society decided it didn't have enough money to support the program. NIST's Radiation Research Center provided financial support for the study conducted by the AAPM committee evaluating the network.

Intellectual Property [Not applicable.]

Technology Transfer Mechanisms Through the subcommittee, it was proposed that three, and later five, RCLs be certified by the AAPM according to AAPM-developed accreditation criteria. While national standards are considered primary standards, this system of five RCLs would be deemed the "secondary standards" level. The five RCLs would ship their x-ray dosimetry instruments to NBS to be calibrated according to the national standards. For those instruments to be suitable for that kind of transportation, the beams would be measured and checked against NBS's instruments and then calibrated. It was agreed that the measurements should agree to within a half percent of the NBS measurements. Then, at the next AAPM meeting, the results would be reported for the committee (now called the Regional Calibration Laboratory Accreditation Subcommittee) to review, and the source of any problems could be identified and taken care of. All other hospitals and medical physicists would have their instruments calibrated at the RCLs. This next level down is called tertiary standards, or the "field instruments" level. This level includes, for example, hospitals and radiation therapists in doctors' offices grouped together to form a clinic. Some field instruments are moved around and used at more than one institution.

Before each RCL was officially accepted, a visiting AAPM committee conducted site visits.[81] Although the committee members would change, there was always someone from NIST involved. They consulted with laboratory staff, examined the instruments and equipment, watched the procedures as practiced at that site, and observed the staff doing calibrations. Also, the committee would bring along "test case" instruments and have them calibrated. If accepted, that site would receive the official seal of approval from AAPM as an RCL.

The five RCLs certified by the AAPM include large private hospitals, a state university hospital, a Veterans Administration hospital, and a small privately-run group of physicists who calibrate machines. More specifically, they are:

- Memorial Sloan Kettering Cancer Center in New York,
- M.D. Anderson Hospital in Texas,
- University of Wisconsin in Wisconsin,
- Allegheny General Hospital in Pennsylvania,
- K&S Associates, Inc. in Tennessee, a relatively newly-formed commercial standards laboratory.

Decisions as to how often the machines need to be calibrated are made by the committee and depend upon time and money available. They are not legally binding. At first, the committee continued to require that the RCLs' instruments be calibrated by NIST once a year via mail.[82] However, NIST's charges were high enough that this was eventually changed to every second year. During the in-between year, the RCLs cross-check in a round-robin fashion among themselves; that is, each sends their instruments to one or more other RCLs and the results are compared. This is less expensive than doing it through NIST. The results should continue to be in close agreement, about a half to one percent level of error. If it becomes apparent that a machine is not calibrated accurately, it would then be checked with NIST. Meanwhile, the site visits are repeated every three to five years. When the RCLs perform their calibration service for the tertiary level hospitals and institutions, they know an instrument's calibration needs to be accurate to one percent in order to conform with NIST's standards. At the tertiary level, it varies as to how often a piece of equipment is calibrated. For example, some machines are calibrated more than once a year, and others are only calibrated when they are first received at the hospital or clinic.

NIST is not allowed to close its doors to individual hospitals, so it still performs the calibration service for some hospitals and other government institutions, particularly Veterans Administration hospitals. However, it is now primarily interested in performing the service for RCLs.

User Groups   The calibration involves x-ray machines and accelerators used for radiation therapy, almost exclusively for cancer treatment. They are not used for treating non-malignant disorders because they are too dangerous. In the United States, approximately 600,000 people undergo radiation therapy for cancer each year. It is estimated that one-third of the population will get cancer; about half are treated with radiation, and the other half undergo surgery. Cancer therapy involves a large portion of the population.

Barriers to Commercialization   The technology was not commercialized, but transferred into general use. It was moved into general use "quickly and painlessly" according to the FLC Winner's publication, so it appears that there were few barriers to implementation and that the program must have been sorely needed at the time it was initiated.

Other Factors   The NIST program staff publishes a regularly updated annual catalog of measurement facilities and other types of instruments available to outside groups, along with the cost of using each device. It also describes several hundred types of services offered[83] including its x-ray calibration service, describes the calibration protocols, and lists the fee schedule. The charge for calibrating x-ray instruments averages $500. Dr. Loevinger[84] said he didn't want to charge outside institutions for that service, but NIST rules required it.

User Benefits/Economic Impact/Outcomes   In the late 1970s, NIST funded a study to determine the adequacy of the RCL system at that time. The study determined that the system as established[85] was working well, but that two additional RCLs would be needed. However, no systematic studies have been conducted since then, so there are no quantitative measures of the impact of the system. The current chair of the AAPM committee, Dr. Geoffrey Ibbott,[86] director of the University of Kentucky's Department of Radiation Medicine, indicated it would be difficult to do a quantitative comparison between the process of twenty to thirty years ago with the current system.[87]

However, anecdotal evidence indicates that this quality assurance system has been successful and will probably continue to be successful. The direct beneficiaries of the system are the nation's hospitals. They benefit financially because it is less expensive to obtain a secondary standard than to obtain a primary standard from NIST. Before this, NBS was the only source in the United States of authoritative calibration of radiation equipment, which made certified calibration costly. Also, calibrations through NIST were more difficult to arrange for many health clinics and laboratories. A regional laboratory is usually closer, so there is less risk in transporting the equipment. As a result, more calibrations are being done by private sector laboratories, so there is better local service to the end users, the hospital patients. This has been an excellent network, serving as a model for other such networks. (See "Government Gains," below.)

The system ultimately is important because people's lives are at stake. People can be injured in radiation therapy due to faulty calibrations. The program works not only to ensure public health and safety, but also to reduce the probability of lawsuits, thereby holding down health costs. Virtually all the lawsuits related to a patient being injured have resulted from a medical physician or physicist being careless or ignorant.

International Activity   In all countries, a government laboratory such as NIST (or group of government laboratories) maintains the national standards that are considered primary standards. Virtually every country's national (or primary) measurement standards are the same because they can be traced to international measurement standards which are provided by the "Bureau International des Poids et Mesures" (BIPM)[88] in Paris. All industrialized countries are members of this organization so that measures are done consistently around the world. If not for this international measurement system, for example, it would not be efficient for American automobile parts to be manufactured in different countries.

Every two years in Paris, the United States and other countries come together and compare each country's primary standards with those of the international system. The international measurement system is independent of politics; it includes countries in both the East and West. The only time the system did not work and all countries did not meet was during World War II.

National or primary standards and BIPM's standards, referred to as "independent primary" standards, are considered co-equal for all practical purposes because BIPM compares differences in the instruments to plus or minus a half percent standard deviation in the level of error from the mean. In the area of ionizing radiation, it is rare for a national primary standard to be more than one percent different from BIPM's.

Government Gains The accreditation system set up with AAPM served as a model for another program established through NIST, which is also a spinoff of Dr. Loevinger's original quality assurance system. This program established standards and "certified" laboratories among the federal laboratories for calibrations for the x-ray *protection* instruments used by radiation workers. This involves workers primarily in federal laboratories such as NIST, but also some private institutions. Dr. Loevinger said that although he did "bug" the radiation protection community to get started in this area, he was not involved in setting it up.[89]

Occupational exposure does not need to be measured to such a high level of accuracy as direct x-ray beams aimed at a patient because this type of exposure involves a much lower level of intensity. A ten to twenty percent level of accuracy is acceptable. In these situations, the measurement system is more of an ongoing process. It is also a bigger operation, because 1.3 million radiation workers are affected.[90] As part of the protective gear for radiation workers, they all wear film badges or carry pocket chambers which measure their exposure over a period of time.

The system for calibrating radiation protection measurement equipment involves the radiation workers themselves as users, rather than patients. Also, a different bureaucracy checks the equipment. For example, NIST might receive a batch of 500 to 1,000 film badges at a time from, say, Brookhaven National Laboratory. Nevertheless, this system involved establishing a similar measurement quality assurance network with industrial and federal laboratories.

In this case, links were established with three U.S. Department of Energy laboratories involved in radiation work—Argonne National Laboratories in Chicago, Pacific Northwest National Laboratory in the State of Washington, and Lawrence Livermore National Laboratory in California; a Food and Drug Administration (FDA) laboratory, the Center for Devices and Radiological Health, in Rockville, Maryland; a private laboratory called Radcal in California; and another private laboratory called Eberline located in both South Carolina and New Mexico.

Elapsed Time Dr. Loevinger joined NIST in 1968, and began working with the AAPM subcommittee around 1970. The first three RCLs took form in 1972. An AAPM committee was established to conduct an independent study of the adequacy of the RCL network beginning in 1976, and subsequently the two additional RCLs were added to it.

The RCL Accreditation Subcommittee was chaired by the same individual, Martin Rozenfeld of the Department of Therapeutic Radiology, Rush-Presbyterian–St. Luke's Medical Center in Chicago for almost two decades, from 1975 until 1995. It is now chaired by Dr. Geoffrey Ibbott.

The system for the radiation protection community was put into place about ten years after the radiation therapy system was established.

Pre-legislation Findings Summary

Appendix D summarizes the findings for each of the selected pre-legislation cases organized according to the questions addressed in the interviews. Whereas, the earlier Level II analysis (appearing in this chapter as the introduction) addressed only certain key questions for the eight FLC awardees for the years 1985 and 1986, the appendix addresses all of the questions for the selected cases documented in this chapter. The summary combines all the cases and presents them in the same topic order as the selected cases except the topic, "Other Factors," which is incorporated into the concluding section of this study. In the appendix, the cases appear in the following order:
a. Penetrometer
b. Thermoplastic Polymer
c. Substance Tracer
d. Alginate Herbicide
e. Root-Control Barrier
f. Radiation Measurement Standards.

The next chapter presents the cases and findings summary from the 1992 to 1993 (post-legislation) time frame.

Notes

1   Confidence in the results was confirmed by having two laboratory researchers read their cases. They were generally pleased and made some comments and minor corrections.

2 1994 FLC "Winners' Document."
3 J. N. Suhayda and C. Ingram, Field Testing and Evaluation of an Expendable Bottom Penetrometer System for Automatic Sea Bed Classification and Estimate of Sediment Shear Strength, Prepared for Naval Oceanographic Office, Marine Geological Laboratory Report No. 523, December 1984.
4 Interview with Mr. Ingram, September 4, 1996.
5 Interview with Dr. Suyhayda, September 23, 1996.
6 Interview with Mr. Ingram, September 4, 1996.
7 Interview with Dr. St. Clair, August 8, 1996.
8 Ibid.
9 Interview with Dr. St. Clair, August 8, 1996.
10 Interview with Mr. Evans, September 16, 1996.
11 Ibid.
12 James L. Haggerty, *NASA Spinoff 1992*, National Aeronautics and Space Administration, Office of Commercial Programs, Technology Transfer Division, ISBN 0-16-038211-4, 1992, p. 109.
13 Interview with Dr. St. Clair, August 8, 1996.
14 Interview with Mr. Evans, September, 16, 1996.
15 Ibid.
16 Ibid.
17 Interview with Dr. St. Clair, August 8, 1996.
18 R. N. Dietz, *Commercial Applications of Perfluorcarbon Tracer (PFT) Technology*, Presented at the Department of Commerce Environmental Conference, "Protecting the Environment: New Technologies—New Markets," Reston, Virginia, September 5-6, 1991, BNL 46265, June 1991, revised September 1991.
19 Ibid.
20 Interviews with Dr. Dietz, August 16, 1996 and September 16, 1996.
21 William J. Connick, Jr., "Controlled Release of the Herbicides 2,4-D and Dichlobenil from Alginate Gels," *Journal of Applied Polymer Science* 27: 3341-3348, 1982.
22 H. Lynn Walker and William J. Connick, Jr., "Sodium Alginate for Production and Formulation of Mycoherbicides," *Weed Science* 31: 333-338, 1983.
23 D. R. Fravel, J. J. Marois, R. D. Lumsden, and W. J. Connick, Jr., "Encapsulation of Potential Biocontrol Agents in an Alginate-Clay Matrix," *Phytopathology* 75: 774-777, 1985.
24 William J. Connick, Jr., "Formulation of Living Biological Control Agents with Alginate," Chapter 19 in *Pesticide Formulations: Innovations and Developments*, Washington, D.C.: American Chemical Society, 1988.
25 Interviews with Mr. Connick, August 28, 1996 and September 6, 1998.
26 R. C. Axtell and D. R. Guzman, "Encapsulation of the Mosquito Fungal

Pathogen *Lagenidium Giganteum* (Oomycetes:Lagenidiales) in Calcium Alginate," *Journal of the American Mosquito Control Association* 3 (3/ September): 450-459, 1987.

27   G. R. Knudsen, J. B. Johnson and D. J. Eschen, "Alginate Pellet Formulation of a *Beauveria bassiana* (Fungi: Hyphomycetes) Isolate Pathogenic to Cereal Apids,*" Journal of Economic Entomology* 83 (6): 2225-2228, 1990.

28   D. J. Daigle and P. J. Cotty, "Formulating Atoxigenic *Aspergillus flavus* for Field Release," *Biocontrol Science and Technology* 5: 175-184, 1995.

29   G. J. Weidemann and G. E. Templeton, "Efficacy and Soil Persistence of *Fusarium solani* f. Sp. *cucurbitae* for Control of Texas Gourd (*Cucurbita texana*)," *Plant Disease* 72 (1): 36-38, 1988.

30   D. G. Kim and R. D. Riggs, "Efficacy of the Nematophagous Fungus ARF18 in Alginate-clay Pellet Formulations Against *Heterodera glycines*," *Journal of Nematology* 27 (45): 602-608, 1993.

31   J. W. Bennett, A. J. Turner, A. K. Loomis, and W. J. Connick, Jr., "Comparison of Alginate and 'Pesta' for Formulation of *Phanerochaete Chrysosporium*," *Biotechnology Techniques* 10 (1/ January): 7-12, 1996.

32   Interviews with Mr. Connick, August 28, 1996 and September 6, 1998.

33   Uma Verma and R. Charudattan, "Host Range of *Mycoleptodiscus terrestris*, a Microbial Herbicide Candidate for Eurasian Watermilfoil, *Myriophyllum spicatum*," *Biological Control* 3: 271-280, 1993.

34   Richard M. Johnson and Armand B. Pepperman, "Mobility of Atrazine from Alginate Controlled Release Formulations," *Journal of Environmental Health Sciences* 30 (1): 27-47, 1995.

35   Interview with Dr. Georgis, August 28, 1996 and fax communication from him, September 23, 1996.

36   Interviews with Mr. Connick, August 28, 1996, and September 6, 1996.

37   Interview with Dr. Walter, September 12, 1996.

38   Interviews with Mr. Connick, August 28, 1996 and September 6, 1996.

39   Interview with Dr. Walter, September 12, 1996.

40   Interview with Dr. Walter, September 12, 1996.

41   R. Georgis, D. B. Dunlop, and P. S. Grewal, "Formulation of Entomopathogenic Nematodes," Chapter 13 in *Biorational Pest Control Agents*, Washington, D.C.: American Chemical Society, 1995.

42   Interview with Dr. Georgis, August 28, 1996, and fax communication from him, September 23, 1996.

43   Interview with Dr. Georgis, August 28, 1996, and fax communication from him, September 23, 1996.

44   Interviews with Mr. Connick, August 28, 1996, and September 6, 1996.

45   Dr. Raulston recently received a 1996 FLC Award of Excellence "for discovery, development, and commercialization of the Riobravis nematode for control of

soil-inhabiting insect pests of major economic importance."

46 Interview with Dr. Georgis, August 28, 1996, and fax communication from him, September 23, 1996.

47 Interviews with Mr. Connick, August 28, 1996, and September 6, 1996.

48 Jianying Gan, Manzoor Hussain and Nasir M. Rathor, "Behaviour of an Alginate-Kaolin Based Controlled-Release Formulation of the Herbicide Thiobencarb in Simulated Ecosystems," *Pesticide Sciences* 42: 265-272, 1994.

49 Debongnie Ph. and Pussemier L., "Encapsulation de Carbofuran ou D'Aldicarbe Dans des Billes D'Alginate: Liberation Retardee, en Solution et sur Colonne de Sol," *Med. Fac. Landbouww. Univ. Gent*, 1992.

50 Qi Mengwen, Wang Fujan and Wang Huaguo, "Study on Release Dynamics of C-Labelled Herbicides from Controlled-Release Formulation into Water," *Acta Agriculturae Nucleatae Sinica* 8 (4): 240-246, 1994.

51 Later, a DuPont scientist developed a more sophisticated injection-molding process for handling this.

52 Robert Cassidy, "The Blood, Sweat, and Years of Developing a Product: How Researchers at Battelle's Northwest Lab Waged War to Bring a Seemingly Hopeless Idea to Market," *Research & Development* 31 (3/ September): 58-64, 1989.

53 Interview with Dr. Van Voris, September 19, 1996.

54 F. G. Burton, W. E. Skiens, J. F. Cline, D. A. Cataldo and P. Van Voris, "A Controlled-Release Herbicide Device for Multiple-Year Control of Roots at Waste Burial Sites," *Journal of Controlled Release* 3: 47-54, 1986.

55 J. F. Cline, D. A. Cataldo, W. E Skiens and F. G. Burton, "Biobarriers Used in Shallow Burial Ground Stabilization," *Nuclear Technology* 58: 150-153, 1982.

56 P. Van Voris, D. A. Cataldo et al, "Stopping Root Intrusion in Sewer Systems," *Water Engineering and Management* (July): 6-9, 1986.

57 J. F. Cline, F. G. Burton, D. A. Cataldo, W. E. Skiens and K. A. Gano, *Long-Term Biobarriers to Plant and Animal Intrusions of Uranium Tailings*, Richland, Washington: Pacific Northwest Laboratory, PNL-4340, 1982; see also, D. A. Cataldo and P. Van Voris, "A Study of Product and Environmental Concerns with Irrigation Devices Containing Treflan," Richland, Washington: Battelle Pacific Northwest Laboratories, Laboratory Project ID 23113207292, unpublished.

58 F. G. Burton, D. A. Cataldo, J. F. Cline and W. E. Skiens, "The Use of Controlled Release Herbicides in Waste Burial Sites," in *Controlled Release Delivery Systems*, T. J. Roseman and S. Z. Mansdorf, editors, New York: Marcel Dekker, Inc., 1983.

59 P. Van Voris, D. A. Cataldo and R. Ruskin, "Protection of Buried Drip Irrigation Devices from Root Intrusion Through Slow-Release Herbicides," *Fourth International Micro-Irrigation Congress, Congress Proceedings,*

Volume 2, Albury-Wodonga, Australia, October 23-28, 1988.

60 R. Ruskin, "Reclained Water and Subsurface Irrigation," Presented at *ASAE International Winter Meeting, December 15-18, 1992,* Paper No. 922578, St. Joseph, Michigan: American Society of Agricultural Engineers, 1992; see also, A. Sanjines and R. Ruskin, "Root Intrusion Protection for Subsurface Drip Emitters," Presented at *ASAE International Summer Meeting, June 23-26, 1991,* Paper No. 912047, St. Joseph, Michigan: American Society of Agricultural Engineers, 1991; and R. Ruskin, D. A. Cataldo and P. Van Voris, "Root Intrusion Protection of Buried Drip Irrigation Devices with Slow Release Herbicides," *Proceedings of the Third National ASAE Irrigation Symposium, Phoenix, Arizona, October 1990,* p. 211-216.

61 Rodney Ruskin, "Subsurface Drip Irrigation in Vineyards," American Society for Enology and Viticulture, Presentation at the Annual General Meeting, Sacramento, California, June 23-25, 1993.

62 Rodney Ruskin, "Underground Irrigation," *Agricultural Engineering* (March): 9-11, 1993.

63 Interview with Mr. Barnes, September 4, 1996.

64 Interview with Dr. Van Voris, September 19, 1996.

65 Interview with Dr. Van Voris, September 19, 1996.

66 Interview with Mr. Barnes, September 4, 1996.

67 Ibid.

68 Until recently, DOE provided funding for specific laboratory technology transfer projects and CRADA work.

69 Interview with Mr. Barnes, September 4, 1996.

70 Battelle, *Putting Science and Technology to Work: A Casebook of Transferred Technologies,* Prepared for DOE Pacific Northwest Laboratory, PNL-SA-16279, October 1988, p. 4.

71 The 1989 NIST Authorization Act changed NBS to the National Institute of Standards and Technology (NIST). The 1988 Omnibus Trade and Competitiveness Act established the new Technology Administration (TA) at Department of Commerce headquarters, with the Gaithersburg laboratory reporting to DOC/TA.

72 The article, which described his analysis, received the journal's "best-paper-of-the-year" award.

73 Interviews with Dr. Loevinger, August 27, 1996, November 12, 1997, and December 9, 1997.

74 Called "TG-3."

75 "Guidelines for Use of the Modernized Metric System," *Dimensions/NBS* (December): 13-19, 1979.

76 See R. Loevinger and T. P. Loftus, "Uncertainty in the Delivery of Absorbed Dose," *Ionizing Radiation Metrology: Collected Papers from the International*

Course, Varenna, Italy, 1974, E. Casnati, editor, Sponsored by CEC, CNEN and CNR, 1977.

77 Another way to measure absorbed dose is to measure the rise in temperature in energy imparted per unit mass using a calorimeter; although this method is more reliable, it is also very difficult and expensive and involves much equipment.

78 The latter is more common.

79 A level of accuracy of more than that is not considered medically important.

80 Lawrence H. Lanzi, Martin Rozenfeld and Peter Wootton, "The Radiation Therapy Dosimetry Network in the United States," *Medical Physics* 8 (1): 49-53, 1981.

81 Generally comprised of Dr. Loevinger and a couple of members of the "mother committee."

82 There are AAPM-established shipping protocols.

83 Related to calibrating scales, weights, etc.

84 Interviews with Dr. Loevinger, August 27, 1996, November 12, 1997, and December 9, 1997.

85 There were only three RCLs at the time.

86 Interview with Dr. Ibbott, September 23, 1996.

87 A federally funded ongoing study, called the "Patterns of Care Study," of the care provided to radiation-treated cancer patients in this country was a possible source of data. But when the investigators involved in this study at the American College of Radiology in Philadelphia were contacted, it was determined that their inquiry did not include data concerning the accuracy of calibration equipment measurements. (Interview with Dr. Jean Owens, October 1, 1996.)

88 International Bureau of Weights and Measures.

89 Interviews with Dr. Loevinger, August 27, 1996, November 12, 1997, and December 9, 1997.

90 Radiation workers handle not only patients undergoing radiation therapy, such as for cancer, but also those patients x-rayed for diagnostic purposes or several million patients each year.

4　Later Winners

This chapter describes post-legislation cases from 1992 and 1993. It introduces the cases using the "Level II Analysis" described in Chapter 1's section on research design. Using information from the 1994 FLC Winner's Document, all of the FLC 1992-93 awards are organized by federal department or agency and topic area. The topics are based on the interview topics defined in Chapter 1's section, "Core Elements of the Government Technology Transfer Process." The following topics are included in the introductory section: technology applications, role of the laboratory researchers and other personnel, technology transfer mechanisms, intellectual property, user benefits/economic impact/outcomes, government gains, elapsed time, and other factors. In order to avoid excessive duplication, topics are addressed primarily with illustrative examples rather than a comprehensive survey of all the cases.

Following the Level II analysis of the 1992-93 group, eight selected cases are examined in greater detail.[1] Because of this more extensive examination, examples for the introductory Level II analysis are largely drawn from the cases not selected. After presentation of the eight cases, the final section of the chapter groups the key data from the eight cases according to the topics.

Introduction – Level II Analysis, Post-legislation Awards

Departments/Agencies and Technology Applications

The seventeen 1992-93 awards were distributed between laboratories of departments and agencies as follows:
- Agriculture (two technologies) – a technology for testing product quality in the paper industry; and a fat substitute for the food industry,
- Air Force (one technology) – a multiband digital speech processor,
- Army (one technology) – establishment of a geographic information system,
- Commerce (two technologies) – an improved gravity measurement

device and optical fiber sensors for use by utilities and manufacturing firms,

- Energy (eight technologies) – a laser method to light up and detect biological samples like genes; a microwave furnace with variable frequencies; an artificial heart pump flow tracking technique; laser lithography technology to etch semiconductor chips; an advanced process for manufacturing metals; establishment of a microelectronics quality reliability center; a nondestructive ultrasound testing technique; and circuits to measure superconductivity for a variety of uses— medical, geological, laboratory instruments, etc.
- Navy (three technologies) – glow-in-the-dark light sticks for fishing industry and military operations; technical assistance provided by a laboratory researcher on paints and coatings; and an advanced polymer resin material.

As with the 1985-86 awards, there were interesting examples of technologies from one field of science being applied to totally different fields with successful results. For example, a team of seven researchers developed a software program to meet military needs, then worked to identify civilian applications. They targeted three users and set up a series of technology transfer mechanisms.

Roles of the Laboratory Researchers and Other Personnel

The seventeen 1992-93 awards involved a total of 31 researchers (25 males, three females, and three persons whose initials or foreign names were not identifiable as male or female).

As with the 1985-86 awards, the role of these laboratory researchers varied. However, the 1992-93 researchers appeared to be more proactive early-on in reaching out to users and industry. There are numerous examples, including the following:

- A researcher identified as having an "entrepreneurial spirit"[2] established a center to transfer reliability testing technology. He visited companies across the country to learn about their needs and convince them government laboratories had something to offer.
- Researchers at DOE's Lawrence Livermore National Laboratory developed a laser for x-ray lithography and "cooperated closely" with other researchers at AT&T, IBM, MIT, University of California– Berkeley, the National Institute of Standards and Technology (NIST), and Naval Research Laboratory (NRL) so they could review it.

- Another group of DOE researchers built a new type of circuitry for a superconducting device co-developed by the laboratory and a company.
- A team of DOE researchers explored industry's need for a particular measuring device and then when companies (e.g., General Motors) indicated interest, demonstrated the device.
- A scientist at the Naval Research Laboratory developed a material and helped a licensing company develop manufacturing procedures. He then helped a CRADA partner identify additional applications and refine the manufacturing procedures.
- A DOE scientist developed a metal manufacturing process which avoided metal fatigue and produced a lighter weight material. He "sparked" the interest of potential producers and users by mailing surveys and then focused on working with the interested respondents.
- A NIST researcher developed an advanced optical sensors technology and solicited interest from companies. Many companies responded, and a relationship developed between the laboratory and 3M Company.

Intellectual Property

Only one patent application was noted in the information on the 1992-93 awards. Many more patents were uncovered in the interviews.

Technology Transfer Mechanisms

In the 1992-93 awards, at least thirteen CRADAs were signed or were in the process of being negotiated. CRADAs are new mechanisms allowed by the 1986 and 1989 legislation. For example, six CRADAs allowed various companies interested in a laboratory's metal manufacturing process to test the process for their products. Creation of a Microelectronics Quality/Reliability Center at Sandia National Laboratories led to that laboratory's first CRADA in 1991. The CRADA was so successful that Phillips Semiconductor and National Semiconductor negotiated follow-on agreements. The new center also led to partnerships with Hewlett-Packard, LSI Logic Corporation, and Olin Hunt Specialty Products, Inc. which resulted in new products on the market.

The laboratories and their partnering companies signed nine license agreements in the 1992-93 awards. For example, Quatro Corporation licensed the non-destructive testing system from Los Alamos National Laboratory. The company commissioned a market survey and developed a business plan that balanced company and laboratory interests.

Some of the 1992-93 awards involved combinations of transfer mechanisms. For example, the Cardolite Corporation licensed NRL's advanced resin technology and manufactured and sold resins to the composite materials industry. The Thiokol Corporation signed a CRADA with NRL to work on aerospace applications for the resins and to improve the manufacturing process.

The laboratories used other mechanisms besides CRADAs and licenses in the 1992-93 awards. In order to codevelop the superconductive device, Conductus officially participated in a Laboratory/Industry Exchange Program with the laboratory. Although this program was a DOE program, all laboratories were allowed to engage in personnel exchange programs. In 1992, Conductus and the laboratory entered into a CRADA to share researchers and further refine the technology.

Agency demonstration programs also transfer technologies. DOD uses demonstration programs to help boost the military industrial base. The transfer mechanism used for the lithography laser was a DOD-sponsored demonstration program conducted through the DOE laboratory with Hampshire Instruments, a small company that developed the concept. Hampshire Instruments developed a high-resolution machine, but it lacked the necessary power for industrial applications. So the laboratory's researchers were seeking industrial partners to improve the process for producing circuit boards.

NIST signed an agreement with 3M Company where 3M contributed funding and sent a guest scientist to the laboratory for a year in order to develop a prototype optical fiber sensing instrument to show potential customers. In negotiating the agreement, 3M became aware of NIST's capabilities, which led to other agreements with the laboratory.

Once it became apparent that the military's Geographic Resources Analysis Support System (GRASS) was beneficial to a variety of users, DOD established a "GRASS Inter-Agency Coordinating Committee" and an "Office of GRASS Integration." To reach private sector users, a non-profit corporation called the "Open GRASS Foundation" was formed. All three mechanisms were intended to reach educational institutions. CRADAs were signed to develop the geographic system for specific user needs.

A researcher at the Naval Civil Engineering Laboratory, an expert in paints and coatings, responded to over sixty technical assistance requests from outside organizations in four years, an average of fifteen a year, at no cost. The scientist responded to questions from companies and state and local governments about surface preparation, failure analysis, and product selection. In responding, the researcher directed them to solutions developed by the Navy and other sources.

User Benefits/Economic Impact/Outcomes

Several companies implemented new product lines and enjoyed product sales and revenues. Several nondestructive testing systems based upon ultrasound technology were sold by Quatro Corporation. In lithography, the United States led the world in producing equipment for manufacturing integrated circuits in the early 1980s, but that lead eroded by 1990. Hampshire Instruments, which was involved in the DOD lithography demonstration, subsequently announced the sale of two systems to major U.S. manufacturers. Cardolite, which licensed NRL's advanced resin, estimated its sales of these materials would be $5 to $15 million within a few years because of the growing number of potential applications.

Other companies have plans or are entering new commercial markets based upon the 1992-93 technologies. One of the companies with a DOE metal processing CRADA was planning to base a new product line on the technology. Conductus based several products on the magnetometer technology that it was marketing. Potential customers included medical diagnostic firms, petroleum surveying companies, electronics manufacturers, and research and educational organizations. The work between NIST and 3M led to new contracts and markets for 3M because the optical fiber technologies offered increased instrument speed and sensitivity.

Several companies experienced cost savings because of new technologies. For example, 3M expected reduced cost because bulk manufacturing costs were anticipated to be quite low.

The clients of the Navy paints expert who provided technical assistance estimated they saved at least $600,000 as a result of the expert advice. This was a result of businesses being able to remove layers of lead-based paint from buildings being remodeled.

The 1992-93 awards included an example of ensuring better quality products. Because of Sandia's Reliability Center, U.S. integrated circuit manufacturers were able to produce circuits with higher quality and reliability. For example, the Center characterized the electromigration tolerance of chemical vapor deposition, performed contrast-induced voltage analysis, and provided other services for industry.

Government Gains

There are several examples of unanticipated government gains in the 1992-93 awards. NASA heard about the optical fiber current sensor that NIST was developing with 3M and decided to sign an interagency agreement with NIST for a small lightweight prototype for NASA spacecraft.

In response to industry demand, the Microelectronics Center at Sandia became a broader "Electronics Center" and the same group of researchers are establishing a "National Center for Ultra-Reliability Engineering."

The Navy researcher's expert advice allowed the Navy to establish unexpected partnerships with private sector companies such as Disneyland (which was able to recoat the channels of the "It's a Small World" ride without draining out the water).

Finally, certain technologies also provided good examples of dual use. The geographic information system was being used by not only the Army, but also the Soil Conservation Service; U.S. Geological Survey; National Park Service; and many other federal, state, and local organizations. It was also used by more than a hundred developing countries.

Elapsed Time

In five of the seventeen awards, the award was made because of the speed in transfering the technology. The awards honored the "rapidity," "swiftness," and "dispatch"[3] with which researchers moved the technologies to the marketplace. The DOE laser lithography method was awarded in 1992. The demonstration for circuit manufacturers was begun in 1990. By 1991, a preproduction prototype was developed for commercial use and sold shortly thereafter. The developer of the metal manufacturing process was awarded in 1992 because he put into place (or began negotiating) six CRADAs in only two years. Developers of the geographic software package were awarded in 1993. Within seven years of the system's debut, land managers around the world were using it.

Other awards provided useful background information on timing. The company co-developing the superconductive device (awarded in 1992) began collaborating with the laboratory in 1989. By 1991, they had produced the first practical device. The NIST researcher developing the new optical sensor was awarded in 1992. He began studying this field in the 1980s, and resolved most of the technical issues by 1989.

On the other hand, one of the awards did not have quick timing. A Navy researcher was awarded in 1992 for developing a polymer. The first patent was issued in 1980. The first license was issued in 1991. Eleven years is not a speedy trip to the market.

Other Factors

This level of analysis uncovered two additional pieces of information which indicated the interconnections between government laboratories, state programs, and legislative representatives. First, the Conductus collaboration with the California DOE laboratory to develop superconducting circuits received grants from the California Competitive Technology Program. Second, Los Alamos National Laboratory and its Quatro teammates met with New Mexico's U.S. senators to work out details for transferring a technology, since the new legislative requirements were not well understood. The transfer was successful and several systems were sold.

Selected Post-legislation Cases

Table 4.1 displays basic data for the eight awards selected for further research beyond the 1993 data highlighted above. The eight cases are presented in the next section. They are:

- Laser-based method to light up biological samples,
- Speech processing coder for telecommunications,
- Paper and plastic quality tester,
- Variable-frequency microwave furnace,
- Gravity meter,
- "Oatrim" fat substitute,
- Chemiluminescent light sticks, and
- Flow diagnostics for artificial hearts.

Case 1 (1993) – Laser-Based Method to Light Up Biological Samples

Role of Laboratory Researchers and Other Personnel   Dr. Edward S. Yeung, principal investigator and developer of this technology, is a widely-acknowledged expert on innovative laser-based technologies for chemical analysis. He won two "R&D 100" awards (annual award by R&D magazine) within three years, including one for this technology in 1991. He was committed to developing a technology that met both scientific and practical

Table 4.1 Selected Post-legislation Case Characteristics

#	Year	Technology	Agency/Lab	Researcher(s)	Partner(s)
1	'93	Laser-Based Method to Light Up Biological Samples	DOE – Ames Laboratory (Iowa State University, Ames, Iowa)	Dr. Edward S. Yeung	Mr. Craig Ranger (Lachat Instruments, Inc.)
2	'93	Voice Coder for Telecommun-ications	Air Force – Rome Laboratory (Hanscom Air Force Base, Massachusetts)	Mr. Luigi Spagnuolo, Mr. Terrence Champion	Dr. John Hardwick (Digital Voice Systems, Inc.)
3	'92	Paper Quality Tester	Agriculture – Forest Products Laboratory (Madison, Wisconsin)	Mr. Theodore Laufenberg, Mr. John Considine, Mr. Dennis Gunderson	Mr. Peter Davis (Isthmus Engineering and Manuf. Co-op), Mr. Curtis Wilson (Bureau of Engraving and Printing)
4	'93	Variable-Frequency Microwave Oven	DOE – Oak Ridge National Laboratory	Dr. Robert Lauf, Mr. Don Bible	Microwave Laboratories, Inc., Mr. Richard Gerard (Lambda Technologies)
5	'92	Gravity Meter	DOC – National Institute of Standards and Technology, Boulder (Colorado) site	Dr. James Faller (Technology Transfer: Dr. Steve ONeil, University of Colorado)	Axis Instruments, Dr. Tim Niebauer (Micro-g Solutions), Dr. Mike Winters (Winters Electro-Optics)
6	'92	"Oatrim" Fat Substitute	Agriculture – National Center for Agricultural Utilization Research	Dr. George Inglett	Mr. Stephen Grisamore (Mountain Lake Specialty Ingredients Company), Mr.

			(Peoria, Illinois)		Mark Freeland (Rhone-Poulenc, Inc.), Mr. Lanny Babbitt (Quaker Oats Company)
7	'93	Chemilumi-nescent "Light Sticks"	Naval Warfare Center – Weapons Division (China Lake, California)	Dr. Herbert Richter, Dr. Ronald Henry, Mr. Joseph Johnson (Technology Transfer: Ms. Martha Harrington)	Mr. Fred Kaplan (Omniglow Corporation)
8	'92	Artificial Heart Flow Diagnostics	DOE – Pittsburgh (Pennsylvania) Energy Technology Center	Mr. Franklin Shaffer, Dr. Mahendra Mathur (Technology Transfer: Ms. Kay Downey)	Dr. Harvey Borovetz (Presbyterian-University Hospital), Ms. Linda Strauss (Baxter Healthcare)

user needs. To do this, he worked with the scientists and facilities at six companies and outside institutions. These were: Lachat Instruments, Inc. in Milwaukee, Wisconsin; the Medical Chemistry Department at Northeastern University in Boston; Roswell Park Hospital Division in Buffalo, New York; Sterling Drug, Inc., which is a subsidiary of Eastman Kodak in Malvern, Pennsylvania; Bio-Rad Laboratories, Inc., in Richmond, California; and Pfizer Central Research, Ltd., in England.

Dr. Yeung provided a range of services to each of these organizations so that they could evaluate the technology for different applications and test its ease of use in anticipation of eventually licensing the technology. For example, Dr. Yeung demonstrated the new technology to a representative of Lachat Instruments, reviewing the technology's advantages, applications, key design criteria, and commercial possibilities. After this, he specified the necessary components of a detector for their particular needs. He received those components from the company at his home, assembled them at his laboratory into a working prototype, and then integrated the prototype on-site with instruments the company was using. He shared his expertise on the prototype system with the Lachat employees and assisted the company in all phases of further developing the prototype into a device the company could manufacture and commercialize. With some of the other organizations, his interaction was not as extensive. For example, Dr. Yeung interacted with Pfizer Research by mail.

In addition, Dr. Yeung and his staff publicized the technology through presentations at scientific conferences around the world and through publishing in scientific journals. Related publications include two early articles in *Analytical Chemistry*, dated 1988 and 1991, co-authored with W.G. Kuhr, a post-doctoral associate.[4] There have been additional articles since then.

After the early business arrangements between the laboratory and its industry partners and other users were finalized, Dr. Yeung continued developing the technology. At this point, he solved a fundamental obstacle to further developing and applying the technology: the micro-manipulation of test samples.

The Technology and Applications   This technology uses a laser-based method to "light up" biological structures (such as blood capillary sections or microbores) fluorescently. This way, researchers can detect and monitor biological processes in fine detail. Because the technology is laser-based, a smaller volume of material is needed for detection and measurements than with previous biological detectors. Thus, the technology helps improve analysis and quantification of even very small volumes of materials.

The technology is easy to use; it can even be used by nonspecialists. Also, it requires relatively little set-up time and can be used in a lighted room. (All previous detectors were based upon conventional light sources.)

The Laboratory   Dr. Yeung directs the Environmental Sciences program at DOE's Ames Laboratory, located at and operated by Iowa State University in Ames, Iowa. Ames Laboratory was founded not long after World War II, following successful development of a uranium production process used for the Manhattan project. It currently has a $38 million budget per year and five hundred employees.

Many of the laboratory's senior scientists also hold joint appointments as faculty members; in turn, the university's graduate students serve on the laboratory's scientific staff. As program director, Dr. Yeung oversees the work of five scientists, seven graduate students, and one postdoctoral fellow. He is also a distinguished chemistry professor at the university.

University Involvement   Dr. Yeung designed and constructed two laser-based fluorescent detectors for Northeastern University's research program in High-Performance Capillary Electrophoresis. To do this, he served as an independent consultant to the university for 20 months. Ames Laboratory's agreement with Northeastern University made Dr. Yeung available for follow-up questions or problems (similar to the laboratory's agreement with Lachat Instruments).

Funding, Financing   DOE and Ames Laboratory funded the basic and applied research that enabled Dr. Yeung to develop the technology. The Iowa State University Research Foundation provided funding for the patenting process.

Lachat Instruments and Northeastern University reimbursed Ames Laboratory for Dr. Yeung's time, travel, and laboratory expenses. For some of the other company interactions, he did independent consulting, which is an activity commonly performed by university professors.

Intellectual Property   After the invention disclosure at the laboratory, a U.S. patent was issued to Iowa State University in April 1991. There were some extensions to the original patent at later dates.

Technology Transfer Mechanisms   Dr. Yeung's interactions with Lachat, a laboratory instrumentation company founded in 1980, produced a licensing agreement for the laser-based method to light samples. In addition to its corporate headquarters in Milwaukee, Wisconsin, Lachat has offices in England, the Netherlands, and Australia. The company is a world leader in manufacturing quality flow injection analysis systems, automated ion analyzers, and mercury analysis systems. The mercury analysis systems are used to perform real-time quality control measures and U.S. Environmental Protection Agency (EPA) compliance monitoring in laboratories. The company's QuikChem^R 1000, 4000, and 8000 series incorporates hardware, Data Quality Management^{TM} expert system software, and chemical formulas. Lachat also markets a distillation system called Micro Dist^R and various other pieces of laboratory equipment. To support its customers, the company offers training packages, service contracts, free telephone support, regional users' group meetings, and newsletters.

Dr. Yeung went to Lachat's location in Milwaukee to finish developing the prototype for them. Clauses in the laboratory's original agreement with the company called for Dr. Yeung to be available by phone for sixty days after delivery of the prototype to Lachat and to provide, as necessary, additional services pursuant to developing the prototype into a manufacturable commercial device. Dr. Yeung made follow-up arrangements with the company to help employees learn how to use the prototype and solve any problems.

The technology was also transferred to both public and private research facilities, including Roswell Park Hospital, Sterling Drug, and Bio-Rad. To transfer the technology, Dr. Yeung pursued both short- and long-term funding through contracts with these outside organizations.

User Groups   Almost any industrial, clinical, pharmaceutical, or university laboratory could benefit from this laser technology. For example, for the pharmaceuticals industry, the technology has been used to analyze blood cells. In the near future, it may allow testing of drug efficacy on a cellular basis or even help doctors detect cancer. It will also contribute to the effort to map the human genome. Furthermore, it will help improve monitoring standards for the environmental field.

Barriers to Commercialization   Many people were skeptical about the technical proficiency of this technology because it was so easy to operate. As a result, Dr. Yeung had to work even harder to make outsiders aware of the extensive engineering devoted to making the technology user-friendly.

Other Factors   Since Ames Laboratory is contractor-operated, the University of Iowa negotiated the license with Lachat. Dr. Yeung says this was back in the days when the university arbitrarily granted licenses without requiring any royalty payments. In fact, usually only an up-front fee was requested as a good faith gesture of reimbursement for the cost of operating the license. However, the university did not even stipulate this requirement. So, the company expected quite a lot with no corresponding input.

User Benefits/Economic Impact/Outcomes   Lachat intended to market a spinoff product to the major environmental research laboratories that monitor the level of contamination in drinking water and wastewater. However, Lachat expected the prototype to be more market-ready and was not willing to put extra effort into developing it further. Dr. Yeung said[5] that if Ames Laboratory could produce an immediately marketable product, they wouldn't have bothered looking for licensees. They would have patented and sold it themselves. The laboratory's license with Lachat died, for all practical purposes. Had Lachat's plans been successful, this technology would have been the first of its kind to be commercially available.

The other institutions are still using the technology for research, not commercial purposes. For example, Northeastern University and Roswell Park are using the detector to study biochemical changes associated with the onset of cancer. The illumination concept developed by Dr. Yeung is best implemented with the prototype instrument he built; thus, it is not necessary to reinvent the instrument. As a result of working with Dr. Yeung's technology, it is possible that these institutions may come up with future patentable developments. However, commercialization of their technology would require use of his instrument in the process.

International Activity   As noted, one of the research facilities using and testing the technology for research purposes, Pfizer, is located in England.

Government Gains   Ames Laboratory is known for its accomplishments in fields like materials science, metallurgy, and superconductivity, although the laboratory also has research programs in biochemistry and the environmental sciences, which are two of the fastest growing areas of R&D. Dr. Yeung's work with these outside organizations and his visibility, internationally, has improved the laboratory's industrial and scientific contacts in those research areas.

Economic Development, Technical Assistance   None.

Elapsed Time   Dr. Yeung began transferring his technology in November 1989. The patent was issued in April 1991, and in that same month he first met with Lachat Instruments. The work with Lachat began in September 1991 and ended in 1993. His work arrangements with Northeastern University spanned from June 1990 to April 1991 and from February 1991 to February 1992.

Case 2 (1993) – Voice Coder for Telecommunications

Role of Laboratory Researchers and Other Personnel   The voice coder technology of this case started as dissertation research at MIT. Researchers at one of the funding agencies' laboratories recognized the potential of the technology early-on and worked with the MIT researchers to develop the concept for digital voice processing. Recognizing the advantages of the MIT-developed technology, Rome Laboratory program managers Mr. Spagnuolo and Mr. Champion were responsible for acquiring funding from the Air Force. They later convinced the Air Force to continue funding the technology, even though it was competing with the recently-established federal standard in this technology area and there were no validated Air Force requirements for this technology. Despite the obstacles, the efforts of the laboratory personnel resulted in this new technology replacing the national standard and the MIT researchers started their own company, Digital Voice Systems, Inc., to commercialize the technology.

Technology and Applications   The technology in this case falls within the realm of telecommunications. It is a voice processor or coder that compresses speech patterns digitally so that they can be sent long distances, transmits the signal, and reassembles at the receiving end. It is a multi-band coder that provides reliable high-quality communications while increasing the capacity of narrowband communications channels.

Conventional technologies that process speech patterns for communication purposes are based upon linear prediction coding (LPC) techniques. Examples of these coding systems are LPC-10, Codebook Excited LPC (or CELP), RELP, VSELP, and so on. Unlike these speech coders, the new "multi-band excitation" coder depends upon sinewave-based strategies. The strategy involves dividing segments of speech into distinct frequency bands and then making a decision as to whether that segment is noise or voice. Traditional coders make a single noise/voice determination for all bandwidths. The voice quality of these coders is low unless they use a "prediction residual" or error signal that helps eliminate the harsh mechanical quality in the speech. (The residual requires a complex process of division into smaller vectors while the computer searches through its codes to find a match.)

This technology is mostly used in digital land mobile radios and mobile satellite telephones. It is also used for voice storage, such as in digital voice answering machines; desktop video conferencing with computers, with no voice lag; and secure communications. An example of the latter application is its use in the secure lines connecting the Pentagon to the White House; for this special application, the coder is used in conjunction with a classified scrambler technology.

The Laboratory Rome Laboratory was established in 1951 at Griffiss Air Force Base near Rome, New York as the Rome Air Development Center. It evolved from the U.S. Army Signal Corps laboratories established in the early 1900s. By 1990, it had become one of the Air Force's four "super" laboratories. The laboratory is the Air Force's center of excellence in the areas of command, control, communications, computers, and intelligence. It actually comprises more than seventy laboratories all over the country, valued altogether at over a third of a billion dollars.

The Electromagnetics and Reliability branch of Rome Laboratory is located at Hanscom Air Force Base twenty miles west of Boston along Route 128, the well-known technology corridor. This branch of the laboratory develops equipment for telecommunications security and conducts research on antennas and electromagnetic energy from targets and terrain. The branch is designated the lead laboratory for the entire U.S. Department of Defense (DOD) for microelectronics compatibility and maintainability, and has a wide array of unique facilities. For example, the Verona Research Facility has a precision antenna measurement system and a radar system evaluation facility.

DOD laboratories have all undergone reorganization in the past half decade, and Rome Laboratory is no exception. As a result, Mr. Spagnuolo is the Acting Chief of the Applied Electronics Division of the INFOSEC Technology Office within the Electromagnetics and Reliability branch of Rome Laboratory at Hanscom Air Force Base. Rome Labortory is still headquartered near Rome, New York, even though Griffiss Air Force Base has been closed as a result of DOD's Base Realignment and Closure program (BRAC).[6]

University Involvement The multi-band excitation coder was implemented on portable, real-time hardware by researchers in a laboratory at the Massachusetts Institute of Technology (MIT). The MIT research that produced a new speech model was the thesis project of several students including John C. Hardwick, Daniel W. Griffin, and S.W. Wong. MIT professor Dr. Jae S. Lim was their advisor.

The MIT team published at least four technical papers in the mid- to late-1980s and early 1990s. At least four of these appeared in the proceedings of the International Conference on Acoustic Speech and Signal Processing, an annual conference sponsored by the Institute of Electrical and Electronics Engineers (IEEE).[7] Another paper was published in an IEEE workshop proceedings on Speech Coding for Telecommunications,[8] the IEEE Transactions on ASSP,[9] and the proceedings from a Digital Signal Processor workshop.[10]

Funding, Financing   The early speech research at MIT was supported by university research assistantships, fellowships, and grants and contracts with federal agencies. Federal funding came from the National Security Agency, the Air Force's Electronic Security Command[11] Cryptologic Support Center, Rome Laboratories, and other agencies. The spinoff company, Digital Voice Systems, Inc. (DVSI), has not received any federal funding since its founding.

Intellectual Property   No patents were filed by MIT on the original research. DVSI currently has at least four patents on the coder and several patents pending, both foreign and domestic.

Technology Transfer Mechanisms   Because of the odds against adoption of a competing technology to the existing standard within both DOD and industry, the Rome Laboratory team developed a strategy to push the technology into the marketplace. The team entered the technology in commercial competitions and established a strong presence at standards meetings, presenting favorable and supportive arguments for the technology before relevant committees.

The Association of Public-Safety Communications Officers (APCO) along with the National Association of State Telecommunications Directors (NASTD) stated they were leaning toward adopting the LPC technology as the new digital standard for land mobile public safety radio equipment in North America (at 4.8 kilobits per second). However, Mr. Spagnuolo and Mr. Champion persuaded the APCO/NASTD committee that this technology would not provide acceptable performance in the harsh operational environments in which public safety equipment would be fielded. They encouraged the committee to consider the multi-band excitation coder. To facilitate the evaluation, they volunteered Rome Laboratory as the host laboratory to test the candidate coders under realistic operational conditions. This effort might be considered outside the scope of a federal laboratory, but was consistent with laboratory efforts to transfer helpful technologies to state and local governments. Evaluating the technology at a federal facility improved the evaluation, which was a service to APCO and NASTD.

In addition to the CELP and multi-band excitation coder, other technologies competing in the evaluation included another LPC-based coder called VSELP and another sinewave-based coder developed by Rome Laboratory, the Sinusoidal Transform Coder (STC). The results proved the multi-band excitation coder to be superior, so APCO/NASTD was able to find a better product. The multi-band excitation coder was named the land-mobile radio standard for public safety communication equipment. The standard was known as "APCO/NASTD Federal Project 25." The evaluation was implemented by the Telecommunications Industry Association.

In addition, the new coding technology was represented in the Telecommunications Industry Association's half-rate and full-rate digital cellular evaluations. In the half-rate evaluations, both the multi-band excitation coder and STC placed in the top grouping of coders. In the end, the multi-band excitation coder performed better than the existing national standards technology and was selected as the new standard.

User Groups   Users of this technology are manufacturers of telecommunications equipment, a multi-billion dollar equipment industry. The new coder technology provides the telecommunications industry with major advances in digital speech processing. Communication is reliable and has increased channel capacity within a fixed spectrum. The technology affords increased error protection against acoustic background noise, channel errors, etc., for harsh environments such as in military and land-mobile radio use.

The end users of land-mobile radios are police and fire departments, other emergency service personnel, trucking and delivery fleets, and taxicabs. Many of these users, in both the public and commercial sectors, are in the process of upgrading from analog to digital radios, which require a speech coder such as the multi-band excitation coder. The radios are either hand-held walkie talkies or vehicle-based units. The technology provides four times the channel capacity within the allocated spectrum for these users while maintaining voice quality.

Mobile satellite phones are used in place of cellular phones where there is no cellular service available. They are found in vehicles, trucks, ships of all types, and aircraft. They are used by journalists and reporters, business travelers, and disaster relief and emergency services personnel in rural areas.

Barriers to Commercialization Traditionally, the communications research communities both within DOD and industry focused on LPC. Rome Laboratory was the only research laboratory developing or supporting sinewave-based techniques. At the time, Mr. Spagnuolo was acting chief of Rome Laboratory's Technology Applications branch in the Boston area. Mr. Champion was director of secure voice research. Both researchers were in the laboratory's Electromagnetics and Reliability Directorate.

Other Factors   The multi-band excitation coder won six out of six of the last standards competitions and/or independent evaluations. These evaluations resulted in adoption by national and international standards groups. Dr. Hardwick, one of the original MIT researchers, noted that negotiations with these standards organizations was "like buying a car." The technology was transferred to the commercial standards organization, and the commercial terms were good. The standards organizations, in turn, collect money from their customers and turn it around to the suppliers.

User Benefits/Economic Impact/Outcomes   The ultimate outcome of the MIT research was a spinoff company, Digital Voice Systems, Inc. (DVSI). DVSI specializes in high-performance speech compression systems based on the multi-band excitation technology. The company sells both hardware and software to other private sector customers.

DVSI developed a software product called the "Improved Multi-Band Excitation" or IMBE™ Speech Compression System. In 1992, *Ocean Voice* magazine noted that this system provided better quality than cellular telephones. DVSI also produces voice coding hardware modules which implement the IMBE speech coding software. DVSI's hardware includes the lower cost IMBE™ VC-20 Voice Codec Module in the 5-inch-square size and the IMBE™ VC-100 Voice Codec Module, the full-duplex version, which is 2 x 2.4-inches. The VC-100 is used in INMARSAT's new Mini-M mobile digital satellite telephone system. DVSI also sells the software by itself, modified for use with other digital signal processing hardware available on the market.

DVSI owns the commercial rights to the voice coding technology. The advantage of DVSI's voice coder is that they offer superior speech quality at lower data rates, which are more desirable. Also, the coders' algorithms are less complex, they require less memory, and can be implemented cost-effectively.

DVSI continues to further develop and commercialize the technology. In fact, the company has recently introduced a new "Advanced Multi-Band Excitation" AMBE^R Speech Compression System implemented by an AMBE-1000™ Coder. According to the product literature, the AMBE-1000 has the highest performance voice coder on the market. The AMBE-1000 exhibited better overall performance than the existing standard in the full-rate digital cellular standard evaluations called "IS-54."[12] The AMBE is interesting because the user can select the amount of error correction and can adjust data rates. Also, the coder is available on a single computer chip and is less than an inch square compared with the IMBE VC-20's five square inches. In 1996, an AMBE-1000 cost $99. An order of at least 360 units reduces the price to $65, and an order of 10,000 reduces it further to $38. To order over 100,000 units, a manufacturer would need to contact DVSI for availability and price.

DVSI, being privately held, is not willing to divulge sales revenues. Dr. Hardwick stated,[13] however, that the company is profitable and has "done rather well" with a "reasonable amount of commercial activity." The company averages ten employees. Dr. Lim, the faculty advisory at MIT, is the Chairman of DVSI, and his former students, Dr. Hardwick and Dr. Griffin both hold management positions in the company, as well. In 1992, *Land Mobile Radio News* said "DVSI is the quiet company whose Vocoder [voice coder] will be heard around the world."[14]

International Activity   The MIT researchers entered the technology in the International Maritime Satellite competition, called INMARSAT, which was held in Australia in 1992. INMARSAT is a commercial organization chartered by the United Nations. At the INMARSAT competition, the multi-band excitation coder competed against all LPC-based coders. The test results showed it was superior, and so it was selected as the INMARSAT-M satellite communication voice coding standard. The Australian Satellite group (called AUSSAT) was so impressed, they, too, selected it as their standard. Other organizations such as OPTUS, another global mobile satellite-based service, followed INMARSAT's lead, as well.

Government Gains   An advantage of this new standard is that a manufacturing base is being established that can provide DOD with commercially available, state-of-the-art speech communications equipment demonstrated to be superior in military applications. Also, the skill-base for using the technology is increasing, both in state and local governments and in industry.

Economic Development, Technical Assistance   DVSI received no economic development or technical assistance services.

Elapsed Time   The MIT work in this area began in the early 1980s. Many of the research papers were published while the authors were still affiliated with MIT. Key papers were published in 1984, 1985, 1988, 1989, and 1991. DVSI was founded in 1988. The technology was commercially available from 1988 to 1991, but was not easy to use. APCO/ NASTD Federal Project 25 announced its selection of the multi-band excitation coder technology in 1992. Negotiations with INMARSAT were completed by mid-1992. DVSI was issued its first patent in 1993.

Case 3 (1992) – Paper Quality Tester

Role of Laboratory Researchers and Other Personnel   During the time frame of this case, Mr. Dennis Gunderson, and Mr. John M. Considine were researchers with the Forest Products Laboratory (FPL) in Madison, Wisconsin, in the Fiber Product Design Criteria Research Work Unit headed by Mr. Theodore L. ("Ted") Laufenberg. This group addresses technological barriers to the use of fiber and paper products. (Mr. Gunderson is now retired, and Mr. Considine does not work at the laboratory any more. Ted Laufenberg is now the laboratory's technology transfer officer, in addition to continuing his work in technical areas as assigned.)

In order to transfer the laboratory prototype version of this paper quality testing technology, the team of three scientists put together a technology transfer plan, including an assessment of the potential market. They contacted over a hundred potential users in all sectors about the commercialization initiative. During this process, three potential equipment manufacturers were identified. Since federal patents were involved, the laboratory solicited CRADA proposals from the three firms on a competitive basis. A panel of technical and administrative staff from the U.S. Forest Service and the University of Wisconsin screened the three proposals based upon pre-determined criteria. Once a CRADA was in place, the research team devoted up to three hundred hours of its expertise to the design and development effort.

FPL scientists are known to publish in a wide variety of publications, including over 100 scholarly journals. Publications involving this technology were published in both the journal[15] and conference proceedings[16] of the Technical Association of the Pulp and Paper Industry (TAPPI), as well as proceedings of the Technical Association of the Graphic Arts.[17] The authors included the team of three scientists, and joint authors from partnering organizations such as the U.S. Bureau of Engraving and Printing and the Swedish Pulp and Paper Research Institute.

Technology and Applications   The properties or behavioral characteristics of paper, and similar materials such as wood, change and deteriorate over time due to environmental conditions, such as when they are stored in a warehouse. An analogy would be the way a barn roof eventually may become sway-backed from exposure to moisture. The technology developed in this case allows a paper product such as a sheet or roll of paper to be tested for those performance characteristics like strength, stiffness, and durability. The test for these characteristics is referred to as the paper restraint testing system. The results of the testing are a quantitative assessment of these properties so that the paper industry can conserve fiber (reduce sheet weight) and reduce the use of expensive coatings which enhance printability and stability, as well as fight humidity.

To test sheets of paper or plastic materials, each sheet is held in a controlled environment by a vacuum that holds it flat. The paper is placed under pressure using a compression technique that uses rigid supports. Air is pulled through and around the material, so that differential pressure is applied to each side of the sheet. The machine that does this must also periodically check for moisture-related "creep" (either stretching or, conversely, shrinking or compressing that is time-dependent and load-dependent). This creeping movement can be measured while the material is flat; if it were wrinkled, it wouldn't be possible to measure it.

The Laboratory   FPL is one of eight government-owned, government-operated Forest Service laboratories within the U.S. Department of Agriculture (USDA). About a hundred scientists and technical professionals work at the laboratory, including foresters, botanists, plant pathologists, microbiologists, chemists, and engineers. In addition, the laboratory employs economists and conducts a great deal of market research for its products. Altogether, 350 people are employed by the laboratory. The laboratory also hosts hundreds of consulting visitors every year. The laboratory's users include not only state and private foresters and landowners, but also regulatory agencies, a variety of industrial sectors, educators, legislative bodies, and the general public.

Broad areas being pursued at FPL include research, technology transfer, and cooperative partnerships related to wood and international forest products. Specialty facilities and groups at the laboratory include: a Pulp and

Paper Pilot Plant, a Conservation and Recycling Technology Marketing Unit, a Fire Testing Laboratory, and an Institute for Microbial and Biochemical Technology.

University Involvement   Technical personnel from the University of Wisconsin (UW) in Madison, where FPL is based, participated in screening the CRADA proposals submitted by industry. Universities were also part of the network established by the laboratory to assess the technology. For example, UW-Madison received a $25,000 cooperative research agreement for evaluation purposes.

Funding, Financing   FPL receives funds from Isthmus Engineering for the joint work under the CRADA. Also as a result of contacts developed during the CRADA/license solicitation process, the laboratory put into place ten cooperative research agreements totaling almost $1 million. These agreements allowed outside organizations to evaluate the technology and equipment for their applications. In the process of evaluating the technology, paper industry associations, private companies, and other federal agencies identified the type of machine configurations they would require in order to use the technology, and some requested information on availability and pricing.

Examples of the cooperative research agreements included: an agreement where FPL awarded the National Starch and Chemical Corporation in New Jersey a small amount of money to evaluate the "cyclic creep" in linerboards. And several cooperative research agreements totaling $430,000 with the American Forest and Paper Association, particularly the Containerboard and Kraft Paper Groups. Additional cooperative research agreements were granted to seven companies in amounts of less than $10,000 each.

Intellectual Property   There was no invention disclosure. For this technology, patents issued to USDA cover the method for conducting the testing and a portion of the equipment used. However, certain aspects of the machine are not patented, specifically, the design of the machine that simultaneously applies pressure to test for paper creep and to condition the paper cyclically.

Two patents were issued, both entitled, "Method and apparatus for edgewise compression testing of flat sheets," for more than one application, dated August 1982 and May 1984, with Mr. Dennis E. Gunderson listed as the inventor. At least one additional patent was issued to the laboratory's company partner described below.

Technology Transfer Mechanisms Isthmus Engineering and Manufacturing Co-op is a small designer and manufacturer of automated machines. The company employs about forty employees and is southeast of Madison, Wisconsin. It was selected based on having the best development plan in its CRADA proposal.

Isthmus was founded in 1980. It averages $10 million in yearly sales developing and manufacturing a variety of specialty equipment and instruments, many of which are one-of-a-kind systems. The company has worked with hundreds of customers on thousands of projects ranging from million-dollar dedicated machining centers to hundred-dollar assembly tools. This includes equipment for plastic packaging, processing equipment for the medical industry, and metal weighing and removal equipment for large machining centers. Additional examples are robotic equipment for John Deere tractors, the pump apparatus in hair spray cans, and the machines for manufacturing Rayovac batteries. The company has also been building special machines for the high-volume automotive industry since its inception. So, the company's products run the gamut from foundry equipment to laboratory equipment.

Since 1988, the company has had an 18,000-square-foot facility that houses its equipment. That equipment includes eight computer-aided design (CAD) stations and a computer-aided manufacturing (CAM) system that links the design computers to three computer-numerically controlled (CNC) vertical milling machines. One of those machines is a 21-tool, four-axis machining center with 30 x 60 inches of "travel." The facility also houses eight manual drafting stations and a coordinate-measuring machine capable of one ten-thousandths of an inch accuracy within a 20 x 40 x 20-inch volume, housed in a 68-degree temperature-controlled room. There is an electrical controls department with full programming, fabricating and documentation facilities. The facility also has other necessary milling, turning, grinding, assembly, inspection, painting, and testing equipment.

According to company board member Mr. Peter Davis, the company does relatively little marketing, and gets most of its work through word-of-mouth. When the company quotes a price for building a machine, it includes everything from design to project schedules, operator's manuals, debug, run-off and service after the sale. When a machine is designed, it includes planning for everything including plumbing, wiring, control boxes and switches, coolant, etc. The company also sells machine-related services such as inspection services on a contract basis.

The company is a workers' cooperative (as opposed to the more common users' cooperatives like grocery stores). It was modeled after cooperatives found in the Basque Madrigon region of Spain, as adapted to Wisconsin laws, of course. Potential employees apply to become self-employed members of the co-op. The relatively low entry fee is based upon a stock equalization basis, free of outside influences. Once a year, the performance of each employee is critiqued by all the other employees. This review results in the entire company being involved in setting the yearly earnings for each employee. So, each employee's earnings result from his or her value to the company, as opposed to being set by outside market rates. After two years of attending board meetings, etc., the member can be voted in as a "partner" or board member, and re-elected every two years. As in a participatory democracy, any employee is eligible to run and be elected as a board member or officer. These board members and officers meet every other week to discuss the company's business and make decisions.

Isthmus' one-year CRADA with FPL was focused on gaining machine design and paper physics expertise so the technology could be commercialized. Also, an exclusive license was granted to Isthmus Engineering based upon both patents. The time frame for the license is equivalent to the life of the patents, and its field of use is paper products. The laboratory team has been particularly protective of the company's wishes to maintain confidentiality regarding the terms of their legal agreements with the laboratory.

User Groups The primary user of this paper testing technology is the paper industry. The paper industry is a major segment of the forest products industry. It produces ninety million tons of paper products a year including boxes, labels, stationary, and raw products like cardboard rolls. In order to avoid defects, most paper products are over-designed or over-processed rather than quality-checked. This inefficiency is compounded as the ingredients for making paper change. Older paper "recipes" involved higher-quality cedar wood fiber. Now that cedar is less available than woods like ash, this presents additional quality issues. When quality testing *is* performed, the equipment is usually supplied by Scandinavian companies.

Quality testing offers paper manufacturers a better understanding of the effect of moisture, in particular, on their products. This technology compresses the window of time required for testing, because it allows testing before the paper leaves the manufacturing facility. As a result, producers can offer competitively priced products through the use of less-expensive paper coatings, a reduction in the use of wood fiber, and savings in the disposal of scrap paper. The savings amount to possibly tens of millions of dollars overall. Also, the testing will allow the use of different combinations of fibers and increased recycling, so that forests can be saved from destruction.

Barriers to Commercialization   The laboratory team had to deal with several obstacles in attempting to promote the commercialization of this technology. First of all, a paper testing market did not exist at the time this technology was developed, which meant there was no demand for a product in this area. There were no U.S. companies that produced paper testing equipment; the laboratory's technology represented the state-of-the art.

The cooperative research agreements for evaluation purposes helped to "spread the word" about the technology. However, each organization involved in evaluating the technology required a different machine configuration and, consequently, required different methods to analyze the data and interpret the findings in order to be able to characterize the behavioral properties of their paper products. This made comparisons and overall technology assessment difficult.

Other Factors   Mr. Laufenberg notes that technology transfer is akin to giving up your children,[18] so it behooves a laboratory to keep the inventors at arms length from the cooperative research efforts particularly now that they share in the royalties because this could possibly create a conflict of interest. Creating a buffer between the actual researchers and cooperative research efforts allows more flexibility in a laboratory's technology transfer program. For example, as a technology gets closer to commercialization the scientists become less involved or even ask to be moved off the project, because at what point would the laboratory be able to say that enough effort has been put into their understandably eager marketing efforts? An intermediary can push the technology with less bias toward that particular technology.

User Benefits/Economic Impact/Outcomes   FPL's partnership with Isthmus Engineering and Manufacturing Co-op opened up a new market for the company. Within eight months after the one-year CRADA was signed, Isthmus delivered to FPL a commercial version of the technology called a

"Vacuum Compression Apparatus" which was an updated and more refined version of the original prototype. This provided the company visibility which attracted an early order for another machine from New Zealand's Pulp and Paper Research Organization. Isthmus also produced a machine for the U.S. Bureau of Engraving and Printing (BEP) in Washington, D.C. The first machine was provided to FPL at cost in return for the laboratory's technical assistance. The other two machines totalled $250,000 in sales for the company. FPL is still using Isthmus' machine in the laboratory's research projects.

The CRADA with Isthmus Engineering was hailed as a first step toward application of this technology in the paper industry. It "set the stage . . .to take a competitive role in supplanting millions of dollars of paper test instrument sales from their Scandinavian counterparts," according to the FLC nomination form.[19] Initial impacts were apparent in the corrugated container and the fine-paper printing industries. However, the paper testing technology was eventually supplanted in the market by other methods that were less expensive, simpler, and easier to implement. The technology is still being used, for example, in university laboratories doing specialty research that requires sophisticated approaches. But, in terms of the commercial marketplace, the technology was a leading-edge product that created the market need for the type of information generated by this testing technique.

In August 1996, Isthmus delivered a fourth machine for $165,000 to a customer at the University of Maryland working with a consortium of companies in the electronics industry. The Maryland consortium will be using the machine to study the long-term effects of cyclic humidity and changing atmospheric conditions on the corrosion and breakage of computer circuits. Computer circuits were an application that had not occurred to Isthmus Engineering. Silicon chips have joints for chemical and heat dissipation. But because the materials that make up the components of a chip are different, the thin films that are deposited on chips can work loose over time. So, this application for Isthmus' testing machine could open up an entirely new market for the company. Peter Davis says[20] the new machine, called a "Thin Film Analyzer," helps to make the company even more diverse. The company was contacted by the University of Maryland through a researcher schooled at the University of Wisconsin who was aware of the Isthmus paper testing machines from his dissertation research.

Meanwhile, Isthmus developed the third generation of the Thin Film Analyzer through advances in computer software and electronics. Its capabilities allow environmentally controlled testing of not only paper and corrugated products, but also many types of materials such as plastics, composites, elastomers, and textiles. This unique testing machine provides information on the material properties, particularly humidity effects. Custom specifications and configurations can be requested to fit user test requirements and facilities.

International Activity   The laboratory also signed a cooperative research agreement that involved in-kind sharing of data with the Swedish Pulp and Paper Research Institute (known as "STFI") in Stockholm. As a result of this relationship, a Cyclic Humidity Conference hosted by FPL in 1992 went international in 1994 and was co-sponsored by both FPL and STFI. The third international conference is slated for New Zealand in February 1997.

Government Gains   Improved testing techniques for paper products will provide savings to American taxpayers in better public-sector methods for printing postage stamps and other government paper products. In fact, the U.S. Treasury Department's Bureau of Engraving and Printing provided FPL with two interagency agreements to evaluate the technology—$200,000 initially with a follow-on agreement for over $400,000.

Economic Development, Technical Assistance   None applied.

Elapsed Time   The technology was patented in 1982 and 1984. The original solicitation for potential industry partners and licensees began in October 1988. The CRADA and license agreements were executed in November 1990, and the CRADA was completed in January 1992. The related cooperative research agreements for assessing the technology were implemented from 1991 through 1993. The jointly sponsored international technical conference was held in 1994. Additional CRADA activity demonstrating test effectiveness for a variety of end uses continued through 1995.

Case 4 (1993) – Variable-Frequency Microwave Oven

Role of Laboratory Researchers and Other Personnel   Dr. Robert J. ("Bob") Lauf is a senior development staff member in the Metals and Ceramics Division at Oak Ridge National Laboratory (ORNL). His expertise is in materials science. Mr. Don W. Bible is a development staff member in the laboratory's Instrumentation and Controls Division. Mr. Bible's expertise is in electrical engineering. Their work required using a microwave furnace for firing ceramics related to the fusion energy program. The microwave tube they used for heating plasma material in a reactor was the size of a person.

Eventually, Dr. Lauf and Mr. Bible conceived the idea of a variable-frequency, as opposed to fixed-frequency, microwave furnace, which would eliminate the hot and cold spots in the furnace. These are similar to the dead spots common to kitchen-sized microwave ovens that necessitate rotating food in the oven. The idea was that they could map the location of the dead spots and program the microwave furnace to change back and forth to different frequencies, a technique called "sweeping." Sweeping was used in radar jamming: a radar screen will lighten up when a blizzard of various frequencies is applied.

When they shopped for a traveling microwave tube that could sweep, they found the power level on most of the commercially available traveling microwave tubes was too low. In their networking, a colleague mentioned a small company in North Carolina producing high-power traveling microwave tubes, so Lauf and Bible tracked down Microwave Laboratories, Inc., a defense contractor in Raleigh, North Carolina. The military initially funded construction of the company's facilities because they needed a particular type of traveling microwave tube. Thus, the company's product line centered around high-power traveling microwave tubes and other broadband products. The company had defense contracts during the Gulf War related to electronic warfare.

When Dr. Lauf and Mr. Bible contacted the CEO of Microwave Laboratories, it happened to be after the Gulf War when the company's business was down. The CEO, Carl Everleigh, immediately invited the researchers over even though he was initially skeptical of their idea. (He is quoted in a journal article as saying he thought they were crazy!) However, since he had previously worked with microwave heating applications, the laboratory researchers were able to convince him that the concept of a variable-frequency microwave furnace was feasible.

So Microwave Laboratories "donated" a $50,000 traveling microwave tube to ORNL (a piece of hardware about the size of an arm), and ORNL bought the "refrigerator-sized" $60,000 power supply. Subsequently, Dr. Lauf and Mr. Bible built a large prototype of the first variable-frequency microwave system. Microwave Laboratories designed and built a smaller bench-top version. ORNL and the company worked closely together, achieving synergism in their skills and expertise.

The two researchers continue to take initiative in identifying industrial and scientific processes that might benefit from this invention. Their knowledge of chemistry and microwave diagnostics allowed them to explore the processing of diamond films. They can explain benefits to potential users in many industries and refer them to the company.

The researchers presented their research results related to the prototype machine at a Materials Research Society symposium on microwave processing of materials.[21] Most of the major microwave processing groups were represented at the symposium. Dr. Lauf said[22] they were practically laughed out of the symposium, since it was thought that traveling microwave tubes were unaffordable.

Technology and Applications   In-roads are being made using the tunable (or programmable) variable-frequency furnace to process coatings and diamond films for semiconductors. Using the furnace, very uniform plasma etching can be obtained on a circuit board. This level of uniformity is difficult to achieve with other such processes. Also, the furnace is effective in applying synthetic diamond films on industrial saw blades, for example.

The technology is also currently being used for heat treatment (or "sintering") of ceramics, as well as the curing of resins. Potential longer-range applications include better cooking effectiveness in food processing and analytical chemistry applications.

Also, there are a variety of industrial applications for this technology in other fields, particularly high-energy nuclear physics. In this area, it would be used to improve ion sources on particle accelerators.

The Laboratory   ORNL was built in 1943 on an almost 3,000-acre site in Oak Ridge, Tennessee. The laboratory was part of the World War II Manhattan Project to produce the atomic bomb. The laboratory's original mission was to produce plutonium for that effort. Today, ORNL is a major DOE multi-program R&D laboratory with an operating budget of over $530 million and a staff of over 5,000. About 4,400 guest researchers spend time at the laboratory every year; one-third of them are from industry.

University Involvement   There was no university involvement in this case.

Funding, Financing   The ORNL researchers obtained funding from three sources to enable them to work on the microwave project. First, conceptual development and design was supported by the Advanced Industrial Concepts Materials Program out of DOE headquarters (now called the Office of Industrial Technologies or OIT). OIT helps industry work with seventeen of the thirty DOE laboratories. OIT has supported more than sixty-five R&D projects that resulted in new technologies being used by industry.

Second, ongoing R&D and funding for the ORNL researchers to be involved in the company-funded CRADA were provided by the laboratory technology transfer program of the Office of Energy Research at DOE headquarters. Created in 1992, this program maintains "quick-response centers" at non-defense DOE laboratories such as ORNL. The program provides technical assistance to small businesses, and promotes CRADAs and personnel exchanges, among other activities. (For fiscal year 1997, DOE's overall funding for this program will be about $20 million. Not only is this funding less than previous years, but it is also restricted to high-risk CRADAs at five DOE laboratories, including ORNL, in three specific technology areas: intelligent manufacturing, tailored materials, and sustainable environments. However, as a result of this cutback, some of the other DOE offices are providing funds for DOE laboratory researchers to be involved in CRADAs. For example, in this case, the DOE Office of Industrial Technologies picked up the laboratory's involvement.)

Third, capital funds to build the large prototype variable-frequency microwave furnace were provided by the Advanced Manufacturing Program at DOE's Y-12 Plant. The plant is located about five miles from ORNL and began operating in 1943 to produce uranium for the Manhattan Project. The plant has an annual budget of over $600 million and a staff of approximately 5,000. The Y-12 Plant is part of DOE's defense programs, as opposed to its civilian energy conservation programs. However, as the government has ended its weapon manufacturing effort during peacetime, the plant's mission has changed. Today, it dismantles nuclear weapon components returned from the national arsenal, maintains a stockpile of nuclear materials, and produces "special" nuclear materials. Most programs at the Y-12 Plant work closely with the nearby ORNL researchers. The plant's capabilities encompass all phases of the design process: from conceptualization to specifications, prototype construction, and an integrated manufacturing process configuration. Another manufacturing program located on the Y-12 Plant site is DOE's Centers for Manufacturing Technology. Facilities available to outside users—industry, universities, and other government agencies—include everything from machine tools to gear and thread technologies. A toll-free telephone line is available for those interested in accessing the plant's facilities or arranging technical assistance or consulting projects.

Intellectual Property The first patent for the "Variable-frequency microwave furnace system" was filed by Martin Marietta Energy Systems, now called Lockheed-Martin Corporation and was issued for a polymer curing application of the furnace. Lockheed-Martin operates ORNL. Dr. Lauf and Mr. Bible are the registered inventors.

Another patent was issued with the co-inventors listed as Dr. Lauf, Mr. Bible, Arvid Johnson (a company scientist), and Dr. Robert J. ("Bob") Markunas, Associate Director of the Center for Semiconductor Research at Research Triangle Institute (RTI). RTI is an independent non-profit organization in Research Triangle Park, North Carolina. RTI had worked with Microwave Laboratories to test the processing of diamond films for semiconductors via plasma-deposited methods. This was reported at a 1993 conference on diamond synthetics in Heidelberg, Germany. Dr. Markunas and the company scientists also worked together to come up with ideas for improved control systems.

Three additional patents relating to the CRADA work with the company are still pending. Dr. Lauf has a portfolio of twenty patents and fifteen pending. On the role of patenting, Dr. Lauf said[23] scientists should get into the practice of patenting first and then publishing. Once an article is published, the scientist still has a year grace period when a patent can be filed without losing the opportunity. Also, the scientists should find customers before they patent because it is impossible, he said, to get a customer without telling them what you have. He added, however, that if technology transfer officers do that, the patent rights will be lost to the company. Dr. Lauf said it is important to let the market indicate the applications. He says few good ideas are transparently obvious from the beginning. For example, the concept of getting better uniformity in a microwave furnace by sweeping the frequencies seems obvious "once you know it works, of course."[24]

Dr. Lauf also remarked that the young scientists employed by his partnering companies are always anxious to show him their latest advances and to get advice regarding their patent portfolio. They talk by phone at least once a week and write many joint publications and patents. He said[25] it is important to get them involved in the technical work early-on so they can be included on the patent, which helps to make it broader.

Technology Transfer Mechanisms   ORNL asked the Microwave Laboratories CEO to sign a proprietary information agreement (the same type of agreement as a non-disclosure agreement, or confidentiality agreement). The laboratory eventually granted a non-exclusive license to the company, the first license related to microwave processing to be signed at ORNL. Ultimately, a three-year CRADA was initiated to further develop the technology and demonstrate new applications. On the three pending patents related to the CRADA work, the laboratory licensed the technology to others and split the royalties with the company.

Both Dr. Lauf and Mr. Bible commented on the technology transfer process and personnel. Dr. Lauf said technology transfer officers worry about the time and money invested in "loser" technologies, but that they end up spending more money trying to assess and identify winners. For licenses, laboratories receive an up-front fee from a company and running royalties based upon a percentage of the company's sales related to that technology. Dr. Lauf commented[26] that technology transfer personnel don't consider it their business to monitor the laboratory's licenses, consequently the laboratory scientist has no way of knowing whether he is getting his "fair cut" of the royalties. A similar problem, according to Mr. Bible,[27] is that laboratory scientists are not allowed to be involved in legal negotiations related to their technologies, so they have no idea about the details of the agreements that are signed. On a related note, he added that in their negotiations the legal staff often commit the technical staff to things that can't be done.

Dr. Lauf felt that consideration of the royalty payments to the laboratory scientists should be secondary to the overall legal agreements between laboratories and companies. However, Mr. Gerard of Lambda Technologies (described below) commented that it is important that a percentage of the royalties go to the laboratory inventors, so that they have an investment in his company's success.[28]

User Groups   The ultimate customers of the company's products will include users for whom conventional (fixed-frequency) microwave sources are ineffective, inefficient, or difficult to control. For example, an athletic shoe company is making plans to use the furnace to cure resin adhesives that attach soles to the bottom of shoes.

The largest potential market is probably diamond film processing. This area includes an estimated $10 billion market for the electronics industry alone. There is a projected $10 billion market for other mass production uses.

Barriers to Commercialization   Both Dr. Lauf and Mr. Gerard commented[29] on the need for exclusive licensing, probably because the company was only granted a non-exclusive license. Dr. Lauf noted that a laboratory's first mission was to get the technology out with fairness of opportunity since they are taxpayer-supported. But a project also has to make business sense. A company can't raise capital unless it gets an exclusive license at least for the first several years or at least one that is limited to fields of use (even with different applications). Mr. Gerard also made the comment that a company needs a window of exclusivity. Even up to the late 1980s, in his experience, a company couldn't get this with DOE laboratories. Now, an ORNL license (through Lockheed-Martin) can be exclusive for one or two years, after which the laboratory will usually retain the right to license the technology to other companies.

Dr. Lauf also says that Congress is getting the message that these issues are not "corporate welfare," but the leveraging of federal R&D money. He added that Lambda Technologies has communicated with its congressional representatives about this.

Other Factors   Mr. Gerard feels the key to their success so far is that they have maintained a close relationship with the DOE inventors, fostering an interpersonal relationship and consulting with them on new ideas. He commented[30] that, even if no new technology results from the CRADA, it keeps them in the loop and bridges a gap. (Other technology is likely as the technology continues to mature through the CRADA.) Also, the relationship with the laboratory researchers helps to streamline the process of working with ORNL.

Dr. Lauf separately commented that he believes the lessons learned in this case are for laboratory researchers to be in good communication with their company CRADA partners. Dr. Lauf says that the business and legal people can do the paperwork involved in CRADAs, but they can't do the necessary technical hand-holding. So, it's the scientists that end up driving the commercial process. It is apparent from his stories that Dr. Lauf serves as a mentor to others in both the laboratory and the company, such as the company's younger, very technical employees, who are recently out of graduate school. Dr. Lauf said these hard-working group of "kids" are at the right place at the right time because "the machine and market are ready" in spite of Microwave Laboratories going out of business. Working together, they encourage each other to work long hours, and as Dr. Lauf believes there is no substitute for hard work.[31] Also, they are always on the lookout for new customers, and they go after awards that might contribute some visibility to their effort.

User Benefits/ Economic Impact/ Outcomes   Microwave Laboratories developed and marketed a new product line based upon the variable-frequency microwave, which included its traveling microwave tube as a component. The products included furnaces, applicators, processing systems, and traveling microwave tube amplifiers. The power levels were high, ranging from 200 watts to two kilowatts, and the frequencies ranged from 900 megahertz to 16 gigahertz. These products were cited in the company's 1992 Annual Report as the cornerstone of a strategy to reduce its dependence on defense-related product lines. After the Gulf War, however, the military's need for Microwave Laboratories' products declined drastically. To make matters worse, the company experienced a quality control problem on one of its Navy contracts. As a result, the company's inventory built up; eventually, the company became insolvent. The company's CEO is now with another company in the microwave business.

Four people from the microwave heating part of Microwave Laboratories spun off to become Lambda Technologies, a privately-owned company in Raleigh, North Carolina. Dr. Lauf described Lambda as "a phoenix rising out of the ashes." (Certain DOE publications relate that, upon the successful development of the new technology, Microwave Laboratories "formed" the new company.) Lambda Technologies acquired Microwave Laboratories' inventory of parts and its jointly developed patents and licenses providing rights to the special tubes. However, all of the intellectual property had to be renegotiated with Lockheed Martin since it was not transferrable. (One of the reasons for this, according to Dr. Lauf, was that the laboratories often give better terms to smaller firms than to larger ones. He added that this often causes venture capitalists difficulty, because it means large firms often can't assume intellectual property rights when they acquire a small firm.[32])

As opposed to Microwave Laboratories, the new company's only product is the variable-frequency microwave oven. After delivering a prototype system to DOE, the company introduced its first standard product in April 1996. They are producing two to three units a month of this research equipment (mostly for university laboratories). The company anticipates that this line of products will grow to significant numbers, possibly 30,000 to

50,000 units. Microwave ovens are often used in scientific laboratories to heat and dry, but conventional microwave ovens do not perform these jobs well because they don't have either uniform energy distribution or controllable energy distribution for selective heating. Lambda Technologies' product is called "Vari-Wave," and it is a multi-functional microwave oven that can serve as a research tool, an analytical tool, or a measurement tool. It can be used for drying, selective or rapid heating, composite and epoxy curing, polymerization, organic synthesis, electronic bonding, moisture and gravimetric analysis, and reaction kinetics. There are three standard Vari-Wave models, each of which offer varying power wattages, gigahertz frequency ranges, and sizes (basic dimensions are 24 x 28 x 21 inches, and weight is less than 100 pounds). Also, a single-function version of the basic model is available. The oven's microprocessor is interactive; the desired parameters can be set up using a keypad and monitored on a control panel. Each model is optionally available with "Vari-Data," which is an off-line PC-based software/hardware interface capability. With Vari-Data, temperature and time can be pre-programmed and operator observations are menu-prompted and recorded.

Lambda Technologies employs a couple dozen employees, including engineers with backgrounds in materials science, chemical engineering, and microwave design. By the year 2000, the company hopes to do in excess of $50 million a year, although the business could be projected to be as much as $100 million. The company now has two production floor sites and plans to expand its product lines. It intends to open a new 12,000 square foot facility in January 1997 that will employ about twenty people. At the main Lambda facility, the company maintains a materials laboratory and a wide range of Vari-Wave and other high-power variable-frequency microwave systems and associated equipment.

Lambda Technologies was re-capitalized with a bank as a major creditor. A venture capital group brought in an entrepreneur from the Northwest, Mr. Richard S. ("Dick") Gerard as President/CEO. Mr. Gerard had a previous track record of building three other high-tech start-ups. Dr. Lauf attributes the success of Lambda's start up to Mr. Gerard's coming on board when he did. A new PhD graduate was scheduled to begin work with Microwave Laboratories on the day the company closed its doors. The venture capital firm that invested in Lambda Technologies temporarily hired the graduate as their employee. He wrote R&D funding proposals for Lambda Technologies that resulted in a DOE Small Business Innovation Research grant that involves an adjacent technology to the variable-frequency microwave technology. The new company was also able to secure some other contract R&D funds.

Lambda Technologies' had a CRADA with DOE that was a hold-over from Microwave Laboratories that was closed out. They signed another CRADA with DOE, initiated in their second year.

During Lambda Technologies' first year and a half, they solidified the base technology and built the intellectual property into a variety of applications. Lambda Technologies' next phase focuses on pursuing new and longer-term applications and developing additional types of customers and market niches. In addition to the standard research equipment they are already selling, the company is marketing equipment which would be used in a production environment, and they are gearing up to respond to customer orders. Mr. Gerard talked about applications like gluing components onto a computer circuit board or heating plastic and rubber products like shoe soles. Adhesive curing and working with polymers doesn't require as much power as the heating of ceramics. There is also the possibility of reviving the military hardware and radar applications. Mr. Gerard feels it is important to fund both research *and* production applications, as well as something in between. If requested, the company will develop custom-designed ovens to specification. He noted that it is important to coordinate the transition of the technology as it is perfected from concept to commercialization.

International Activity   At least two foreign patents have been filed in more than one application area.

Government Gains   A technology developed exclusively for military hardware, the traveling microwave tube, was adapted for civilian use in the new variable-frequency microwave furnace. It's successful commercialization is contributing to the economy.

Economic Development, Technical Assistance   Another aspect of Lambda Technologies' next phase of growth is the development of the company's own financing.[33] Mr. Gerard did not mention connections with locally available business assistance.

Elapsed Time   The license agreement was signed, and initial license fees paid, in February 1992, less than three months after filing the patent application in November 1991. The ORNL researchers presented the Materials Research Society paper in 1992 and published in 1993. In 1994,

Microwave Laboratories closed and Lambda Technologies was founded.

Case 5 (1992) – Gravity Meter

Role of Laboratory Researchers and Other Personnel Dr. James ("Jim") Faller is chief of the Quantum Physics Division and chairman of the Joint Institute of Laboratory Astrophysics (JILA) in Boulder, Colorado. At the time of the transfer, he was senior scientist at NIST and a fellow at JILA. Dr. Faller is one of the world's leading experts in measuring gravity. It was the subject of his dissertation at Princeton, and he soon developed an international reputation in the field. The researchers led by Dr. Faller at JILA are highly regarded by their colleagues in laboratories in Austria, Germany, England, Italy, France, Canada, Japan, the former Soviet Union, and Finland. Dr. Faller is the principal investigator of the group's projects. He has published more than ninety scientific papers (particularly on the gravity meter technology).[34] He also designed and built one of the first instruments deployed on the Moon by Apollo 11. Dr. Faller has received many honors and awards, including the Department of Commerce's Gold Medal and election to Fellowship in the American Physical Society.

Dr. Faller developed the mechanical, electronic, and optical (laser) technology incorporated into the JILA gravity meter. He was the primary technical expert throughout the development of the technical specifications and work statement of the original competitive solicitation. Once the contract was in place, he advised the NIST contracting officer's technical representative (COTR) who, at the time, was the NIST division chief. The FLC nomination form stated that the COTR "relied heavily" on Dr. Faller's technical advice. Dr. Faller's participation in the project was considered "critically important" when the sudden unavailability of a unique material necessary for a component required major redesign. And, he had primary responsibility for testing and evaluation of the instruments once they were delivered by the contractor.

Technology and Applications Traditional gravity measurement devices are based upon relative measures rather than absolute measures of changes in local height relative to the center of the Earth. NIST/ JILA developed an absolute gravity meter that makes absolute measurements of the Earth's gravitational field five to ten times more accurately than previously possible. In measuring the Earth's crust and density, it can detect changes smaller than a centimeter in relation to the center of the Earth. A unique portable first-generation device with ultra-high accuracy, national decisions are based upon its measurements.

The Laboratory   NIST works with industry and other outside organizations to develop and apply technology, measurements, and standards. NIST's technical work is performed in eight operating units or laboratories (focused on physics, chemistry, electronics, materials, manufacturing, buildings/fire research, computing, mathematics, and technology services) located in Gaithersburg, Maryland, and Boulder, Colorado. For these laboratories, NIST has a $740 million annual operating budget, and for congressionally-appropriated programs run by NIST, it has $260 million in funding.

NIST equipment and facilities available to industry focus on fire research, large-scale structures testing, computer network security, polymer composite fabrication, and transverse electromagnetic cells. The institute has a number of highly specialized facilities such as a ground screen antenna range and mechanical behavior facility.

The overall NIST staff numbers some 3,000 at both the Gaithersburg and Boulder sites. NIST is proud of the fact that it now implements an average of one CRADA for every three researchers on its staff. Since 1988, NIST has implemented 582 CRADAs.

University Involvement   The NIST Boulder site and a laboratory of the National Oceanic and Atmospheric Administration (NOAA), also a part of the U.S. Department of Commerce (DOC), are co-located in the same building in Boulder, Colorado. The research in this case, however, did not take place at NIST's Boulder site. It occurred at JILA, a cooperative research venture of NIST and the University of Colorado located on the university's Boulder campus. The laboratory is sponsored and run jointly by the university and NIST. JILA's research interests relate to gravitational physics, among other areas. One of its missions is to develop state-of-the-art precision instruments to measure the Earth's gravity or "g," a fundamental measurement in physics.

The NIST employees at JILA report to NIST's Physics Laboratory in Gaithersburg, which conducts research on quantum, electron, optical, atomic, and molecular physics. JILA's research is also centered on ionizing radiation used in medicine, the focus of an earlier NIST Physics Laboratory case study in this series of cases. The Physics Laboratory has highly specialized equipment, such as polarized electron microscopes, scanning tunneling microscopes, and a synchrotron radiation source.

Employees of both NIST and the University of Colorado freely intermingle at JILA; the chain of command is invisible. JILA is unique in terms of its structure. Probably the only analogy to JILA is the federal-university biotechnology collaboration at the University of Maryland called CARB, Center for the Advancement of Research on Biotechnology. The difference between JILA and DOE's university-operated laboratories is that in the case of JILA, the University of Colorado is an equal partner, not a contractor.

JILA has a total of about 250 employees. This includes eight full-time NIST professionals and fifteen university professors. Each organization allocates funding for faculty salaries, overhead, and direct expenses, with NIST's contribution to the university laboratory being in the form of block grants. As part of the arrangement between NIST and JILA, NIST pays sixty percent of Dr. Faller's salary; the other forty percent comes from other sources such as grants that he negotiates through the university. Dr. Faller can earn up to 100 percent of what his federal salary would be.

Funding, Financing   The funding for the absolute gravity meter came from different organizations at different points in its process. Funding for most of the fundamental R&D underlying the technology transfer was provided by NIST and the Defense Mapping Agency. NOAA provided the bulk of the funding for the 1990 procurement through its Advanced Technology branch. Each of the two initially procured instruments cost NOAA $250,000. NOAA also provided partial support for Dr. Faller's time spent on development activities related to the procurement. In addition, two gravity meters were purchased by the military for about $300,000 each. So, in total government agencies purchased instruments for approximately $1.4 million.

On the commercial side, two of the three principals of the company JILA selected to transfer the technology to invested almost $1 million of their own money to develop the gravity meter. Other follow-on investment for development of a more compact version of the technology was funded with $250,000 Phase I and II SBIR grants through NOAA.

Intellectual Property   There is no patent on this technology. Dr. Faller said that until about five years ago, NIST urged its researchers to publish, giving the laboratory up to one year to file for a patent. Apparently, during those years, NIST did not provide incentives to its researchers to initiate patent applications within the deadline. If Dr. Faller had patented the gravity meter, he said he would have received $2,500 plus fifteen percent of the royalties returned to the government. To make matters worse, when Dr. Faller received his FLC award, he learned at the awards presentation that his counterpart award recipients at DOE each received $10,000 just for getting the award.[35]

For follow-on commercialization efforts, the absolute gravity meter was a proprietary instrument.

Technology Transfer Mechanisms   To transfer the gravimeter, NIST initiated a competitive public solicitation through *Commerce Business Daily* and subsequently signed a five-year procurement contract with Axis Instruments Company, a small high-tech start-up in Boulder, Colorado. The contract stipulated that Dr. Faller and his group at NIST/JILA would design and develop the gravity meter technology, while Axis would manufacture the instrument.

Axis' agreement with NIST was a procurement contract mechanism where the product was to be built to the customer's specifications. Dr. Faller said, in hindsight, they should have used a CRADA rather than a traditional production contract, but they didn't know about CRADAs in 1992.[36] Among other requirements, the contract specified that the first two field instruments be delivered to NIST within two years. As part of its original contract, NIST had retained an option to purchase as many as six additional instruments for its own laboratory use three years beyond the initial production period at $250,000 each. NIST certified the gravity meters and turned them over to NOAA. NOAA provided most of the funding for the contract and participated in developing the solicitation and selecting the contractor, while NIST supervised the contractor's performance.

User Groups   The Axis principals were former instrument makers with JILA. Axis undertook aggressive marketing efforts. According to the company's marketing brochure, Axis Instruments was formed:

> "To design, manufacture and market the finest quality, high-precision scientific instruments—mechanical, optical, electronic, and vacuum. We offer cost effective, 'one-stop' design, prototype development, manufacturing and marketing service."

NIST encouraged the formation of the company, and encouraged it to market and sell gravity meters to other customers. The NIST contract was the company's first major contract, but it subsequently won contracts to build more "FG5 Gravimeters" for the Canadian, English, and French governments. In exchange for a gravity meter, Axis acquired the prototype for a new type of iodine laser from the Bureau International des Poids et Mesures (BIPM) or International Bureau of Weights and Measures in Paris, France. Before that, Axis used a less accurate laser in its gravity meter, and the French laser was used as a laboratory physicist's instrument only. The French measurement bureau gave the company circuit diagrams describing how to make the laser as well as international manufacturing and marketing rights. In return, Axis agreed to pay a royalty to BIPM on each one sold. The laser is a component of the gravimeter; it is also a stand-alone device available to both commercial and private laboratories.

In the meantime, the company appeared to have high expenses with salaries, rent, and purchased equipment and parts. The president of the company attempted to sell the company to large manufacturers like Texas Instruments, but they were not interested in Axis' relatively small market. In mid-August 1993, Axis had to let go of its technical employees and was left with management only. They made it through the initial stages, but eventually, Axis went out of business.

Subsequently, Dr. Tim Niebauer, the Axis chief scientist who had developed the original prototype for Axis (and Dr. Faller's graduate student at the university), bought the rights to the gravity meter from Axis and the Axis equipment inventory. He agreed to pay back the Axis founders a certain amount for each instrument sold. He moved the company to a "low-rent" area and hired back only four of the Axis employees. The new company, Micro-g Solutions, began selling instruments and eventually moved to a better location. They bought a former supermarket/dance hall on the main street of Erie, Colorado. Being eager to attract a high-tech firm, the Erie municipal officials granted the company an exception to city zoning regulations so that they could manufacture in the middle of town.

Another of the original Axis principals, Dr. Mike Winters, obtained the rights to the Paris iodine laser and he and his wife formed a separate company called Winters Electro-Optics, Inc. Since the laser is a component in the gravimeter, Micro-g Solutions buys the part from Winters Electro-Optics. Dr. Winters was the fourth person brought in to Axis Instruments, hired when Dr. Faller convinced the Axis founders they needed a physicist to build the laser used in the gravity meter. Dr. Winters did the development work required to turn the French iodine laser into a product and integrate it into the gravity meter. So, Winters Electro-Optics now pays what was previously Axis' royalty responsibility to BIPM. They have the same agreement as Axis had with BIPM, and they can use as much of the design as they want. Winters Electro-Optics also currently has a two-year CRADA with NIST.

Winters Electro-Optics was "financed" through an agreement with Axis whereby they would handle the warranties in return for the Axis laser inventory and related equipment such as oscilloscopes. In responding to the warrantees, it was easier for Dr. Winters to replace the existing lasers with his new one for free. So, basically giving away about ten lasers comprised Dr. Winters' start-up costs. Dr. Winters is pleased to report that the only outside money used in the course of starting up his business was $5,000 to $10,000 borrowed from relatives.

Dr. Winters says the laser is a perfect product for a very small company, although the market will eventually saturate. Their major competition is the National Physical Laboratory (NPL) in Great Britain, which has recently been privatized. However, NPL has access to markets in Malaysia and Singapore that Winters Electro-Optics does not have. Dr. Winters says it is not feasible to use agents overseas because they don't know his product, and distributors often raise their price in those locations. On the other hand, being smaller, the company has to go outside for machining work unlike Micro-g which has its own machine shop.

A gravity meter can be used to measure, for example, a rising sea level due to global warming. Also, exploring for oil reserves requires tools that measure gravity, among other things. For oil prospecting, absolute gravity measurement saves time and labor over relative gravity measurement. Relative instruments involve repeating measurements over several days. The instrument goes back and forth, aimed at a series of distant sites, a labor-intensive and time-consuming process. An absolute gravity measurement tool would require only one visit to the same site. In addition, NOAA requested two absolute gravity meters for field use by its Climate and Global Change Program at the same time that other organizations were also expressing an interest.

Barriers to Commercialization   When the contract was issued with Axis, NOAA's requirements included a new level of compactness and robustness for several critical parts within the gravimeter. Therefore, NIST was in the process of designing certain improvements and preparing the necessary engineering documents. This made negotiating the contract difficult because the design of the instrument was evolving while it was being negotiated.

Other Factors   In this type of joint scenario with NIST and the University of Colorado, it is common to have inventors from each organization involved in one invention. When this happens, an invention disclosure is filed with each organization. Then, the two technology manager officers cooperate to move the technology forward.[37] The procedure is to select a lead party from one of the organizations in order to file and market the invention, negotiate the license, collect the licensing fees, and share them with the other party. Who is chosen as the lead depends upon such criteria as how many researchers are from each organization, their seniority, and existing institutional portfolios. If there are two inventors, they each receive half of the royalties.

In negotiating the license, the university's motivation is to get funds into the university. The university will usually entertain several companies. The company most likely to commercialize and that offers the highest royalties is awarded the license. On the other hand NIST, as a high-profile public institution, doesn't like to favor one company with selection over another (in the case of exclusive licensing) unless it was the only way to get the technology commercialized.

NIST is not as adamant about patenting either. NIST will allow the university to patent, but reserves the right to says it may still make the technology available for free. In this case, it becomes useless for the cooperating university to patent.

In the case of a CRADA that involves solely NIST personnel on the non-company side, it would be put into place by NIST rather than the university. So it is apparent that the fluidity between the two organizations creates both challenges and opportunities due to the differences in their philosophical approaches, procedures, and cultures. Dr. Faller's former graduate student[38] noted that NIST employees are still reluctant, even in 1996, to talk with or work with companies. They are afraid that, as government employees, they would create a problem in fairness of access. The technology transfer officer for the university, Dr. Steve ONeil, says NIST takes very seriously its position as a "high-profile public institution" in the state and the nation and impresses this upon its employees.[39]

The four University of Colorado campuses made $1.6 million in royalties last year, yet Dr. ONeil says they almost broke even on their technology transfer efforts. This is because twenty-five percent of this return went to the inventors, twenty-five percent went to the relevant department, and twenty-five percent went to the university's general research account. The remaining $400,000 was used to pay for expenses involved in technology transfer activities such as marketing and administrative staff, patent searches, attorneys, etc. So the issue of how to keep incentives high and still recoup enough to keep the technology transfer function operating is problematic. Dr. ONeil added that out of several thousand faculty and employees at the university, only a handful understand technology transfer.

User Benefits/ Economic Impact/ Outcomes Axis's customers included the University of Colorado, local businesses in Colorado, and private individuals such as independent inventors as well as international governments. The company built up to revenues of about $3 million a year, and twenty employees.

Micro-g grew to seven or eight employees and became successful, selling four instruments in 1995, and four as of mid-1996. The projected revenues for an anticipated product from SBIR-funded research are $10 million per year. Also, NIST and Micro-g Solutions have a CRADA to continue development work on the absolute gravimeter. JILA is trying to be more aggressive in getting the gravity meter in a form that outsiders can use. The laboratory is becoming more industry-oriented, bringing in corporate executives to communicate user requirements to the laboratory personnel. Micro-g Solutions' president, Dr. Niebauer,[40] noted that the gravity meter market was not a bad market, but also not such a huge market that any one company would have spent ten years as NIST did in developing the gravity meter to where it would be commercially viable. He also noted that a CRADA was not necessarily an appealing business deal for companies. Unless the CRADA is structured in a certain way, it won't make money for the company because there is no specific goal (like a product that makes money) and no deadline. Furthermore, the company can't provide direction to the government employees working on the CRADA: they can only access information and ask specific questions. The CRADA is based upon the good faith of both parties that useful research will ultimately result from the CRADA work. Companies don't normally do business that way, he said. So CRADAs are useful for ideas, but not necessarily for products.

In summary, including the NIST 1990 contract, eighteen gravity meters were sold, representing approximately $5.4 million in sales at an average of $300,000 each. (Axis sold eight; Micro-g sold ten). Overall, the government has recovered about a third of this, $1.8 million, in taxes from the two companies. On average, over a period of five years, ten people were employed by Axis and Micro-g Solutions. Micro-g Solutions has eight or nine employees, and Winters Electro-Optics has two employees. Winters

Electro-Optics provided the first ten lasers for free and donated one to an institution. So the company actually sold roughly forty lasers at $32,000 each, totaling $1,280,000 over three years, providing a little over $425,000 a year in gross revenues.

International Activity   In the early 1990s, before being transferred to the private sector for mass production, six of the NIST instruments were being used at various sites around the world and more were in demand. Early on, gravity meters were sold by Axis Instruments to BIPM, Canada's Geological Survey, and the National Environmental Research Council of England.

Micro-g Solutions sold four gravity meters to the Japanese government. Having experienced several major earthquakes recently, the Japanese have devoted large amounts of funding to their earthquake science programs, which purchased the instruments. Also, Italy purchased an instrument to monitor Mount Vesuvius. Mount Vesuvius is dormant but not dead and, unlike many other volcanoes, has many people living on it. Belgium and Australia also bought gravity meters. A gravity meter was even sold to China; for this sale, there was a great deal of export-related paperwork. Transfer of the gravity meter to all of these countries has helped the United States' balance of trade payments.

Government Gains   The original NIST contract stipulated the delivery of the first two gravity meters to NIST (for NOAA). Since then NIST has purchased one additional instrument for use in its Gaithersburg, Maryland, laboratory in anticipation of a new mass standard for kilogram replacement. This area of the standards community is becoming more active because the absolute gravity meter makes it possible for a new standard to be adopted. A device less expensive and not quite as precise would allow wider audiences to avail themselves of this level of accuracy. It is predicted that the standards in this area will change between 1998 and 2002.

Economic Development, Technical Assistance   There were state-funded incubators available in the area, but neither Axis, Micro-g, nor Winters Electro-Optics chose to be affiliated with these facilities. Dr. Tim Niebauer, president of Micro-g, noted that economic development services and banking institutions were not that helpful. On the other hand, he referred to the useful federal government research funding programs as an "engine for the future."

Elapsed Time   NIST has invested in gravity measurement research at JILA since the early 1960s. In the mid-1980s, Dr. Faller and the NIST researchers began developing a portable absolute gravimeter. In 1989, NOAA requested the two gravity meters which set in motion the 1990 procurement process. The gravity meter technology was transferred to the firm in late September 1990 through the contract. The first two instruments were delivered to NIST in the Spring of 1992. At that time, NIST undertook a six-month testing and evaluation phase. So the actual technology development phase for those particular instruments took about eighteen months, which is quite fast!

Axis was actively in business from April 1990 to the summer of 1993. Winters Electro-Optics was founded in September 1993.

Case 6 (1992) – "Oatrim" Fat Substitute

Role of Laboratory Researchers and Other Personnel   Dr. George E. Inglett, a specialist in biopolymer materials at the National Center for Agricultural Utilization Research (NCAUR), invented the technology to manufacture a natural fat substitute and food additive called Oatrim. Early in his research, Dr. Inglett provided information on the technology to interested companies and responded to specific questions. Papers about the technology were published in *Food Technology*, the journal of the Institute of Food Technologists[41]; USDA's *Science of Food and Agriculture*[42]; and conference proceedings of the American Chemical Society.[43]

After Oatrim's announcement at the American Chemical Society conference, Dr. Inglett received thousands of industrial inquiries. Consequently, he arranged a technology transfer conference at the laboratory, which was attended by approximately seventy industry representatives. More than half of the seventy companies attending the conference applied for licenses to the technology. Three were granted non-exclusive licenses.

Once Oatrim was licensed, Dr. Inglett assembled an information packet on the technology that included the names and addresses of the licensees producing the material. At this point, he hosted corporate visits to the laboratory to review equipment and processing needs.

Dr. Inglett invited an engineer from ConAgra, the first company to start a pilot plant, to spend several days at the USDA Peoria laboratory to participate, first-hand, in Oatrim processing. Dr. Inglett also visited the pilot plant site in Mountain Lake, Minnesota, to consult with plant personnel and corporate management and analyze the initial product. Like ConAgra,

Rhone-Poulenc Food Ingredients relied heavily on Dr. Inglett and maintained close contact with him during the product development phase.

In addition, Dr. Inglett initiated "human studies" of Oatrim. The first such study was carried out by nutritionists at the USDA/ARS center at Beltsville, Maryland, using two dozen volunteers. The results of the five-week study were published in the December 1993 issue of ARS' *Agricultural Research*. As expected, the results showed a decrease in "bad" LDL without a decrease in "good" HDL cholesterol. Volunteers improved in their ability to process sugar from their meals, an indication of their "glucose tolerance" that is linked to diabetes. Also, surprisingly, most of the volunteers lost an average of 4.5 pounds during the study, despite increases in their caloric intake. And, a 1994 article in *Science News* noted that "nobody in the Oatrim research test complained about being hungry."

Subsequently, Dr. Inglett conducted additional research to improve processing of the product and to address problems associated with scale-up from laboratory to industrial production. For example, he experimented with various food products and created an ice cream substitute that fared well with the taste panels. Also, he searched for varieties of oats containing higher levels of cholesterol-fighting beta glucan.

Technology and Applications   Oatrim is a bland, rich, creamy gel derived from oat starch that has less than one calorie per gram compared to nine calories per gram for fat. It remains stable when frozen, cooked, or left at room temperature. It causes no change in flavor or loss in textural quality (that is, it "has the feel of fat on the tongue") upon incorporation into many foods including dairy products of all types, dressings and sauces, meats and cereals.

Oatrim has the double benefit of lowering cholesterol levels in the blood while it reduces fat intake because it decreases artery-clogging LDL cholesterol and increases levels of beneficial HDL cholesterol. This is because Oatrim is the only fat replacer on the market that contains beta glucan, the soluble fiber in oats.

The Laboratory   NCAUR is one of the four major centers in USDA's Agricultural Research Service. It is located in Peoria, Illinois. Within NCAUR, there are a number of "individual laboratories" for bioscience, food and fiber, industrial products science, crop protection, and natural toxins.

Funding, Financing   One of the licensees, the Rhone-Poulenc/Quaker Oats partnership found the difference in scaling up from small "bench-top" levels to full-scale production to be "world's apart."[44] Mark Freeland[45] of Rhone-Poulenc said Dr. Inglett's original work was good science, but was done on such a small scale that the partnership found it necessary to invest millions of dollars and a great deal of time into formulating the Oatrim ingredients for incorporation into mass-production levels. The partnership produced over twenty million pounds every year as far back as 1993. Mr. Freeland said the companies probably should have better understood that they were getting into a "laboratory process." They didn't appreciate how difficult scaling up to full production would be. The companies still believe in the technology and are not abandoning it, but they are at the end of their investment in product development. The only feasible new investments at this point would be made to increase capacity or tweek the production process or re-launch the product.

The companies also experienced problems and devoted a great deal of money toward technology development. They noted that any grain product has an inherent cereal flavor, and they found that the original Oatrim had a weakness as a fat replacer because it had an oat flavor that didn't work well with cheese and ice cream products. Particularly with higher levels of production and use, the oat flavor crept in.

Intellectual Property   So far, Dr. Inglett has produced three types of Oatrim from oat flour or bran. The U.S. patent listed in the FLC's midwest regional directory as available for license is: "A method for making a soluble dietary fiber composition from oats."

The technology is also easily applied to other crops such as barley and corn. Thus, a second patent involves barley, as follows: "A method for making soluble dietary fiber compositions from cereals." Dr. Inglett is listed as the inventor on both patents.

Technology Transfer Mechanisms   Non-exclusive licenses were granted to three companies in the early 1990s, described below. With the keen competition for the technology, it is presumed that these licenses were probably somewhat above average in fees and royalties paid to USDA and Dr. Inglett.

ConAgra Specialty Grain Products Company: The first license was with ConAgra Specialty Grain Products Company in Omaha, Nebraska. ConAgra, Inc., also headquartered in Omaha, is a $25-billion company with over 83,000 employees, the second largest food manufacturer in the United States behind Kraft Foods.[46] ConAgra is comprised of a number of independent operating companies which it owns, including: Armour[R], Banquet, Butterball[R], Chun King, Eckrich[TM], Hebrew National[R], Hunt's, Knott's Berry Farm[R], La Choy, Mamacita's, Marie Callender's[R], Orville Redenbacher's[R], Peter Pan[R], Reddi Wip[R], Sargeant's, Steak Express, Swift Premium[R], Swiss Miss[R], Taste o'Sea, Van Camp's, Wesson[R], and many others.

Rhone-Poulenc, Inc.: A license was also granted to Rhone-Poulenc, Inc., a $16 billion French company based in Cranbury, New Jersey, with more than 140 offices throughout the world including major facilities in Canada and Mexico. The company's primary North American laboratory facilities produce meat, poultry, and seafood in Washington, Pennsylvania; brew and make wine in Chicago, Illinois; provide dairy culture in Madison, Wisconsin; and, perform general food research at their headquarters in Cranbury, New Jersey.

Rhone-Poulenc Food Ingredients, a world leader in R&D on food applications, has annual sales in excess of $300 million. Its core products and typical applications include: phosphate blends used as leavening in baked products and other uses; "performance ingredients," including fat replacers, for salad dressings and sauces; stabilizer systems for dairy products; and starter cultures and and Direct Vat[TM] inoculants for cheeses, yogurt, etc. Some of their well-known brand ingredients include: MicroGARD[R] to extend shelf life, Assur-Rinse[TM] Safety Wash to remove bacteria during beef and poultry processing, and Rhodigel[R] to maintain sauce viscosity. The company's tag line is, appropriately, "Your Food-Tech Partner[SM]."

Quaker Oats Company: A license was also granted to Quaker Oats Company headquartered in Chicago, Illinois. Quaker Oats is a $6 billion retail-oriented food company with sales based largely upon a wide variety of oat-based products as well as Gatorade[R] and Snapple to supermarkets. The Quaker Oats Industrial Cereal Group sells large quantities of wheat flour products to bakeries and grocery stores. This amounts to $40 million in business for the company, but it is less than one percent of the overall company business.

Rhone-Poulenc/ Quaker Oats: Rhone-Poulenc Food Ingredients and Quaker Oats Company's Industrial Cereal Group formed a partnership in April 1992, splitting expenses and profits fifty/fifty. Instead of building a plant right away, the companies reduced their risk by partnering and proving the validity of the technology first. Rhone-Poulenc contributed process development, engineering, and applications. In addition to technical services, Rhone-Poulenc brought a sales organization to the partnership, since it can call on General Mills, Pillsbury, Kellogg, Kraft, and other food retailers that could be considered competitors to Quaker Oats. Quaker Oats' contribution was their specific expertise related to oats, since the company represents more than a century of experience with oat technology. The partnership's development work related to the process of scaling up the production is being carried out at the Quaker Oats' operation in Cedar Rapids, Iowa, which specializes in flour and oat production. The scaling up process produced reasonable quantities for analytical testing and screening of applications.

User Groups   Food companies are the immediate users as they incorporate the product into their processed food products. For example, Oatrim is used in dozens of varieties of no-fat cheeses.

The ultimate end users are adult customers who are concerned about their health. According to the National Cholesterol Education Program, more than fifty percent of all adults in the United States have an increased risk of heart attack because their blood cholesterol levels are above the desirable range. Americans derive about thirty-seven percent of their calories from fat, as opposed to the thirty percent maximum recommended by nutrition experts.

Back in 1990, the market for non-fat foods was a mere $750 million. In 1995, a record 1,914 fat-modified products were introduced to the commercial market, according to the editor and publisher of *New Product News*, Lynn Dornblaser. By the year 2000, experts are projecting a multi-billion-dollar industry, although introductions of new products may level off by then. Oatrim is expected to share in at least $100 million of this market and create new markets in combination with wheat and meat products in particular.

Barriers to Commercialization   Because Oatrim is a food product, it required approval from the U.S. Food and Drug Administration (FDA). But since it is made from all-natural fiber ingredients with natural enzymes, it did not face major regulatory hurdles and years of safety tests to get to the market. ConAgra filed a petition for FDA approval by "self-affirmation" through an FDA method known as GRAS ("Generally Recognized as Safe"), which takes months instead of years. Using this process, the company convened a panel of paid experts to review the product information. Once all criteria and panel recommendations were met, the product was submitted to the FDA as information for the file. At that point, ConAgra assumed the liability. By comparison, Procter & Gamble's highly publicized Olestra took seventeen years to obtain FDA release for limited distribution only, using strict production standards, and just for certain applications like snack foods—potato, corn, and tortilla chips and crackers.

As with other cases in this series, one reason the product has been slow to influence the market is that it is new to industry. Therefore, it requires education; for example, potential customers need to get product samples.

Finally, the first-hand experience of the partners indicated that joint ventures and alliances of this type depend upon a number of factors for success in the long run. Apparently, they are easy to describe and conceive, but hard to manage because they are heavily affected by the differences in relationships, personalities, businesses cultures, and even nationalities. All of this makes decision-making difficult. For long-run success, the individuals involved have to be motivated to avoid the risk of alienation.

Other Factors   The Mountain Lake partnership and USDA at ARS-Peoria formed a CRADA. Dr. Inglett, a Peoria biochemist Richard Greene, and researchers from the Mountain Lake partnership did some additional processing to Oatrim which resulted in a new fat substitute called Z-trim which has zero calories per gram.[47] Under that CRADA, any inventors involved, including company inventors, are given first right of refusal to a new invention. Dr. Inglett introduced Z-trim at an American Chemical Society meeting in August 1996 and held a laboratory-hosted announcement of Z-trim in early October 1996.[48] Like Oatrim, Z-trim is made from all-natural ingredients. It replaces fat (removing calories) and adds fiber. It is flavorless, made from the hulls of soybeams, oats, peas, rice, or bran from corn or wheat. The hulls are processed into microscopic pieces, dried, and milled into powder. In powder form, Z-trim absorbs up to twenty-four times its weight in water: when cool, it forms a fat-like gel that supplies texture qualities such as moistness and smoothness. USDA taste test panelists rated highly some brownies made with Z-trim.

A 1996 edition of *Time*[49] said the following: "Just three weeks ago, a U.S. Department of Agriculture researcher presented a fat substitute made from the hulls of oats, corn and soybeams. If it is remotely palatable, it is sure to sell. History suggests, however, that it won't make much difference. Despite the arrival of Olestra last winter, despite NutraSweet and one percent milk, despite an estimated $33 billion spent every year on diet books, over-the-counter medications, health-club memberships and low-calorie foods, the flab still remains, entrenched solidly on waists, hips and thighs."

Regardless of *Time* magazine's cynicism, the Mountain Lake partnership has the first right of refusal on a potential exclusive license to the Z-trim invention. The patent for Z-trim was filed by USDA in November 1995. In the meantime, USDA is not sharing information on Z-trim.

User Benefits/ Economic Impact/ Outcomes   The users and partners are discussed below in turn.

ConAgra: ConAgra's brand name for Oatrim is TrimChoice. When Oatrim was first introduced, then-CEO Mike Harper leveraged the benefits of TrimChoice for its independent operating companies manufacturing Healthy Choice™ products. These ConAgra companies were strongly encouraged to incorporate TrimChoice into their products, and ConAgra was soon picking up orders from its food manufacturers on a regular basis. The original pilot plant was in place for only a few months in order to define the process parameters needed for large-scale production. Widespread product marketing began in 1991, while the actual TrimChoice commercial plant in Mountain Lake, Minnesota, was being completed. The company began full-scale production there in the Fall of 1991. The manufacturing facility in Mountain Lake was very automated, employing only nine people, which kept ConAgra's operating costs low. ConAgra's first Oatrim-related food, an extra lean (96 percent fat free) ground beef product, sold in the Midwest under the Healthy Choice brand name.

With a change in top management in 1992, ConAgra began offering TrimChoice to other companies outside the ConAgra family, and, according to Steve Grisamore,[50] then sales manager for new product development, the company sold a lot of TrimChoice to other companies. In February 1993, ConAgra formed a fifty/fifty profit-splitting joint venture with A.E. Staley Manufacturing Company of Decatur, Illinois. Staley is a publicly traded Fortune 500 company, like ConAgra, owned by a foreign sugar company called Tate and Lyle. This partnership created a stand-alone company called Mountain Lake Manufacturing. (Mr. Grisamore became General Manager of Mountain Lake, although he is based at corporate headquarters in Omaha.) Con Agra contributed the R&D and manufacturing expertise, and Staley originally contributed the sales and marketing expertise to the partnership.

Healthy Choice is an overall $1.2 billion brand for ConAgra. It had its own Internet site before this became commonplace. The ConAgra 1996 annual report stated that its Healthy Choice products were important contributors to earnings growth for the year, particularly the soups and fat-free cheeses. Healthy Choice products that enjoyed successful debuts in 1996 were hot sandwiches, ice cream, and processed deli luncheon meats. The Oatrim ingredient is listed on the labels of hundreds of products as "hydrolized oat bran." According to the company's yearly financial reports, ConAgra's gross margin increased in 1996 due to margin improvements in various of its businesses. This included specialty food ingredients, which also contributed to the company's pre-tax earnings increase.

In July 1996, some changes were made to the Mountain Lake partnership, and the name was changed to Mountain Lake Specialty Ingredients Company. Media articles indicated that Staley is not actively handling TrimChoice any more. The product competes with its own starch-based fat replacers. However, Staley is still a half-partner in the Mountain Lake venture.

Rhone-Poulenc/Quaker Oats: The partnership introduced its product in national markets and sells to fifty to sixty companies. Quaker Oats had hopes of high sales within a short period of time. Lanny Babbitt of Quaker Oats' Industrial Cereal Group said[51] that for a company this size, "significant sales" would translate into $10 - 100 million, a level they have not yet reached, with an approximate one-year commercialization period. Mark Freeland of Rhone Poulenc Food Ingredients stated[52] that Rhone-Poulenc's average time frame for a product going from "ground zero" to being a commercial success is five to seven years. Rhone-Poulenc has also been disappointed in the growth of the business. Given the heavy interest in low-fat products, the company would have liked to have seen more growth by now. At four years of development, the company is within the framework for new product development and still has good expectations.

However, representatives of both companies noted they are already improving their market position with a new version of the technology that was launched in September 1996 based upon revisions to the original Quaker™ brand Oatrim. Their product was re-named Beta-Trim, related to the fact that it contains five percent beta-glucan. They enhanced their technical process to separate the soluble from the insoluble fiber, which makes the ingredient virtually flavorless. This expands the market to the more delicate food products such as processed cheeses with low- or no-fat content (versus, say, Mexican hot sauces). The product material states that ... "Beta-Trim gives reduced fat foods the mouthfeel, texture, and taste consumers expect from full-fat products . . .and provides lubricity and creaminess of foods that starches and maltodextrin cannot provide." As a result, they feel their partnership could soon experience some breakthroughs. For example, a Rhone-Poulenc newsletter[53] notes:

* "New production techniques that remove the 'oaty' taste from some Oatrim products have led to stunning successes in both hard-pack and soft-serve ice cream mixes."
* Also, "Working with one of our Oatrim family of fat-replacers, a well-known producer of high-quality franks has successfully marketed a totally fat-free hot dog that has all the characteristics of its full-fat counterparts. Taste, texture, bite, juiciness—they're all there. Only the fat is missing."

At the same time, the partnership is experimenting with various combinations of Beta-Trim blended with their other products such as Raftiline[R]/Raftilose[R], hydrocolloids, phosphates, and dairy cultures to develop more versatile food ingredient systems. Next month,[54] the Rhone-Poulenc/Quaker Oats partnership will bring together a professional taste panel at the Quaker Oats facility in Barrington, Illinois, to compare their "old" and "new" Oatrim. They intend to identify and measure the results quantitatively using human taste buds instead of machinery.

International Activity   Since the USDA waived its right to foreign filing, Dr. Inglett chose to patent Oatrim outside of the United States and, based upon that foreign patenting, he granted ConAgra an exclusive worldwide license outside the United States.

The Rhone-Poulenc/Quaker Oats partnership subsequently negotiated a sub-license from ConAgra, Inc. for rights in fourteen countries. In other words, wherever ConAgra could sell, the Rhone-Poulenc/Quaker Oats partnership could also sell in those countries by paying royalties to Con Agra. However, Con Agra "let go" of some of their international patents by not paying the annual fee in certain countries, so the patents are no longer in place in eight of the fourteen countries. Therefore, each organization is now selling the product in only six foreign countries.

Government Gains   The conversion of oats, which is a raw agricultural farm commodity, into a value-added product has helped the United States' balance of trade in this area. So, the government has indirectly realized a benefit from Oatrim.

Economic Development   No economic development services were provided.

Elapsed Time   Dr. Inglett recognized the potential for Oatrim in 1988. A patent application was filed in early 1990. Subsequently, in April 1990, the development of Oatrim was announced at a trade association conference while the patent was still pending. A conference about the technology was held at the Peoria laboratory in May 1990. In June 1990, the first company, ConAgra, signed a license with the laboratory for rights to the technology. (Apparently, the U.S. Patent and Trademark Office must have issued a statement that the technology had passed initial screening, and was a likely candidate for a patent.) The first Oatrim product reached the market in November 1990 as part of a product sample distribution of pilot plant quantities about six months after the license was granted. Widespread product marketing began in 1991 while the pilot plant was being completed. Overall scale-up from bench-top to commercial production was accomplished in record time by ConAgra.

By the end of 1991, Oatrim was licensed to the two additional companies. All three license negotiations were based merely upon the patent application being filed. The patent was issued in February 1991. The Rhone-Poulenc pilot plant arrangements were completed in 1992 to 1993. In the meantime, Rhone-Poulenc partnered with the third licensee, Quaker Oats. So, all three companies moved quickly to get the technology into the marketplace. "Human" tests were conducted in 1993.

An article in *Science*[55] noted that all-natural fat replacers take four years to get to the market; apparently, the journal was on the mark vis-a-vis Oatrim. Dr. Inglett's award from the FLC noted that it "honored the dispatch with which he moved the technology into the commercial mainstream." Steve Grisamore, (then) sales manager for new product development for ConAgra, was quoted in the Fall of 1992 as "attributing the speed [of commercialization] to a concerted corporate effort to get [the product] to market, and to excellent cooperation from Dr. Inglett."[56] He said[57] the amount of time involved in going from bench-top to commercialization, eleven months, was "unheard of" and would normally take eighteen to twenty-four months.

Case 7 (1993) – Chemiluminescent "Light Sticks"

Role of Laboratory Researchers and Other Personnel   Dr. Herbert Richter, supervisory research chemist, has directed a group of chemists in the

Weapons Division of the Naval Warfare Center at China Lake, California. Dr. Richter's team has been developing new "energetic materials," propellants, and explosives. Members of the team receiving the FLC award for technology transfer were Dr. Ronald Henry (now deceased), senior scientist among the research chemists, and Joseph Johnson (retired), physical science technician, who loaded and evaluated the experimental formulations.

Over the years, the team performed a series of tests for luminescence on a variety of chemical compounds. The compounds selected for this application exhibited the most sustained, temperature-resistant, and non-flammable luminescence. All of the mixing of compounds was performed using various glass laboratory vials. Two smaller vials contained the chemicals; these vials were enclosed within a larger vial before they were broken to create the mixture. Field tests were sometimes performed under adverse conditions; at one point, the team did night testing in the cold in Alaska, which included a harrowing helicopter ride.

In order to create user awareness, Dr. Richter worked with George Linsteadt, the first ORTA officer at China Lake. Mr. Linsteadt was also one of the FLC's founders and later its administrator upon retirement from Navy work. Mr. Richter cooperated with the subsequent ORTA officers as well.

Dr. Richter created a sample kit containing products in a variety of sizes and chemicals. The team performed countless demonstrations and provided data for potential manufacturers and for government agencies (Department of Transportation (DOT), the Forest Service, and state and local safety agencies). Some of the demonstrations were dramatic: for example, a DOT demo involved dropping a light stick from a third story window to the street below. The interest created by the glow stopped traffic for several blocks. They also briefed many military officials both within and outside of China Lake.

Also, many papers (including Navy technical notes, monographs, and publications) covered this technology over the years, and it has been discussed at many conferences. As noted, the papers and data were shared with interested parties from the beginning, and licensees resulted from these early contacts.

One of the contacts was interested in having Dr. Richter serve as a consultant to them in the early 1980s. For a number of years, DOE and other departments and agencies have had provisions where their laboratory scientists could be hired out. (DOE's program, for example, is called "Work-for-Others.") However, the military services, for the most part, do not take part in this practice.

Technology and Applications   "Light sticks," like those novelty items that glow in the dark are created when chemical compounds, each containing a dye and a catalyst, are mixed together. The fluorescent dyes determine the color of the item. The light produced by this chemiluminescent mixture is long-lasting and resistant to environmental extremes such as heat, sand, or snow.

The technology was developed at the China Lake site of the Naval Warfare Center as a marker to locate downed pilots and to illuminate potential targets for airplane bombs. The devices are visible from the air, produce little heat, and can be scattered to form a very large visible signal *without* starting a fire. (Putting together certain chemical compounds often starts a fire.)

The Laboratory   The China Lake laboratory covers over a million acres of the Mojave Desert in Southern California. It was founded in the early 1940s as a facility to test rockets developed at the California Institute of Technology (CalTech) in Pasadena, California, and as a Navy proving ground. It was known as the Naval Weapons Center for several decades, then as the Naval Air Warfare Center, and now as the Naval Warfare Center. The facility employs some five thousand civilians, including one thousand scientists and engineers doing R&D, and about five hundred military personnel. The China Lake facility has an estimated annual operating budget of about $800 million per year.

University Involvement   CalTech does not have a role with the China Lake facility at this point. The Jet Propulsion Laboratory at CalTech became one of the nine official NASA field centers when NASA was created in 1958.

Funding, Financing   Over the years, the research in this case was sponsored by the Navy. Support was received from the Office of Naval Research, the Naval Air and Naval Sea Systems Commands, and the Marine Corps. Support was also received from the Army. Funds to file the patents for the technology came from the Marine Corps Exploratory Development Program.

Intellectual Property   There are two current Navy patents for the technology which cover "The latest advances in chemiluminescent technology offering improved color delineation and longer staying power under adverse conditions." Dr. Richter, Dr. Henry and Mr. Johnson are listed as the inventors.

Omniglow Corporation holds 99 percent of the chemiluminescent patents worldwide. It has patented processes for: multicolor Lite-Rope[R] necklaces; white, red, pink and purple fluorescents; Lightwrap[R] glowing strips; Flex-Stick[R] fishing lures; S.E.E.[TM] Systems; and Speculite[R] lightsticks.

Technology Transfer Mechanisms   An early license was signed some time in the mid-1970s with a company called either Chemical Devices Corporation or Coolight. Later license agreements were signed in May 1989 with American Cyanamid Corporation and in July 1989 with Omniglow, Inc. in Novato, California. Omniglow was the successor to Chemical Devices Corporation. The duration of the license agreements was about 4.5 years for each, extending to the end of 1993. Both licenses were revocable, non-exclusive licenses so the companies could use the inventions to produce light sticks, light wands, and safety lights.

Early on, both licensees indicated they were happy with their licensing arrangements, and one of them indicated a desire to work further with Dr. Richter. Dr. Richter said[58] CRADAs were "talked about" at that time, but still not commonly done at China Lake. For example, a China Lake CRADA announced in the laboratory's weekly "Rocketeer" newsletter in 1996 took almost ten years to sign, from 1986 when the CRADA legislation was enacted. The CRADA involved Thiokol Corporation, a major multinational company and a leader in "energetic" materials. Now, however, the Naval Warfare Center's Weapons Division pursues CRADAs more actively.[59]

In the early 1990s, Omniglow filed a lawsuit against American Cyanamid that alleged that the company had illegally filed a light-stick-related patent excluding the government from rights to the technology which actually belonged to the government. (Presumably, this would apply to the other government licensee, as well) The government decided not to join in prosecuting the case. Yet, Omniglow won the case and in 1993 as a result ended up acquiring the division at American Cyanamid that manufactured light stick products. Since then, the rest of American Cyanamid has been bought out by American Home Products, an $8.5 billion acquisition. As a result of all of this, Omniglow wrote a letter to the Navy cancelling both of its license agreements to pay royalties to the Navy (by now, it had also acquired the American Cyanamid license), saying the licenses were no longer needed.

During the course of the licenses, the companies paid the China Lake laboratory over $86,000 in royalties. Of this, $25,000 was paid to the three inventors. The remainder was used to fund other research at the laboratory. A China Lake "Success Story" sheet states that this "provided an incentive to other [laboratory] inventors with potentially licensable patents."[60] In the Navy laboratories, the inventor receives at least twenty percent of the royalties. If there is more than one inventor, the twenty percent is split ..... g them. Also, having a patent in itself, gives the scientists credit when it comes time for promotions and other benefits. (Dr. Richter noted that if he had had the foresight to file a patent on a frisbee made with the chemiluminescent materials, he'd be a millionaire now![61]) The other eighty percent of the royalty income can be used by the laboratory. Dr. Richter said the Navy often uses the income as seed money to fund exploratory, high-risk ideas.

User Groups   The fields of use for this technology range from novelty items to equipment that search and rescue dogs can wear in night searching. This is a technology that has such wide-ranging fields of use, that the number of potential user groups of products made with light sticks is practically limitless.

Barriers to Commercialization   When serious efforts to transfer the technology began in the late 1980s, the Naval Weapons Center, as it was called then, did not have extensive licensing experience. Also, the interested industry partners had never licensed federal technology. Because of the lack of experience in formal technology transfer, working through the process was time-consuming.

User Benefits/ Economic Impact/ Outcomes   Before it was acquired by Omniglow, American Cyanamid sold a significant number of light sticks to state and local governments and commercial markets including the fishing industry.

Omniglow Corporation was a small start-up business when it signed its original license with China Lake. The company was founded in 1986 with a staff of three, a small laboratory/ production facility in Novato, California, and one basic light-stick product. As a result of the license signed within three years of the company's founding, the company was able to enter into the defense market. Today, the company has a staff of three hundred, over a hundred products, a 20,000 square foot facility in California, and a second 180,000 square foot production plant in West Springfield, Massachusetts. Omniglow is now a publicly traded company and the largest producer of chemical light products in the world. Their markets include not only the defense marketplace, but also markets for novelty and retail products, industrial safety, and biomedical products, and commercial and recreational fishing.

Omniglow sells chemiluminescent products to more than twenty-five military and law enforcement organizations worldwide, including all the major defense departments and law enforcement agencies. Omniglow sells over fifteen million units, amounting to some $150 million worth of lights sticks, to DOD each year. The use of light sticks by the U.S. Government dates back to the Viet Nam conflict. More than 20,000 units were used in that war. At that smaller level of production, and with the lack of an industrial base to manufacture light sticks, some of those light sticks may have been produced at China Lake at that time. Since the early 1990s, the Omniglow quality assurance team has supervised production of more than 250 million chemiluminescent devices overall. Omniglow's non-infra-red military line currently consists of 1.5- to 15-inch CyalumeR and SnaplightR light sticks; PMLR personnel marker lights; S.O.S. light sticks; and LiteshapeR stick-on buttons. The infra-red line is also available in many such versions.

Similarly, police and public safety agencies use light sticks as flares and in place of flashlights. Omniglow's various-sized green and thirty-minute high-intensity red, yellow and white light sticks are used by police for traffic control and by emergency services personnel for hazard identification.

Since 1990, Omniglow has sold novelty items to wholesalers, retailers, and distributors in more than thirty countries. In the United States, Omniglow toy and novelty products can be found on more than 5,000 store shelves including major retailers such as Toys 'R Us, Wal-Mart and K-Mart. This includes products such as 4- and 6-inch Lite-UpR and Glow StickR light sticks, and Magic in the NightTM earrings, bracelets, and eye glasses. In addition, Omniglow sells 22-inch necklaces and other products directly to major amusement parks such as Disneyland, Walt Disney World, Universal Studios, Six Flags, Tokyo Disneyland and Euro-Disney.

Omniglow's first biomedical product, the SpeculiteR light stick, is being developed and marketed in cooperation with The Trylon Corporation. It involves proprietary technology utilizing chemiluminescence as an endoscopic light source. Speculoscopy has received FDA approval, and the

Speculite light stick is being distributed through major women's health care providers both in the United States and overseas.

As of 1993, Omniglow had invested at least $100,000 to develop its non-governmental markets (anticipated annual sales in that area were two to five million units). This includes expansion into cold weather applications and industrial safety. The company is particularly excited about its new industrial safety product called Snaplight[R] Emergency Evacuation System or S.E.E.[TM] system. The S.E.E. system includes the company's unique hexagonal design that increases the light generated by a light stick. Other applications being explored include a new approach to measuring pollution from smokestacks using light stick technology.

Also, Omniglow provides light-stick materials and instructions to high school classrooms so that students can perform chemiluminescent experiments in their science classes.

International Activity In 1990, Omniglow expanded into the commercial fishing fleet market. In the deep sea fishing industry, luminescent baits attract big fish such as swordfish in the dark, low temperatures of ocean waters around the Arctic and Alaska. These are areas where these fish were never caught in the past. (Dr. Richter created and tested some of the original fishing lures, but he said[62] the Navy did not want to spend the $3,000 it would have cost at that time to file a patent for this application.) Tests showed that light sticks could improve a catch by as much as thirty percent. Since 1991, Omniglow has been working with the Japan Tuna Association (JTA) to further develop the chemiluminescent technology for this industry. JTA is the economic and political arm of the largest tuna fishing fleet in the world, representing more than 2,000 members with 750+ vessels. Omniglow now markets 4-and 6-inch Snaplight[R] and Cyalume[R] light sticks and patented Flex-Stick[R] lures to commercial fishing fleets.

By 1994, in addition to a Domestic Sales and Marketing Office in Pompton Plains, New Jersey, Omniglow had established three sales offices in: Toronto (in the province of Ontario) Canada; Portsmouth, England; and Tokyo, Japan. Omniglow also has three joint venture facilities under construction or consideration in the People's Republic of China, Indonesia, and Eastern Europe.

Government Gains Although light sticks were originally developed specifically for military targeting, they ended up providing countless other military benefits. These benefits serendipitously became evident as early as the Viet Nam war, as the following examples illustrate:

- It was discovered that the chemical mixture could be put into aluminum cigar tubes and used by Army troops in place of flashlights which required traditional batteries that corroded in the humid forests of southeast Asia.
- At one point during the war, the military actually documented the number of pilots and jets that were saved because the pilots could hold light sticks in their mouths, shining them onto the dials if the cockpit lights failed. Normally, if this happens, pilots must eject, possibly into enemy territory. In the meantime, the light sticks were attached with rubber bands or velcro to the doors of the aircraft.
- Other uses included: landing helicopters, emergency lighting for Navy divers, arm bands for groups of parachutists close to the ground at night, tips for the wings of Air Force jets, helicopter markings, and man-overboard float lights.

In 1991 and 1992, during Operation Desert Shield and Desert Storm, the invisible infrared light stick was standard military issue for covert signaling and marking to identify allied soldiers and reduce the risk of friendly fire. Omniglow's infra-red product was the industry standard. During the Gulf war, Omniglow provided the government with millions of high-quality light sticks on short deadline. As a result, Omniglow was elected as a member of the U.S. Military Quality Vendor Program.

Economic Development   None.

Elapsed Time   The technology was under development from 1962 to the late-1980s. Over this twenty-five year period of time, light yields from the light stick compounds increased by a factor of 10,000. Because of the lengthy duration of the research, informal transfer to the chemical industry was going on for many years. The light stick was considered "invented" in 1973. Shortly after that, it was patented for the first time. That original patent has since expired. Dr. Richter "went on the road with his show" in the mid-1970s. The technology was first licensed in the 1970s. The early license, also, has long since expired. Two updated patents were issued to the Navy in December 1986 and April 1987. Also, in 1986 Omniglow was founded, and the company licensed an improved version of the light stick that same year. Omniglow was awarded its first contract with the DOD in 1989. Efforts to more formally transfer the technology began in mid-1988. Another license was signed with American Cyanamid in 1989. In 1990,

Omniglow began selling novelty items both in the United States and abroad.

Case 8 (1992) – Artificial Heart Flow Diagnostics

Role of Laboratory Researchers and Other Personnel   Mr. Franklin D. Shaffer, a mechanical engineer, was the principal investigator on several projects to develop a system for flow analysis at the Federal Energy Technology Center (FETC) in Pittsburgh, Pennsylvania. He was the key developer of the system at FETC's Fluid Analysis Laboratory. Dr. Mahendra P. Mathur was a supervisory physical scientist in the Fundamental Combustion Group at the laboratory. These two FETC researchers trained researchers at the medical company that they worked with under a CRADA to use FETC's fluorescent technique and helped them to set up the system. The FETC researchers also visited Stanford's research center, met implant patients, and observed implant surgery in the operating room.

The group of laboratory scientists co-wrote at least ten papers published in journals or conference proceedings, each involving a combination of authors from the FETC laboratory, Presbyterian-University Hospital, and Baxter Healthcare. These included, for example, articles for the IEEE International Conference on Pattern Recognition,[63] the National Fluid Dynamics Congress,[64] a Cardiovascular Science and Technology Conference,[65] the Journal of Applied Optics,[66] and an International Conference on Mechanics of Two-Phase Flows.[67]

Technology and Applications   In order to study the flow of coal-based fuels, FETC scientists developed a Flow Analysis Laboratory. This work was headed by Mr. Shaffer, who eventually patented a flow imaging technique called Fluorescent Image Tracking Velocimetry (FITV). Part of that system is the accompanying computer software. The FITV system uses pulsed-laser light to excite microscopic, fluorescently dyed particles that are added to a fluid. The fluorescent emission of the particles is viewed with digital photography, making the flow visible and measurable. With multiple exposure photography, the particles are tracked, thus revealing the motion of a fluid. This data is then used to measure flow properties such as velocities, streamlines, and shear stresses.

These same techniques, once refined, were applied to blood flow, coded according to pulse and measured in the same way. Low blood flow velocities near surfaces can promote blood clotting, the nemesis of artificial hearts. FITV allows researchers to identify and eliminate areas of low blood flow velocities. Once some initial problems were overcome, it was the first time blood flow was visualized and measured on the internal surface of an artificial heart.

The Laboratory   The Pittsburgh Energy Technology Center consolidated, effective December 1996, with Morgantown Energy Technology Center in West Virginia to become the Federal Energy Technology Center with two "campuses." The subject of this case is the FETC laboratory in Pittsburgh, Pennsylvania, with over three hundred researchers specializing in coal and liquid fuels. It has been a major DOE single-program laboratory for over fifty years. Unlike many of DOE's major laboratories, it is government-operated rather than contractor- or university-operated. FETC is actually located in a suburb south of Pittsburgh, sharing over 200 acres with the Bureau of Mines and the Mine Safety and Health Administration (which is part of the Department of Labor). The site is the nation's largest federal research complex focused on coal. FETC's budget is over $175 million per year.

University Involvement   About ten to twenty organizations in the health care field produce artificial hearts, including Baxter Healthcare, which created an artificial heart pump called the "Left Ventricular Assist Device" (LVAD). Novacor Laboratory in Oakland, California (eventually bought out by Baxter Healthcare) designed the pump. It is now manufactured by the Baxter Healthcare Novacor Division.

The Baxter LVAD was approved by the FDA for clinical testing; therefore, the Baxter laboratory is testing the pump at Presbyterian-University Hospital associated with the University of Pittsburgh in Pennsylvania, at Stanford University's Falk Cardiovascular Research Center in California, and at St. Louis University in Missouri. The Baxter LVAD is also being tested extensively in Europe where regulations are less stringent than in the United States.

In their efforts to perform the FDA clinical trials, Presbyterian-University Hospital and the University of Pittsburgh's School of Medicine felt they did not have the proper blood flow analysis techniques they needed. Therefore, Dr. Harvey Borovetz, professor of surgery and director of biomedical engineering approached the (then) Pittsburgh Energy Technology Center to request assistance with measuring and analyzing the flow of blood through the LVAD heart pumps.

Since FETC has a well-funded Flow Analysis Laboratory, it made sense to try to apply a flow analysis technique, which is normally used to analyze the flow of fuel through pipes, to analyzing the flow of blood through an artificial heart pump. In explaining the technology, FETC ORTA officer Kay Downey noted that, in terms of this application, blood clots are like slurries.[68]

Funding, Financing   Because this was one of DOE/FETC's first CRADAs, the agency was not sure how to handle the receipt of private industry funds when Baxter offered it; therefore, this particular CRADA and the follow-on CRADA did not involve private funds. The equipment from Presbyterian-University Hospital and Baxter set up at FETC was valued at $500,000.

Intellectual Property   A patent was issued to DOE for the blood flow analysis technique. Frank Shaffer is listed as the inventor. Also, Baxter Healthcare has a new patent on its LVAD.

Technology Transfer Mechanisms   In order to cooperatively test the heart pump, the group of pump designers and testers signed a multiple-partner CRADA with FETC. Partners in the CRADA are FETC, Presbyterian-University Hospital, and the University of Pittsburgh's Schools of Medicine and Engineering. Baxter Healthcare was not an official CRADA partner, but worked in the laboratory under the auspices of the University of Pittsburgh. The LVAD and a cardiac simulation flow loop were set up at FETC's facility. Up to six researchers at a time from the hospital and from Baxter worked in the FETC laboratory for about three months. To accommodate Baxter's extensive needs, an FITV system was set up at Baxter's laboratory in California. Eventually, FETC-like systems were set up at each of the LVAD testing centers.

User Groups   The ultimate users of artificial heart pumps are critically ill patients with heart diseases. Using a heart pump bridges the time gap between heart failure and transplant of a donor heart. This allows a certain number of patients to lead productive lives, although many do not survive until a heart is available.

The intermediate users of the heart pumps are the health care organizations that make use of them in their practices. Mr. Shaffer said, however, that the market is small since there are only ten to twenty health care companies in the world making artificial hearts. Although the market is small, most heart pump manufacturers are now using this technology.

Other Factors   The FLC nomination form notes an interesting point, and that is that this case "represents technology transfer in its finest sense, because it embodies the application of technology from one discipline, fundamental engineering in fossil fuels, to a quite different one—medical technology." Every year, some 30,000 to 50,000 patients wait for heart transplants, yet only about 2,000 donor hearts are available each year; therefore, speed of development was critical in this case.

User Benefits/ Economic Impact/ Outcomes   The CRADA proved to be helpful, and certain medical constraints were overcome because of FETC's FITV system. The test data allowed the researchers to experiment with various blood pressures and rates for the artificial heart pump hooked up at FETC. Ultimately, they tested it on patients with artificial hearts at the hospital. Data from the FETC research contributed to patient management, transplant operation techniques, and heart pump redesign. Of the three applications, pump redesign is the longest process.

When this CRADA began, the Baxter LVAD was approved by the FDA for testing under an "investigational device exemption," which involved limited clinical trials at certain clinical centers of excellence on human patients. Through the clinical testing, it was revealed that blood clotting in the LVAD was a serious problem and that the measurement methods being used by Baxter were inadequate. Through the work with FETC and the use of FITV, the Novacor LVAD was redesigned to eliminate the clotting problem. The new design then had to go through eighteen months of animal testing and finally was approved for clinical use in 1995. It was approved for investigational testing again in April 1996.

It was quickly apparent that the application of fuel flow analyses to artificial hearts was successful. In fact, it was so immediate that the CRADA was amended early-on to include measurements of hemodynamics in an artificial lung invented by a University of Pittsburgh heart surgeon, Dr. Brack Hattler. The Hattler artificial lung "oxygenator" consists of a pump surrounded by fibers that release oxygen into the blood stream. Like the artificial heart, the artificial lung was also set up in FETC's laboratory. A progression from the animal testing phase to clinical tests occurred around 1993, and the measurements have been successful.

The Pittsburgh hospital was very pleased with the CRADA work jointly performed with FETC. Dr. Borovetz[69] at the School of Medicine noted that this satisfaction stemmed from several factors. For one thing, they had access to expensive equipment that the university wouldn't have been able to afford. But more importantly, he added, "All that equipment wouldn't have meant anything if it weren't for the Frank Shaffers of the world." He said the project would not have succeeded without Mr. Shaffer at the laboratory helping. As a result of Mr. Shaffer's involvement, it was a "big winner," resulting in a superior pump, which Mr. Borovetz claims has been one of the high points of his career. Mr. Shaffer was willing to be a true partner, not just show them the equipment, he said.

The CRADA-related work in the Baxter laboratory is still active and has a high priority. Baxter Healthcare has made some money on the device overseas, but will not realize revenues in the United States until they can overcome all the regulatory hurdles and sell it on the open market. Each state-of-the-art blood pump costs $80,000. In any case, profits are difficult to measure because the tremendous development effort has cost many millions of dollars.

More recently, the FETC analysis technique was applied to the development of other artificial organs. Now most of the companies around the world researching artificial organs use the flow analysis technique and have facilities that duplicate the original FETC facility. This includes laboratories in Europe, Australia, and Korea.

International Activity   Baxter Healthcare's widest application of their LVAD is now in Europe (England, France, and Germany), where FDA approvals are not relevant. In Europe, the Baxter LVAD is being used not only as a bridge to transplant, but it is also being permanently implanted in older patients who are not good candidates for heart transplants.

Government Gains   The FETC flow analysis technology continues to progress and is benefiting from the input of researchers in the medical field. The imaging technology is also used in research being done in other laboratories within FETC.

Economic Development   None.

Elapsed Time   The early jointly authored scientific papers were published from 1989 to 1992. The original CRADA was signed in January 1991. The term of duration was five years (to January 1996). This CRADA, now expired, was followed up by a new CRADA to extend the FITV technique with a small fiber-optic line that will be used to identify and measure blood and cancer cells.

Post-legislation Findings Summary

Appendix D summarizes the findings for each of the selected post-legislation cases organized according to the questions addressed in the interviews. They appear in Appendix D after the pre-legislation summary. Whereas, the earlier Level II analysis (appearing in this chapter as the introduction) addressed only certain key questions for the seventeen FLC awardees for the years 1992 and 1993, the appendix addresses all of the questions for the selected cases documented in this chapter using the same order as is used for the "Pre-legislation Findings." The responses to the "Other Factors" question are incorporated into the next chapter. In the appendix, the cases are presented under each topic in the following order:

a. Laser Method to Light Samples
b. Voice Coder
c. Paper Quality Tester
d. Variable-Frequency Microwave Oven
e. Gravity Meter
f. Oatrim Fat Substitute
g. Chemiluminescent Light Stick
h. Artificial Heart Blood Flow Analysis.

The next chapter compares the findings from the cases in the pre-legislation and post-legislation time frames.

Notes

1   Three laboratory researchers read drafts of their cases as part of the interview process.
2   1994 FLC Winners' document.
3   1994 FLC Winners' document.
4   E. S. Yeung and W. G. Kuhr, "Indirect Detection of Native Amino Acids in Capillary Zone Electrophoresis," *Analytical Chemistry* 60: 1832-1834, 1988; E. S. Yeung and W. G. Kuhr, "Indirect Detection Methods in Capillary Separations," *Analytical Chemistry* 63: 275A-278A, 1991.
5   Ibid.
6   Rome Laboratory was the only military facility at that location to survive the BRAC initiative.

7 D. W. Griffin and J. S. Lim, "A New Model-Based Speech Analysis/Synthesis System," *Proceedings of the International Conference on Acoustic Speech and Signal Processing (ICASSP) 1985, Tampa, Florida*, p. 513-516, March 1985; J. C. Hardwick and J.S. Lim, "A 4.8 kbits/sec Multi-Band Excitation Speech Coder," *Proceedings of the ICASSP 1988, New York, New York*, p. 374-377, April 1988; J. C. Hardwick and J. S. Lim, "The Application of the IMBE Speech Coder to Mobile Communications," *Proceedings of the ICASSP 1991, Toronto, Canada*, p. 249-252, May 1991; S. W. Wong, "An Evaluation of 6.4 kbits/s Speech Codecs for INMARSAT-M System," *Proceedings of the ICASSP 1991, Toronto, Canada*, p. 629-632, May 1991.

8 J. C. Hardwick and J. S. Lim, "A 4800 bps Improved Multi-Band Excitation Speech Coder," *Proceedings of IEEE Workshop on Speech Coding for Telecommunications, Vancouver, Canada, Institute for Electrical and Electronics Engineers, September 5-8, 1989.*

9 D. W. Griffin and J. S. Lim, "Multiband Excitation Vocoder," *IEEE Transactions on ASSP* 36: (8), August 1988.

10 D. W. Griffin and J. S. Lim, "A Speech Spectral Analysis/Synthesis System," *Proceedings of 1984 Digital Signal Processing Workshop*, p. 5.1.1-5.1.2, October 1984.

11 Now the Intelligence Command.

12 DVSI product technical literature, May 31, 1995.

13 Interview with Dr. Hardwick, September 23, 1996.

14 *Land Mobile Radio News* (September 11): 1992.

15 D. E. Gunderson, "A Method for Compressive Creep Testing of Paperboard," *TAPPI Journal* 64 (11): 67-71, Technical Association of the Pulp and Paper Industry, 1981; J. M. Considine, D. E. Gunderson, Peter Thelin, and Christer Fellers, "Compressive Creep Behavior of Paperboard in a Cyclic Humidity Environment—Exploratory Experiment," *TAPPI Journal*: 131-136, 1989.

16 T. L. Laufenberg, "Characterization of Paperboard, Combined Board, and Container Performance in the Service Moisture Environment," *Proceedings of the TAPPI International Paper Physics Conference*, 1991.

17 J. M. Considine and J. F. Bobalek, "In-plane Hygroexpansivity of Postage Stamp Papers," *Proceedings – Technical Association of the Graphic Arts*, 1991.

18 Interviews with Mr. Laufenberg, September 4, 1996, September 12, 1996, and September 19, 1996.

19 FLC Award for Excellence in Technology Transfer, Statement of Achievement, Submitted by Allen J. Schacht, December 4, 1991.

20 Interview with Mr. Davis, October 2, 1996.

21 D. W. Bible, R. J. Lauf and C. A. Everleigh, "Multikilowatt Variable Frequency Microwave Furnace," *Materials Research Society Symposium Proceedings 1992, San Francisco, California, Volume 269*, 1993.

22 Interview with Dr. Lauf, September 4, 1996.
23 Interview with Dr. Lauf, September 4, 1996.
24 Interview with Dr. Lauf, September 4, 1996.
25 Ibid.
26 Interview with Dr. Lauf, September 4, 1996.
27 Interview with Mr. Bible, August 28, 1996.
28 Interview with Mr. Gerard, September 19, 1996.
29 Interview with Dr. Lauf, September 4, 1996; interview with Mr. Gerard, September 19, 1996.
30 Interview with Mr. Gerard, September 19, 1996.
31 Interview with Dr. Lauf, September 4, 1996.
32 Interview with Dr. Lauf, September 4, 1996.
33 Interview with Mr. Gerard, September 19, 1996.
34 Examples are: J. E. Faller, "Precision Measurement of the Acceleration of Gravity," *Science* 158 (60), 1967; J. E. Faller, M. A. Zumberge and R. L. Rinker, "A Portable Apparatus for Absolute Measurements of the Earth's Gravity," *Metrologia* 18 (145), 1982; J. E. Faller, M. A. Zumberge and J. Gschwind, "Results From an Absolute Gravity Survey in the United States," *Journal of Geophysical Research* 88: 7495, 1983; J. E. Faller and I. Marson, "'g'—the Acceleration of Gravity: Its Measurement and Its Importance," *Journal of Phys. E: Scientific Instruments* 19 (22), 1986; J. E. Faller and I. Marson, "Ballistic Methods of Measuring "g"—The Direct Free-Fall and the Symmetrical Rise-and-Fall Methods Compared," *Metrologia* 25 (49), 1988; J. E. Faller, G. Peter, R. E. Moose, C. W. Wessella and T. M. Niebauer, "High Precision Absolute Gravity Observations in the United States," *Journal of Geophysical Research* 94: 5659, 1989.
35 Interview with Dr. Faller, July 8, 1996.
36 Interview with Dr. Faller, July 8, 1996.
37 Since JILA is part of the NIST Physics Laboratory headquartered in Gaithersburg, it works with the NIST/Gaithersburg technology transfer office rather than the NIST/Boulder office.
38 Interview with Dr. Niebauer, July 25, 1996.
39 Interview with Dr. ONeil, August 29, 1996.
40 Interview with Dr. Niebauer, July 25, 1996.
41 G. E. Inglett and S. B. Grisamore, "Maltodextrin Fat Substitute Lowers Cholesterol," *Food Technology* 45 (6/June): 104, 1991.
42 G. E. Inglett, "OATRIM Cuts Fat, Cholesterol in Ice Cream," *Science of Food and Agriculture* 2 (2): 4-5, 1990.
43 G. E. Inglett, "OATRIM Products With Elevated Beta Glucan Contents as Natural Fat Substitutes," Fourth Chemical Congress of North America, New York, New York, August 25-30, 1991, *Abstr. Pap. American Chemical Society*

202 (1-2), 1991.

44 Interview with Mr. Babbitt, October 4, 1996.

45 Interview with Mr. Freeland, October 11, 1996.

46 Internationally, Nestle is largest.

47 Somewhere in between Oatrim and Z-trim, Dr. Inglett developed "Oatrim-10." A one-ounce chocolate bar, for example, contains a gram of soluble beta-glucan from Oatrim-10, a half gram of Z-trim, oat fiber, corn syrup, milk chocolate, and artificial sweetener.

48 "USDA Develops No-Cal Fat Substitute," *The FLC Newslink* (December): 4, 1996; "USDA Fat Buster," *Technology Transfer Business* (Winter): 39, 1997.

49 *Time* (September 23): 63, 1996.

50 Cheryl Pellerin, "Lowering Cholesterol: An Oat-Based Fat Substitute ConAgra is Speeding to Market," *Technology Transfer Business* (Fall): 25, 1992.

51 Interview with Mr. Babbitt, October 4, 1996.

52 Interview with Mr. Freeland, October 11, 1996.

53 "Losing the Fat, Winning the War," *Focus on Food Ingredients*, Rhone-Poulenc Food Ingredients, 5th Anniversary Issue, p. 3.

54 Interview with Mr. Freeland, October 11, 1996.

55 Constance Holden, Random Samples editor, "Another Fat Substitute," *Science* 273 (September 13): 1495, 1996.

56 Cheryl Pellerin, "Lowering Cholesterol: An Oat-Based Fat Substitute ConAgra is Speeding to Market," *Technology Transfer Business* (Fall): 25, 1992.

57 Interview with Mr. Grisamore, September 20, 1996.

58 Interview with Dr. Richter, August 7, 1996.

59 Some recent Naval Warfare Center - Weapons Division CRADA titles include: "Color-neutral electrically conducting polymers"; "Ring vortex projection for medical applications and law enforcement"; and "Electronically steerable antenna systems for autos, boats, and aircraft."

60 "Chemiluminescence Technology," no date (finalized August 1996), released by the NAWC Weapons Division Technology Transfer Office, p. 2.

61 Interview with Dr. Richter, August 7, 1996.

62 Interview with Dr. Richter, August 7, 1996.

63 R. Srinivasan, R. Singh and F. Shaffer, "Analysis of Pulse-Coded Particle Tracking Velocimetry Data," *IEEE International Conference on Pattern Recognition, Copenhagen, Denmark, September 1991.*

64 F. Shaffer, M. Shahnam, M. Mathur, J. Edmann, H. Borovetz, R. Schaub and J. Woodard, "Particle Image Velocimetry on the Surfaces of a Pericardial Trileaflet Valve," *National Fluid Dynamics Congress, Los Angeles, California, June 22-25, 1992.*

65 J. Woodard, F. Shaffer, R. Schaub, L. Lund, H. Borovetz and J. Antaki, "Flow Visualization of the Novacor Left Ventricular System," *Cardiovascular Science*

and Technology Conference, Bethesda, Maryland, December 2-4, 1991.

66   E. R. Ramer and F. D. Shaffer, "Automated Analysis of Multiple Pulse Particle Image Velocimetry Data," *Journal of Applied Optics* (November): 1991.

67   F. D. Shaffer and E. R. Ramer, "Pulsed-Laser Imaging of Particle-Wall Collisions," *International Conference on the Mechanics of Two-Phase Flows, Taipei, Taiwan, June 1989.*

68   Conversation with Ms. Downey.

69   Interview with Dr. Borovetz, September 16, 1996.

5 Comparing the 1980s and 1990s Cases

To improve our understanding of the effects of the technology transfer legislation, this study compared a 1980s group of pre-legislation cases with a 1990s group of post-legislation cases. A number of findings emerged from that analysis. Table 5.3, "Similarities and Differences between the Two Time Frames" lists the findings according to the interview topics (at the end of the chapter).

In this chapter, the findings are grouped and discussed according to their impact on technology transfer outcomes and process. The first section discusses those findings indicating a positive impact, while the next two sections present findings with no impact or negative impact, respectively.

Government Technology Transfer Improved

The technology transfer legislation has had a positive effect on technology transfer outcomes and process, as summarized below:

- Researchers' Roles – Researchers produced more prototypes and samples.
- University Involvement – The university role became more institutionalized.
- Intellectual Property – Laboratory patenting increased.
- Technology Transfer Mechanisms – Researchers displayed increasingly strong opinions about licensing royalties.
- User Groups – Small-firm involvement increased; the transfer process shifted from technology push to market pull; and users narrowed from broad groups to targeted markets.
- User Benefits/Economic Impacts/Outcomes – Number of products increased; sales revenues increased, including international activity; number of spinoff start-up companies and jobs generated increased; company and product failures decreased; and dual uses, government gains, and spinbacks increased.

• Elapsed Time – The time to market decreased.

Researchers' Roles – Researchers Produced More Prototypes, Laboratory Samples

In the 1990s cases, the researchers produced more laboratory prototypes and laboratory material samples than in the 1980s cases. In the 1980s cases, laboratory material samples were developed in the thermoplastic materials and the chemically-imbedded herbicide/ pesticide.[1]

The majority of the researchers in the 1990s cases developed prototypes and samples as part of their role in the technology transfer process. These cases involved the laser-illuminated biological samples, paper-quality tester, microwave oven, gravity meter, and blood-flow diagnostic system. The scientists developed laboratory samples for the two materials-oriented technologies in the later group, Oatrim and the light sticks.

University Involvement – University Role Became More Institutionalized

In both the 1980s and 1990s cases, universities were involved in government technology transfer. In the 1980s, interactions took place at the individual scientist level, whereas in the 1990s cases, formal inter-institutional ventures characterized the joint government/ university work.

The university work in the 1980s cases involved professors and graduate students on an individual basis at the level of the bench scientist, as follows: penetrometer – prototype development by a university professor; thermoplastic material – on-site graduate student co-inventor; tracer technology – graduate assistants for various demonstrations; and alginate-based herbicide/ pesticide – university professor co-inventor.

The university work in the 1990s cases involved teams of university researchers, and resulted from inter-institutional structures or ventures, as follows: voice coder – university technology development work and spinoff company; paper quality tester – technology testing and screening by universities; gravity meter – joint government/ university laboratory development; and artificial heart pump – university CRADA partner.

Intellectual Property – Laboratory Patenting Increased

Laboratories and their company partners both pursued patents more aggressively in the 1990s cases than in the 1980s cases.[2] For the five patentable technologies in the 1980s cases: The relevant agencies patented the two herbicide/ pesticides several times in each application area.[3] NASA patented the thermoplastic material twice through joint laboratory-university filings. Additional findings are less positive: DOE patented the tracer technology, with its practically limitless applications, only once.[4] Also, the Navy researcher attempted to patent both versions of the penetrometer, but was denied application filings by the Navy.[5]

In the 1990s cases: Iowa State, DOE's contracting laboratory operator, patented the laser-based method for lighting up biological samples. The U.S. Department of Agriculture patented the paper-quality tester and fat substitute. Both the Navy and its corporate licensees patented the light sticks many times.[6] Both DOE and Baxter Healthcare patented the blood-flow technique. Oak Ridge filed several patents jointly with its spinoff company and other research partners for the microwave oven.[7] For the voice coder, the Air Force was not the developer of the technology and couldn't patent it, but the spinoff company patented it several times. In the remaining case, the company currently producing the gravity meter considers the technology proprietary, although neither NIST nor the company patented it.[8]

Mechanisms – Researchers Had Increasingly Strong Opinions About Royalties

In the context of most laboratory budgets, royalty income is not that significant (although it may become more important as budgets decline). For example, China Lake received a little over $86,000 in royalties for its licenses on the light-stick technology. However, at the level of the individual researcher, licensing royalties provide important incentives to transfer technologies. Apparently, these payments are becoming more meaningful because the researchers' opinions about royalties were strong in both the 1980s and 1990s cases, becoming stronger in the 1990s.[9]

In the 1980s cases: The chief scientist in the thermoplastic material case provided detailed information about the royalty sharing arrangements in the case of multiple inventors from different institutions, and opined about what the laboratory should do with its portion of the royalties. He also recommended that mechanisms to handle royalty-sharing disputes be created. The researcher in the alginate-based case commented on the amount of royalties an ARS scientist receives when a USDA patent is licensed.

In the 1990s cases: The inventor of the laser-based method for lighting up samples stated that only one in a hundred products on the market benefits the actual laboratory inventor financially. The chief scientist in the microwave oven case was stated that royalty payments to the laboratory scientists should be a secondary consideration in terms of the overall legal agreements between laboratories and companies. In the same case, the head of the partnering company stated the importance of sharing a percentage of the royalties with the partnering laboratory inventors to give them an investment in his company's success. A USDA scientist in the alginate case referenced the inventor in the Oatrim case, widely known for making the most of the new technology transfer incentives. The lead researcher in the light-stick case provided details about the laboratory's royalty-sharing arrangements and his views about the contribution of patents toward promotions. He said that if he'd had the foresight to file a patent on a frisbee made with the chemiluminescent materials, he'd be a millionaire by now! A "Success Story" sheet from his laboratory stated that the royalties provided an incentive to other inventors in the same laboratory. The chief scientist in the gravity meter case commented on the royalties he missed by not patenting the gravity meter. He also noted that some agency incentives for transferring technology were better than others. As an example, he said that other FLC awardees received bonuses with their technology transfer award while he did not.

User Groups

The user groups and laboratory partners became more small firm-oriented and targeted.

Small-Firm Involvement Increased   In the 1990s cases, more small firms were involved in technology transfer activities than in the 1980s cases. Overall, small firms comprised ten of the 31 technology partnering companies (three from the 1980s cases and eight from the 1990s cases), an indication that small firms have become more involved in technology transfer efforts.

In the 1980s cases, the three small firms included: High Technology Services in the thermoplastic material case, AIM Inc. in the tracer technology case, and Thermo Trilogy Corporation in the alginate-based herbicide/ pesticide case. Large corporations comprised the remainder of the partners in the 1980s cases, including Hoescht-Celanese Corporation, M&T Chemical, Consolidated Edison, Grace-Sierra, Mycogen Corporation, Reemay Inc., and Geoflow Inc. The interviewees did not identify the other

1980s partners as large or small.

In the 1990s cases, the eight small firms included: Digital Voice Systems, Inc. in the voice coder case, Isthmus Engineering and Manufacturing Co-op in the paper-quality tester case, Microwave Laboratories Inc. and Lambda Technologies in the microwave oven case, Axis Instruments, Micro-g Solutions, Winters Electro-Optics in the gravity meter case, and Omniglow Corporation in the light stick case. The 1990s cases involved several large corporations, including ConAgra, A.E. Staley Manufacturing, Quaker Oats, Rhone-Poulenc Inc., and Baxter Healthcare. The interviewees did not identify the remaining 1990s partners as large or small.

Process Shifted from Technology Push to Market Pull   "Market pull" characterized the transfers in the 1990s cases, while "technology push" characterized the transfers in the 1980s cases. Users or partners initiated contacts with the laboratory researchers in the 1990s cases, whereas the laboratory scientists initiated most of the contacts in the 1980s cases.

The 1980s "technology push" cases included the: penetrometer, thermoplastic material, substance tracer, chemically-based herbicide/ pesticide, and radiation therapy. In the case of the alginate-based herbicide/ pesticide, both market pull and technology push were at work. The original Grace-Sierra work stemmed from existing relationships between the company and USDA. Later, Biosys, Inc. and EcoScience contacted the USDA scientist.

Five of the 1990s cases were market pull, while three were a combination of technology push and market pull. In the 1990s paper-quality tester and gravity-meter cases, Isthmus Engineering and Axis Instruments both responded to competitive solicitations from their developing laboratories. Similarly, the Oatrim manufacturers responded to the developing laboratory's request for expressions of interest at a technology transfer conference for industry. The artificial heart testers and manufacturer contacted the DOE laboratory in search of a workable technology. Originally, laboratory researchers sought out the microwave oven manufacturers, but the "replacement" manufacturer sprang up specifically to manufacture the laboratory's technology. The voice coder, light sticks, and laser-based luminescent technologies all involved both technology push and market pull. For example, the university laboratory that developed the voice coder was seeking funding from a federal agency, but once the agency understood the technology's potential, it not only funded it but also pushed it to standard-setting organizations.

Users Narrowed from Broad Groups to Targeted Markets   The 1990s cases had more targeted and specific technology user groups than the 1980s cases. The 1980s cases involved earlier-stage technologies, so the user groups were broad-based and not as market-oriented.[10] For example, both the thermoplastic material and tracer technology had a broad range of existing and/or potential users. The same was true for the two herbicide/ pesticide technologies. Although based in very different materials, each appealed to agricultural, public works, and home owning groups.[11] Addressing requirements for oceanography and waterway projects, the penetrometer was used by government organizations, whether state, local or federal. The radiation therapy system, which expanded to include radiation protection, was intended for medical and industrial arenas.

The 1990s cases had more targeted user groups and markets. The voice coder addressed telecommunications industry needs. The paper testing technology solved storage problems for paper product manufacturers. The gravity meter provided measurements for geological, climate, and environmental programs. The fat substitute was an ingredient used by producers of processed foods and meals. The blood flow diagnostics assisted the heart transplant portion of the medical community. The microwave furnace was employed in laboratories of all types (university research, commercial testing, etc.) and later targeted to manufacturing environments. If commercialized, the laser-based method for illuminating biological samples would appeal not only to research laboratories, but to chemical and pharmaceutical laboratories (eg., clinical, pharmaceutical). The light-stick technology was the most broad-based of all the technologies, being applicable to everything from defense reconnaisance to toys, and from industrial plants to medical clinics, and likewise to both the recreational and commercial fishing industries. Omniglow even provided its chemiluminescent materials to high schools for science experiments.

User Benefits/Economic Impacts/Outcomes

Compared to the 1980s cases, the 1990s cases exhibited greater economic impact in terms of increased number of products, sales revenues, spinoff start-up companies, and jobs generated. Combined with a decline in company and product failures, and an increase in dual uses and government gains, these findings indicate significant improvements in technology transfer results.

Number of Products Increased   The 1980s cases produced sixteen successful products or product lines, whereas the 1990s cases produced 114 successful products. In the 1980s cases, successful commercial products included: Thermoplastic material – M&T-4605 and Techimer 4001; Tracer technology – "COPS" substance tracer/ sniffer; Alginate-based herbicide/ pesticide – GlioGard™ Biological/ Microbial Fungicide, BioSafe^R 20 and BioSafe^R 100 Insect Control, SoilGard™ 12G Microbial Fungicide, Aqua-Fyte™ Bio-herbicide, Vector^R MC for mole crickets, Lesco™ Vector^R MC Insect Control, and BioVector^R 355 for citrus weevils; Chemically-imbedded herbicide/ pesticide – Rootguard^R Subsurface Irrigation Systems, including an entire line of irrigation products, Biobarrier^R Root Control System, Biobarrier^R II Pre-emergence Weed Control System, Root Shield™, and Grow Guard.

In the 1990s group of cases, the following products made it to the commercial marketplace: Voice coder – IMBE™ Speech Compression Systems, IMBE™ VC-20 Voice Codec Module, IMBE™ VC-100 Voice Codec Module, AMBE^R Speech Compression System, and AMBE-1000™ Coder; Paper quality tester – Vacuum Compression Apparatus, Thin Film Analyzer; Microwave oven – Variable-Frequency Microwave Furnace, Applicator, Processing System, Traveling Microwave Tube Amplifier, Vari-Wave multi-functional microwave oven, Vari-Data hardware/software interface; Gravity meter – FG5 "Gravimeter," Gravity Meter, Iodine Laser; Oatrim fat substitute – TrimChoice, Beta-Trim; Light sticks – S.E.E.™ Emergency Evacuation Systems, Flex-Stick^R fishing lures, Speculite^R, Cyalume^R, Snaplight^R, Lite-Up^R and Glow Stick^R light sticks, PML^R stick-on buttons, Lightwrap^R glowing strips, "Flourescers," Lite-Rope^R necklaces, and Magic in the Night™ earrings, bracelets and eye glasses. Most of these products are offered in various sizes and colors, for a total of over 100 products; and Heart-flow diagnostics – Left Ventricular Assist Device (LVAD) artificial heart pump.

Not only did the number of products increase, but also two companies involved in post-legislation cases reported on the specifics of their commercial success. Lambda Technologies received orders for two to three microwave units a month in 1996 compared to none when the company was first spun off. Omniglow produced over 250 million lights sticks since the early 1990s.

Several products were taken off the market and replaced by other products, reflecting changing corporate priorities. In the 1980s group, these products included GlioGard™ Biological/Microbial Fungicide, BioSafe^R 20 and BioSafe^R 100 Insect Control in the alginate-based herbicide/ pesticide case. In the 1990s group, they included the FG5 Gravimeter from the gravity-meter case.

Sales Revenues Increased, Including International   Most of the privately-held companies were reluctant to reveal revenues associated with their products. In the 1980s cases, very little information was made available on the sales revenues of the companies.[12]

In the 1990s cases, four companies provided sales revenues amounting to a total of $7.7 million in revenues. The cooperative producing the paper quality tester realized almost $500,000 in sales from the Forest Products Laboratory technology. The first company producing the gravity meters sold $3 million in gravity meters before going out of business; its spinoff did almost $2.5 in sales of gravity meters after that. Another spinoff manufacturing a major gravity meter component had realized $1.2 in sales or almost $500,000 in gross revenues.

On a related note, both the 1980s and 1990s cases, with one exception (the 1980s penetrometer case), reported international commercial *activity* of some sort; however, the companies involved in the 1990s cases indicated substantially more international *sales* than the companies in the 1980s cases.[13] Only the chemically-imbedded herbicide/ pesticide in the 1980s group of technologies had significant international commercial sales. For example, Agrifim International's marketing materials listed major systems installed in other countries. From the 1990s time period, several companies enjoyed significant international commercial sales. These included companies in the gravity meter, fat substitute, and light stick cases.

Number of Spinoff Start-up Companies, Jobs[14] *Increased*   In contrast to the 1980s cases (none of which led to the creation of spin-off companies), the 1990s cases produced six new companies to commercialize the technology developed in the case. For example, Digital Voice Systems, Inc. was established to commercialize the voice coder, and Omniglow, Inc. was created to advanced the light stick technology. Lambda Technologies, a spin-off of the bankrupt Microwave Laboratories Inc, commercialized the microwave oven.[15] Axis Instruments, founded to commercialize the gravity meter, later spawned Micro-g Solutions and Winters Electro-Optics when it closed its doors. The lead scientists for both the microwave oven and the gravity meter[16] described the resulting spin-off companies as "a phoenix rising out of the ashes."

Although the companies involved in the 1980s cases did not provide any employment data, none was dedicated exclusively to the laboratory technology as occurred in the 1990s cases.

Six companies in the 1990s group of cases provided the number of jobs generated by the laboratory-developed technology; a net total of 348 new jobs were created. The company commercializing the voice coder employed "about ten" people since its founding in the 1980s.[17] In the microwave oven case, Lambda Technologies grew from four to nineteen employees, and projected adding several new positions by the end of 1996. The first company producing gravity meters employed about twenty people before it went out of business, but two new spinoff companies, Micro-g and Winters Electro-Optics, employ eight and two people, respectively. In the light stick case, Omniglow increased from three to three hundred employees since its founding in the late 1980s. Although the 1990s Oatrim fat substitute case involved very large companies, specific divisions or sales groups were dedicated to the laboratory technology. The Mountain Lake "Trimchoice" manufacturing facility employed nine people, and the Quaker partnership included a sales force of about two dozen persons dedicated to its Beta-trim product.

Company, Product Failures Decreased   The companies in the 1990s cases experienced fewer technology transfer and commercialization failures than companies involved in the 1980s cases. Most of the failures in the series of cases came from the 1980s cases, with a couple of exceptions.

In the 1980s time frame, five of the commercializing companies were subject to corporate buy-outs, take-overs, or company failures which, in turn, affected the outcome of the technology. Also, in at least five of the 1980s cases, production costs were prohibitive. As a result, two products stopped short of the commercial marketplace, and of the commercialized products, five technology licenses were inactive although three commercial products resulted from them. The penetrometer technology did not make it to the commercial marketplace, and was eventually supplanted by a more cost-effective and improved technology. The thermoplastic material product, M&T-4605, also failed through two licensees[18] because of changing company strategies and high production costs, although the small company that continues to hold an active license hopes the technology will succeed commercially. There were also several failures in the alginate-based herbicide/ pesticide case. A problem experienced by all of the companies in this case was the high cost of raw materials. One of the original licensees, Mycogen, never commercialized any products, and another company tested its Aqua-Fyte™ Bio-herbicide in EPA field experiments, but did not commercialize.[19] Two additional companies experienced changes in business ownership and resulting changes in product lines that precluded the development of certain of the alginate-based products. In the chemically-imbedded herbicide/ pesticide case, the companies producing Root Shield™ and Grow Guard did not pursue their markets aggressively and the licenses became inactive.

In the 1990s group, there was one *license failure* in the case of the laser-based biological samples. Apparently, the company did not want to invest in the laboratory prototype, in spite of it being customized to the company's equipment, reflecting a syndrome in American industry known as "NIH" or not-invented-here.[20] The only *failed commercial product* in the 1990s group resulted from overall company failure, namely Microwave Laboratories and its line of variable-frequency microwave oven products: the Variable-Frequency Microwave Furnace, Applicator, Processing System, and Traveling Microwave Tube Amplifier. However, the technology survived; the microwave oven was being produced by another company that spun out of the original failed company.

Dual Uses, Government Gains, Spinbacks Increased   Three cases in the 1980s and 1990s resulted in federal government gains and/or dual (military/civilian) uses. Of the 1980s cases: The penetrometer was developed for state economic development purposes, but the Navy became the chief beneficiary, using six holotype versions of the invention. In the case of the tracer technology, the developing laboratory (Brookhaven National Laboratory) received unexpected revenue by providing tracer services. The chemically-imbedded herbicide/ pesticide was developed for use at radioactive military storage sites.

Five 1990s cases provided government benefits. The Forest Service developed the paper quality tester for the paper industry, but the Bureau of Engraving and Printing was able to use the technology in its currency printing activities. Both the voice coder and the microwave oven element were developed for military communication purposes. NIST developed the gravity meter for environmental purposes, but the military services also purchased gravity meters for their use. The Navy developed the light stick technology for military search and rescue purposes.

Elapsed Time – Time to Market Decreased

Development time, that is, the time from laboratory R&D to commercialization or pre-commercialization stages was seven to ten years less in the 1990s cases than in the 1980s cases. In the 1980s cases, the penetrometer technology and the quality assurance program did not involve commercialization. Of the remaining 1980s cases, many of the technologies were under development, as follows: The thermoplastic material had spent fifteen years in the phases between conceptualization and pre-commercialization. The tracer technology was under development for over twenty years. Because each existing application area reached a different stage of development at a different time, the technology needed to be examined on an application-by-application basis. Because not enough data was available to do a mini-case study on each application, development was protracted. The alginate-based herbicide/ pesticide was under development for about twenty years. The chemically-imbedded herbicide/ pesticide was under development for nearly twenty years, with different applications reaching the market at different times. However, the team leader emphasized that the ultimate value of this technology wouldn't become apparent for another fifteen to twenty years. In summary, in the 1980s cases, fifteen to twenty years was the predominant development time frame.

In the 1990s cases, however, all eight technologies reached commercialization within eight to ten years: The laser method had been under development for about ten years, although still not commercialized. Both the voice coder and the paper quality tester were under development for about ten years before being commercialized. The tunable microwave oven and the fat substitute were both market-ready in about eight years. The gravity meter *instrument* reached the market in ten years, while the technology underlying the instrument was under development fifteen years before that. The light-stick technology was first licensed in the late 1970s, less than ten years after its discovery, although it continued under development for fifteen years after that. The blood-flow diagnostics application was under development for about eight years.

In summary, the cases showed that the legislation led to shorter development time frames. To get a technology from R&D to the marketplace took fifteen to twenty years in the 1980s cases and just eight to ten years in the 1990s cases. Table 5.1 summarizes the data on elapsed times.

Table 5.1   Elapsed Times

	Technology	Years under development
Pre-legislation	Penetrometer (no commercialization)	–
	Thermoplastic material	15
	Tracer technology	20+
	Alginate-based herbicide/ pesticide	20
	Chemically-imbedded herbicide/ pesticide	<20
	Radiation program (no commercialization)	–
Post-legislation	Laser method	10
	Voice coder	10
	Paper quality tester	10
	Microwave oven	8
	Fat substitute	8
	Gravity meter	10[a]
	Light sticks	<10[b]
	Blood flow diagnostics	8

[a] Instrument, only, under development 10 years; technology an additional 15 years.
[b] Technology first licensed in 10 years; development continued 15 more years.

Certain Aspects Did Not Change

Certain aspects of the cases did not change, or remained consistent, from the 1980s (pre-legislation) time frame to the 1990s (post-legislation) time frame, as follows:

- Roles, Technologies, Laboratories, Funding – Consistency characterized the "system" in both time frames: many aspects of the researchers' roles were consistent, the technologies continued to represent diverse technology areas, the laboratory groups remained similar, and funding combinations remained similar.
- Technology Transfer Mechanisms – Licenses and CRADAs predominated in both time periods.
- User Benefits/Economic Impacts/Outcomes – Researchers and partners

continued to describe success many ways: they consistently used miscellaneous indicators; and they consistently used intangible measures.

- International Activity – The laboratories remained uninterested in obtaining foreign patent rights.
- Economic Development, Technical Assistance – American-owned firms predominated in both time frames, but did not use available services.

Roles, Technologies, Laboratories, Funding

The government technology transfer system is represented by the roles, technologies, laboratories, and funding sources. Consistency characterized that system in both time frames.

Aspects of the Researchers' Roles Were Consistent   Many aspects of the researchers' roles did not change from the two time frames. The laboratory scientists in both the 1980s and 1990s cases were internationally-recognized experts. All scientists performed the traditional roles of research, publishing in scientific journals, and speaking at technical conferences to communicate their research findings with others in their fields. In both groups of cases, researchers cited a number of scientific and technical articles they wrote for reviewed journals. As noted in the alginate-based herbicide/ pesticide case, USDA researchers "lived off publications" because that is how they were professionally graded.

Most scientists interviewed weren't happy only performing basic or applied research; they also wanted to have their technologies put to practical use. From the laboratory researchers' perspectives, activities in this regard were meaningful whether a business venture failed (as with EcoScience Corporation in 1980s group, or Lachat Instruments' failure to invest in commercialization in the 1990s group). To the extent possible, the laboratory researchers "championed" their technologies by performing special activities to promote them. In the 1980s cases: The inventor of the penetrometer used the prototype instrument to perform the survey work for the local government. The developer of the tracer technology conducted numerous multi-party tests, large-scale demonstrations, and international experiments, and even produced a video explaining the technology. The creator of the alginate-based herbicide/ pesticide explained how his technology worked to industry and university visitors in his laboratory. Developers of the chemically-imbedded herbicide/ pesticide tested their product at a nuclear waste site.

In the 1990s cases: The inventors of the microwave oven sought out a company to manufacture the variable-frequency component and then concentrated on mentoring the eager young scientists at the partnering company. The artificial heart-flow diagnostic researchers spent much time training CRADA partners on the flow diagnostic technique in their own laboratories. Decades before the light sticks were commercialized, Navy researchers conducted numerous demonstrations in all types of settings, temperatures, and times of day. However, once licensing and commercialization ensued, these activities subsided over time.

In spite of the above explicit examples of enthusiasm, an *undercurrent* of caution on the part of laboratory researchers was also evident in both time frames, and the researchers seemed reluctant to pursue technology too aggressively. For example, the 1980s penetrometer case involved a patent filing rejection and a lack of encouragement by the laboratory regarding commercialization. However, the negative attitude on the part of the Navy laboratory may have been related to the classified nature of the technology.[21] In the 1990s cases, in the gravity meter case, it was stated that some NIST employees were reluctant to talk with or work with companies. NIST was referred to as a "high-profile laboratory," implying that it did not want to attract adverse publicity. Similarly, the researchers in both time frames expressed little interest in venturing outside their laboratories to start up businesses or become involved in other entrepreneurial activities.[22]

In both the 1980s and 1990s cases, there were some but not many examples of references to market evaluations or technology assessments by the researchers. In the 1980s group: In the chemically-based herbicide/pesticide case, the head of the research team provided market statistics in anticipation of future applications of the technology. The only other case that with a similar level of market awareness was the thermoplastic material case, in which the laboratories were described as becoming more businesslike regarding technology transfer. The scientist in that case further commented that the NASA technology transfer offices seemed to be employing a "Madison Avenue-type of marketing." However, this description applied to their technology marketing and deal-negotiating activities rather than to market analyses. There were no references to technology evaluations in the 1980s cases.

In the 1990s: The paper-quality tester case provided the only example of a laboratory (the Forest Products Laboratory) making a conscious effort to develop a defined strategy for commercialization, including market assessment. The remainder of the cases in this group ranged in their approaches to market assessments. For example, in the gravity-meter case, NIST had been using a laboratory version of the gravity meter in its scientific work, and it wasn't until other agencies expressed interest in owning a gravity meter that the laboratory transferred the technology. In two of the 1990s cases, technology assessment work was performed: In the paper-quality tester case, the laboratory issued cooperative research agreements to a number of universities and trade associations to evaluation the technology because of the need for a variety of manufacturing configurations. In the microwave oven case, the Oak Ridge technology transfer office "spent money" to identify and assess "winner" technologies, which could mean they contracted out this activity.

Technologies Continued to Represent Diverse Areas   The technologies were diverse because the laboratories have a wide diversity of applicable technologies. Both the 1980s and 1990s cases included a representative range of technology and application areas, but with environmental technologies and biotechnology being especially evident—which is not unexpected since they are two of the fastest growing fields of science.

Technology areas for the six technologies in the 1980s cases were: oceanography, materials, three environmental technologies, and medicine.

The technologies for the 1990s cases also involved a typical group of federal laboratory technology and application areas, including: biotechnology, telecommunications, environmental technology, and medicine. Two technologies were related to industrial equipment, and two were related to consumer goods. Each of these latter four technologies could be categorized in more than one technology area (materials, biotech, medical, etc.).

Laboratory Groups Remained Similar   There was little difference between the groups of laboratories in the 1980s and 1990s cases; both represented the range of federal/ national/ defense laboratories and field centers in this country. The 1980s cases involved USDA, NIST, NASA, Navy, and two DOE laboratories. The 1990s cases involved two USDA, NIST, Air Force, Navy, and three DOE laboratories.

Funding Combinations Remained Similar   In all of the successful 1980s and 1990s cases, the laboratories and companies used a variety of sources to fund technology development, technology transfer, product development, and commercialization. While funding sources included federal, state, international, and private sources, the federal laboratories and agencies financed the bulk of research and technology transfer; DOE was the only department with funding "issues."23 Also, virtually all of the successful cases involved private sector support for product development and commercialization in both large and small firms. CRADAs are assumed to be "funds-in" mechanisms, except in cases where noted otherwise. Leveraged "outside" investments included venture capital funds or angel investments and agency SBIR funding and/or state technology program funding.24

In addition to the federal support, in the 1980s group of cases: The thermoplastic material case involved SBIR funding, state government funding, and company-supported product development. The tracer technology involved a CRADA, consortia-supported demonstrations, international testing grants, and laboratory equipment sales. The alginate-based herbicide/ pesticide case involved CRADAs and company-supported product development. The chemically-imbedded herbicide/ pesticide case involved industry-paid work-for-others and company-supported product development.

In addition to federal funding, in the 1990s group of cases: The laser-based case involved some university-supported consulting work. The voice-coder case involved university funding contributions and outside capital for the spin-off firm. The paper-quality tester involved company support of product development through the CRADA. The microwave oven case involved SBIR funds and substantial venture capital investment. The gravity meter case involved interagency transfers of funds and SBIR funds to develop a more compact instrument. Both spin-off companies in the gravity meter case received angel funding. The following cases relied on company-supported product development: the light sticks, Oatrim fat substitute, and artificial-heart pump. In the latter two cases, millions of dollars were invested by large multinational corporations, including ConAgra, A.E. Staley Manufacturing, Quaker Oats, Rhone-Poulenc, and Baxter Healthcare.

Mechanisms – Licenses, CRADAs Predominated in Both Periods

Licenses continued to be a traditional mechanism for transferring technologies in both the 1980s and 1990s cases. Although CRADAs were not possible legislatively during the early 1980s, eventually they became used in the 1980s cases. In the 1990s cases, CRADAs were the primary mechanism. No exchanges of technical personnel were uncovered in either time frame.

In the 1980s cases, the laboratories used the following technology transfer mechanisms—besides information exchange, addressed above:

- Technology Licenses: Licenses included two non-exclusive licenses for the thermoplastic material, and two exclusive licenses and one non-exclusive license as well as two sub-licenses for the chemically-imbedded herbicide/ pesticide. The tracer technology, which involved a variety of mechanisms, included an exclusive license.
- CRADAs: In the alginate-based herbicide/ pesticide case and the tracer technology case, the laboratories turned to CRADAs after they were legally possible.[25]
- Technical Assistance: Two of the 1980s cases, the alginate-based herbicide/ pesticide case and the radiation therapy case, involved technical assistance.
- Reimbursable Work: Even with five DOE cases spanning the two time frames, only one research team did "work for others." This involved the 1980's chemically-imbedded material case at DOE/PNNL.
- Scientific User Facilities: The use of scientific equipment and facilities would be most likely to arise in the DOE-related cases because of their large and unique hardware. Only one case in the 1980s group, the tracer technology case, involved user facilities. Not surprisingly, it involved a DOE laboratory.
- Standards Promotion: The radiation therapy case involved proactive work by federal personnel with standards-setting communities to transfer calibration technology and responsibility to "regional calibration laboratories."
- Procurement Contracts: The penetrometer case involved two procurement contracts, one with a university to develop a prototype, and another with a company to develop six holotypes for use by the Navy. Also, the tracer technology involved a collaborative procurement project.
- International Tests and Miscellaneous Mechanisms: The tracer technology involved equipment loans and sales by the laboratory, joint public-private testing, and international tests or experiments. The scientist made many attempts to commercialize the instruments and services related to the tracer technology but was not successful.

In the 1990s cases, the laboratories used the following technology transfer mechanisms—other than information exchange, addressed earlier:

- Technology Licenses: License agreements included a license for the laser-illuminated method, a non-exclusive license for the microwave oven, three non-exclusive licenses (one of which evolved into a CRADA on a related technology) for the fat substitute, and two non-exclusive licenses for the light sticks.
- CRADAs: Five of the eight cases involved CRADAs and three of those involved more than one CRADA. These included the paper-quality tester, microwave oven, gravity meter,[26] fat substitute, and blood-flow diagnostics. The blood-flow diagnostics case provided the only example of an unusual CRADA configuration: two multiple-party CRADAs.
- Technical Assistance: One case in this group, the microwave furnace case, involved technical assistance.
- Standards Promotion: In the voice-coder case, the Air Force's technical team funded the voice-coder technology which was actually researched in a university environment. The Air Force used unique technology marketing strategies in promoting the technology to national and international standards organizations.
- Procurement Contracts: The gravity-meter case involved a procurement contract to develop the gravity meter for use by the Department of Commerce and eventually by the Department of Defense.
- Cooperative Research: The paper-quality tester case provided the only example of cooperative research in both sets of cases. The Forest Service granted cooperative research agreements so that outside organizations could test and evaluate the technology in a variety of settings.
- International Tests: The voice coder and artificial-heart diagnostics involved international testing.

Although CRADAs had been possible since the 1980s, laboratory scientists in two 1990s cases were just becoming familiar with them.[27] In the 1992 creation of the gravity meter, the chief scientist would have encouraged use of a CRADA rather than a procurement contract because it would have alleviated certain problems, but he didn't know about CRADAs. Eventually, however, NIST signed two CRADAs with the two firms spinning off from the original gravity-meter contractor. Similarly, the lead scientist for the light stick case said CRADAs were "talked about" when the technology licenses were being signed, but not commonly used at his laboratory.

The technology transfer mechanisms used in both the earlier and later time frames are summarized in Table 5.2.[28]

Table 5.2   Technology Transfer Mechanisms Used

Case / Mechanism	Pre-legislation						Post-legislation							
	1	2	3	4	5	6	1	2	3	4	5	6	7	8
Licenses		X	X		X		X			X			X	X
CRADAs			X	X				X	X	X	X			X
Technical Assistance				X		X				X				
Personnel Exchanges														
Reimbursable Work					X									
User Facilities			X											
Standards Promotion						X		X						
Procurement Contracts	X		X								X			
Cooperative Research									X					
International Tests			X					X						X

User Benefits/Economic Impacts/Outcomes

The researchers and partners continued to describe success in many ways. They consistently used miscellaneous indicators and intangible measures.

Researchers, Partners Consistently Used Miscellaneous Indicators
Researchers in both the 1980s and 1990s cases used numerous miscellaneous indicators of outcomes and success other than the economic impact indicators discussed above. Most of these miscellaneous indicators were quantifiable or "semi-quantitative," or at least measurable on a ranked scale if more than one data point were available. In the 1980s group of

cases, the miscellaneous indicators included:

- Increase in Market Share: In the alginate-based herbicide/ pesticide case, a company was producing three new alginate-based products. The company's market share increased from approximately 25 to 85 percent based upon the success of one of those products.
- Number of Publications and Literature Citations: In the alginate-based herbicide/ pesticide case, the diffusion of information played a major role in transferring the technology. In the interview, the researcher suggested a formal scientific literature citation check as a way to understand the flow of information on that technology.
- Gain of a Unique Patent Position: M&T Chemical gained a unique patent position by licensing the thermoplastic material.
- Number of Awards: The radiation therapy standards program was put into place by a NIST researcher who received a number of awards for his work in that area.

In the 1990s group of cases, miscellaneous indicators included:

- Taxes Paid by the Private Partners: One of the companies in the gravity-meter case provided information on the amount of taxes paid by the company over the course of its several-year existence.
- License Royalty Revenues: The Navy laboratory provided information on the total amount of royalty revenues generated as a result of the two licenses on the light-stick technology.
- New Market Opening: The paper-quality testing technology opened up an unexpected new market for Isthmus Engineering in computer film-quality testing.
- Level of Teamwork: In the case of the microwave furnace, both the lead laboratory scientist and the company head commented on the importance of laboratory/ company relationships, citing this as a key to the company's commercial success. The company head said close interpersonal relationships and good communication with the laboratory scientists helped streamline the process of working with the laboratory generally. The lead scientist commented that the laboratory's business (technology transfer) and legal staff could do the CRADA paperwork, but not the necessary technical hand-holding and mentoring.
- Speed-to-Market Advantage: The Oatrim fat-substitute case provided an example of a company gaining a niche through unusually fast speed to the market.
- User Satisfaction: In the heart-flow diagnostics case, the hospital partner was extremely pleased with the CRADA work with the laboratory and

with its access to equipment it wouldn't have been able to afford. One doctor praised the inspired assistance and attitude of the laboratory researcher as key to the development of a superior heart pump. This is corroborated by the fact that the hospital and laboratory signed follow-on agreements. In the light stick case, both licensees indicated early on that they were pleased with their licensing arrangements, and one indicated a desire to work further with the laboratory.

- Traded or Bartered Goods: In the 1990s group, both the paper quality tester and gravity meter cases provided examples of commercial products being traded or bartered for intellectual property rights or consulting time. This type of data can be converted to dollar amounts and added to the statistics indicating commercial impact.
- Commercial Space Occupied: Two companies showed physical growth: Lambda Technologies expanded to an additional production floor site; Omniglow grew from 20,000 to 180,000 square feet in office and factory floor space.

Researchers, Partners Consistently Used Intangible Measures   Two cases in the 1980s group indicated quality-of-life improvements. Two cases in the 1990s group illustrated improvements in the laboratory or company image. Both of these indicators are intangible factors.

In the 1980s group: The project leader on the chemically-based herbicide/ pesticide emphasized the technology's ability to improve environmental conditions and to serve as an enabling technology for similar breakthroughs. The tracer technology also contributed to a better quality of life through its environmental applications. The radiation therapy quality assurance case referenced a contribution to safety and the avoidance of costs imposed by legal liability. However, in that case the benefits were indirect and consequently even more difficult to measure. A formal study of the impact of the radiation therapy quality assurance system confirmed its success. In that case, the transferred technology improved the nation's medical system rather than having a commercial impact. In cases where there are few or no (economically) measurable variables, official reviews become significant. Since that study, the quality assurance system's success had become so well-accepted that no additional studies have been commissioned.

In the 1990s group: In the case involving the laser-based method, the laboratory experienced an improvement in its image as a result of the scientist's ground-breaking work in the "hot" biotechnology area of science.

In the gravity meter case, Axis Instruments showed that commercialization of the technology was possible, paving the way for two subsequent spinoffs which may not have been launched otherwise.

International Activity – Laboratories Remained Uninterested in Foreign Patent Rights

No laboratories applied for foreign patents either in the 1980s or 1990s cases, in spite of the increase in international activity and sales in the 1990s cases. However, there were instances in the 1990s cases when the scientist or the partnering firms sought foreign rights. In the 1990 case related to the Oatrim fat substitute, the scientist obtained foreign rights when the laboratory chose not to, and the partnering firms licensed and sub-licensed from him. Several of the partnering companies in the 1990s time frame sought foreign patents, including: Digital Voice Systems Inc. from the voice coder case, Lambda Technologies from the microwave oven case, and Omniglow from the light stick case.[29]

Economic Development, Technical Assistance – American-Owned Companies Predominated in Both Time Frames, But Did Not Use Available Services

In all of the cases, only two foreign-owned firms were involved in technology transfer, but they involved U.S.-based manufacturing. These firms were joint venture partners in the 1990s Oatrim case, but both had major manufacturing operations in the United States.

There appeared to be no connection between economic development programs and the American-owned firms involved in technology transfer in either the 1980s or 1990s cases. In neither time frame did the industry partners avail themselves of state, regional, or local economic development or technical assistance programs. One case in the 1980s time period, the thermoplastic material case, involved a small firm receiving state R&D funding, but no technical assistance or economic development services were granted.

Denver, Colorado ranks among the top four or five areas of the country in terms of entrepreneurial development. However, none of the small gravity-meter firms based there chose to be affiliated with any of the incubators or technology transfer networks. The president of one of the Denver area firms noted that economic development services and banking institutions were not that helpful. However, he said federal R&D funding programs were useful and referred to them as an "engine for the future."

Challenges to the Technology Transfer Environment

The findings also show conditions or changes that could be considered threats or challenges to the technology transfer environment, as follows:
- Roles of Other Laboratory Personnel – The technology transfer function and research functions continued to register tension.
- Barriers to Commercialization – Private sector commercialization problems related to technology transfer increased, including scale-up problems, difficulties in marketing highly-technical products, and partnership problems.

Roles – Technology Transfer, Research Registered Tension

Throughout most of the cases in both time frames, a tension was "sensed" between the technology transfer function and the research function, which is not necessarily identifiable by specific data points from throughout the cases. As illustrations, in two of the 1990s cases, there was an effort on the part of the technology transfer personnel to maintain objectivity which was either explicitly or potentially an issue with researchers. This issue did not come up in the pre-legislation cases. In the microwave oven case, the researchers commented that they were not included in legal negotiations on behalf of the technology and, as a result, the legal staff made commitments that the technical staff could not honor. In the paper-quality tester case, the laboratory rotated its researchers periodically to keep them from becoming too attached to particular technologies and over promoting them.

Barriers – Private Sector Commercialization Problems Increased

In the 1990s cases, there was an increase in commercialization-related problems faced by the private sector partners (eg., production and marketing) that were not as evident in the 1980s cases. Most of these problems were fairly typical business problems that were accentuated by the highly-technical nature of the products.

Introducing highly-technical products to the market is especially difficult because they often require educational campaigns, making them more expensive to market than other new or mature products. In the 1980s cases, the chemically-imbedded herbicide/ pesticide case illustrated the difficulty in marketing a product that necessitates a paradigm shift on the part of the customer from a short-term to a long-term orientation, and from the necessity for large quantities to only small quantities. Comments were also made about the difficulties for companies, particularly small firms, in achieving full-scale production of a highly technical new product and in purchasing expensive raw materials. Three of the 1980s cases involved these problems: the thermoplastic material case, alginate-based herbicide/ pesticide case, and chemically-imbedded herbicide-pesticide case. In the thermoplastic material case, High Technology Services manufactured and sold the material in sample-like quantities which did not allow volume discounts. When larger volumes were ordered, HTS negotiated a price and then contracted with a manufacturer to produce the material on a toll basis of two to five times the production costs, which did not provide much profit margin.

From the 1990s time frame, several cases highlighted the problem of new products with no initial demand, and for which the market had to be created and customers educated. These were: the voice coder, paper quality tester, gravity meter, and the Oatrim fat substitute cases. The voice coder, in particular, was very different technologically from existing products. In response to these marketing difficulties, the partnering companies relied on their reputation and networking within their own technical communities to obtain business. For example, the company commercializing the paper quality tester, Isthmus Engineering, relied on word-of-mouth to spread its reputation for quality products. Axis Instruments, the original gravity-meter partner that went out of business, had emphasized formal publicity; in contrast, the successful Axis spin-off company relied more upon contacts within the international technical community.[30] Two 1990s cases also encountered scaling-up problems: the gravity meter and fat substitute cases. In the gravity meter case, the smaller of the two spin-off companies had to "go outside" for machining work because it did not have its own machine shop. Large corporations also experienced problems in scaling up. The large companies in the Oatrim fat substitute case said they should have been more aware they were receiving only laboratory samples from USDA. They discovered that production in larger quantities resulted in an essentially different product. Furthermore, in the 1990s group of cases, in the fat substitute case, the laboratory's licensees, being partnerships and joint venture arrangements, faced the management challenges inherent to these types of arrangements. According to the company representatives, such partnerships were easy to conceive but hard to manage because they involved merging different business cultures and personnel nationalities.[31]

Summary

Table 5.3, "Similarities and Differences between the Two Time Frames," summarizes the details in this chapter. (Not all of the findings listed in the table were discussed in the text above.) The next chapter summarizes the key findings and presents implications and conclusions.

Table 5.3 Similarities and Differences between the Two Time Frames

Topic	Pre-legislation	Post-legislation
Roles of Lab Researchers & Other Personnel	Lab prototyping not done; some sampling	Lab prototyping & samples on the rise
	Champion role by lab researchers focused on tech marketing, demos, etc.	"Champion" role focused on interaction with CRADA partners
	Lab researchers not risk-adverse, but generally careful approach to tech transfer	Pockets of cautiousness still exist, depending upon lab culture
	Labs uneasy with market & tech assessments	Inexperience with market & tech assessments continues
	Tech transfer & research functions had subtle tension	Tech transfer and research functions registered tension
Technologies and Applic.	Incremental improvements & evolutionary tech applications	More revolutionary applications occurred
University Involvement	Joint government/ university inventing prevails at bench scientist level	Joint government/ university inventing results from inter-institutional ventures

Funding, Financing	Combination of funds used to support tech transfer & commercialization; CRADA funding issues only apparent in DOE cases	Combination of funds used to support tech transfer & commercialization; CRADA funding issues only apparent in DOE cases
Intellectual Property	More emphasis on scientific publishing; less emphasis on patenting	More aggressive patenting by both labs & partnering companies
	Royalty-sharing incentives are very important to the lab inventors	Royalty-sharing incentives are even more important to the lab inventors
	Lack of incentives to patent; researchers less "patent-savvy"	Similar lack of incentives (earlier-on); lack of patenting incentives eventually subsided
	Inventing partners = universities	Inventing partners = companies
Tech Transfer Mechanisms	Scientific journals are std. for the scientific community	Scientific journals still important
	Licenses a more traditional mechanism, although it took time to gain expertise in implementation; CRADAs not legislatively possible; when possible, labs turned to them	Licenses widely used, with only minor issues unresolved; CRADAs prevail as a mechanism after long-term unfamiliarity
	Some researchers have negative perceptions of CRADAs	Some company partners have negative perceptions of CRADAs
User Groups, Lab Partners	Small firm involvement in tech transfer not as prominent	More small firm involvement in tech transfer
	Tech push; user contacts initiated by lab researchers	Market pull; contacts were user-initiated
	Earlier-stage technologies, less market-orientation, broad-based user groups	Targeted user groups & markets

Barriers to Commercial-ization	Failures: two products stopped short of commercial marketplace; five licenses inactive, although three products resulted Scaling-up difficulties (particularly small firms)	Failures: one inactive license; one failed line of products (although several of these products taken over by a new company) Scaling-up difficulties, difficulties in marketing highly-technical products, and joint venture & partnership problems among companies
Other Factors	No involvement of foreign firms in tech transfer	Foreign-owned firms were involved, but with U.S.-based manufacturing
User Benefits/ Economic Impacts/ Outcomes	16 products resulted from the technologies in the cases; "$ few thousand" in sales revenues; no spinoff start-ups resulted; no other sales or employment data made available Variety of miscellaneous & intangible indicators used to describe success	114 products resulted from the technologies in the cases; $7.7 million sales revenues; six spinoff start-ups resulted; 348 *net* jobs generated Variety of miscellaneous & intangible indicators used to describe success
Intl. Activity	Moderate international sales No labs applied for foreign patents	More extensive intl. sales No labs applied for foreign patents; inventors & partnering firms exercised foreign rights
Govt. Gains	Unanticipated govt. benefits, dual uses, and spinbacks apparent	Unanticipated govt. benefits, dual uses, and spinbacks apparent
Econ. Dev., Tech Assis.	No connection between econ. dev. programs & tech transfer	No connection between econ. dev. programs & tech transfer
Elapsed Time	15 to 20-year development time	8 to 10 years up to commercialization

Notes

1   The only prototype development work in the 1980s cases was not performed by the laboratories, but was contracted out—to a university engineering department (in the penetrometer case) and to instrument makers (in the tracer technology case).

2   Other aspects of intellectual property remained the same. For example, the laboratory researchers from both the 1980s and 1990s cases did not elaborate or provide details on invention disclosures. None of the cases in this series involved filing a provisional application, an option under the new international patent regime. Also, the issue of software copyrighting did not come up because there were no cases centered around software development; an FLC award in the 1985-1986 time frame was for a geographical information system, but it was not developed into a case for this study. The only aberration in terms of intellectual property issues related to the litigation in the light stick case, which is illustrative of the fact that patent court cases are becoming more common.

3   However, there were still some intellectual property lessons to be learned in one of these 1980s cases where the researcher shared his knowledge about his technology during a corporate visit to his office, only to find the company avoided his patent when commercializing.

4   The inventor of the tracer technology commented that laboratory researchers tended not to patent until recent years.

5   The inventor of the penetrometer said that in the 1980s when he attempted to transfer the technology, a patent counted only about as much as a publication toward professional advancement. Further, Navy scientists received $50 to $100 per patent, which was not much incentive.

6   However, the inventor of the light-stick technology said he was denied patent filing for the fishing lure application because the Navy did not want to spend the money it would have cost to file.

7   Furthermore, the chief scientist in the microwave oven case, who had a portfolio of twenty patents and fifteen pending, was sought out by proteges for his patent-related advice.

8   The inventor of the unpatented gravity meter said that, until about five years ago, NIST urged its researchers to publish rather than patent.

9   It is debatable whether the reseachers' strong opinions regarding royalties are an indication of improvement or of greater awareness. In either case, an assumption underlying the legislation is that such incentives would have an effect, so this change is viewed as an overall positive sign and is, therefore, included in this first section about positive changes.

10 The radiation therapy quality assurance program was standards-oriented rather than market-oriented.

11 With one exception: the alginate-based herbicide/ pesticide could be used by landscape firms to control weed and pests slowly; the chemically-imbedded herbicide/ pesticide was so long-term oriented that landscape outfits saw it as competition in the sense that a major portion of their maintenance services would not be necessary.

12 Because the tracer technology proved to be not viable as a stand-alone business venture, at least three small firms sold the tracer service and/or instruments as "sideline" businesses. However, the laboratory scientist who developed the technology said the firms had not realized more than *a few thousand dollars worth of business* in this area. The laboratory, itself, offered tracer technology services for a fee. Thermo Trilogy from the 1980s alginate-based herbicide/ pesticide case noted that the company would not be able to provide sales figures until it goes public.

13 Consistent with the companies' reluctance to share sales revenues, they did not provide precise international sales figures.

14 State government technology programs are often administered by state economic development agencies, the majority of which use job creation as their key measure of success according to a 1996 survey by the National Association of State Development Agencies. A problem, however, is that they measure jobs on an annual and regional basis rather than a cumulative and per-project basis, making this difficult to apply to longer-term technology transfer and commercialization.

15 DOE publications are not accurate in saying that Microwave Laboratories *formed* the new company.

16 In a presentation at an FLC meeting.

17 The group of university researchers forming Digital Voice Systems, Inc. received funding from other sources besides the Air Force so it is not clear how much of the credit the Air Force can claim for this success.

18 The two original licenses for the thermoplastic material are inactive.

19 The product was abandoned when the company failed financially.

20 However, another factor may have been the fact that the University of Iowa's license with Lachat Instruments was negotiated in the days when licenses were arbitrarily granted without requiring royalty payments; not even an up-front fee was requested as a good faith gesture in return for the cost of operating the license. This lack of value may have also contributed to the company's lack of commercialization effort.

21 In contrast, the 1990s paper testing technology being used by the U.S. Bureau of Engraving and Printing which prints currency, also likely involved classified technology (the Forest Products Laboratory interviewee was reluctant to talk about that application). However, it did not seem to be a deterrent to transferring the technology to other applications in that case.

22 In the 1990s cases, the chief scientist in the gravity meter case expressed such an interest (possibly because one of his graduate students started one of the spin-off companies). He noted that scientists often feel a moral obligation to give back to society. However, he said regulations at his laboratory would not allow him to do so. Meanwhile, the PNNL leave-of-absence program has spawned ten new local businesses in less than two years by laboratory scientists, ranging from medical products to agricultural services; the Sandia National Laboratories entrepreneurial-leave policy involved five employees taking leave within a year to start businesses.

23 In both the 1980s and 1990s cases, federal funding cutbacks became an issue only in cases involving DOE laboratories. This was because of the uniqueness and competitiveness of the DOE R&D and CRADA funding system. Of the 1980s cases: The scientist from the tracer technology case (at Brookhaven National Laboratory) indicated that laboratory user fees were a welcome revenue for the laboratory when it was experiencing downturns. The team leader for the chemically-imbedded herbicide/ pesticide (at PNNL) was frustrated with the DOE system and the lack of funding for his team to continue its development work. From the 1990s time frame: In the microwave oven case (at Oak Ridge), the DOE Office of Industrial Technologies, a DOE headquarters program, supported the laboratory's CRADA involvement when special DOE CRADA funds were no longer available. In the artificial-heart case, the laboratory turned down private funding offered by one of the CRADA partners (Baxter Healthcare) because it was not sure how to handle "funds-in"; however, Baxter Healthcare and its hospital partner set up $500,000 worth of equipment at the laboratory.

24 At least four of the small firms from both time frames (High Technology Services, Lambda Technologies, Micro-g, and Winters Electro-Optics) mentioned the issue of capitalization.

25 The scientist from the alginate herbicide/ pesticide case stated that laboratories were "practically giving away" intellectual property rights to companies under CRADAs. He added that while technology-transfer personnel were in favor of CRADAs, the laboratory inventors stood little to gain by them. The inventor of the penetrometer had some similarly negative views about CRADAs, commenting that CRADAs eased the technology transfer process, but made other laboratory activities more difficult.

26 A company partner from the gravity-meter case noted that a CRADA was not necessarily profitable because there was no specific goal or deadline (unless structured a certain way). He said CRADAs were useful for ideas but not necessarily for commercial products. Further, a company partner couldn't provide direction to the laboratory personnel involved in the CRADA; they could only access government information and ask specific questions. He said

CRADAs are based upon the belief of both parties that useful, industry-oriented research would result from the CRADA effort. Yet, he said, companies don't normally "do business" that way.

27　Technology transfer officers may have been aware of CRADAs as an option early on, but most agencies took years to adopt CRADA regulations; similarly, in the 1990s light stick case, restrictions on outside consulting created problems for one scientist. Although the issue was resolved and laboratory personnel can now consult, the scientist who raised the issue was unaware of it.

28　The cases, as numbered, represent the order of the cases as presented in Chapters 3 and 4.

29　It is presumed that Agrifim International (from the 1980s chemically-imbedded herbicide/ pesticide case) and Baxter Healthcare (from the 1990s heart pump case) also had foreign rights due to the extent of their international activities, although this was not mentioned.

30　This company was also faced with a nearly saturated market for its highly specialized instrument, and with competition from well-funded foreign government agencies producing gravity meters and having greater access to international markets. It was infeasible for the company to use overseas agents because of their lack of familiarity with its highly technical product.

31　This made decision-making difficult and fraught with the risk of alienation and ultimately failure.

6 Lessons Learned

The purpose of this study was to examine the process and impact of government laboratory technology transfer, based upon a qualitative comparative analysis of successful cases before and after passage of related legislation. The two key laws, passed in the mid- and late-1980s, were intended to encourage cooperative research for commercialization purposes.

This last chapter discusses certain of the study's findings presented in the previous chapter. It summarizes the positive findings, and provides implications and issues related to both positive and negative findings. Some of the findings are combined for purposes of discussion and drawing conclusions. The second section of the chapter discusses the lessons learned from this study about technology transfer measurement and evaluation.

No doubt some of the broad societal factors noted in the background section of the literature review chapter (defense conversion, increased attention to global competitiveness, etc.) influenced the outcomes documented through this study. It is impossible to know with total accuracy whether the legislation—as opposed to other factors—produced the study's findings. In order to control for other factors, a control group would need to be built into the research design up front. However, this project involved retrospective research.

Assessing the Effect of the Legislation

Roles of the Laboratory Researchers and Other Personnel

The findings showed that the laboratory researchers' roles continued to involve research, publishing, speaking at technical conferences, producing prototypes and samples, and serving as champions. However, aspects of these roles changed from the pre-legislation period to the post-legislation period.

The findings showed that the laboratory researchers produced more laboratory prototypes and samples. This indicated more proactive technology transfer, but also raised a potential longer-term structural issue

267

related to funding. A higher level of prototyping activities may cause pressure to identify potential sources of funding unless royalty revenues (or other income) support these activities. A dramatic trend toward increased prototyping by many laboratories could signal a move away from basic research towards applied research and even later phases of development.

Although the researchers continued to serve as technology champions in both time frames, closer analysis showed that the champion role changed. In the pre-legislation cases, the champion role took the form of demonstrations, not only for proof-of-concept, but also for technology marketing purposes. In the post-legislation cases, once CRADAs were signed, the researchers roles as champions were funneled towards their close working relationships with their CRADA partners rather than technology marketing.[1]

In addition to this "funneling" effect of CRADAs, the efforts of the technology transfer staff to maintain objectivity in technology transfer may have been downplaying the champion role. The researchers were not involved in deal-making with potential industry partners. This may have been a factor of the complexity of the deals, and the need for the laboratory negotiators to have strategizing skills matching those of their private-sector counterparts. However, a by-product of the researchers being excluded from agreement negotiations is that they may feel that no one is looking out for their individually-vested interests which doesn't help to engender trust.

The implications of a change in the researchers' champion role are that laboratory researchers are doing less marketing and are focusing their energies on their partners, so that the technology transfer staff is losing the marketing assistance of the researchers. Therefore, the technology transfer staff must be prepared to take on more of the role of technology champion and marketing agent.

Neither the pre-legislation nor the post-legislation cases contained many examples of formal market analysis or technology evaluation work with the exception of the Forest Products Laboratory (FPL). In this regard, it is appropriate to view commercialization efforts within the context of laboratory missions. For example, the mission of certain laboratories (like FPL) inherently involves working with industry and other users, so such commercialization initiatives are more "normal" for them than other laboratories (such as the defense laboratories and NASA field centers). As a result, it is difficult to measure degrees of proactivity in this area, but a positive aspect is that the industry-oriented laboratories can serve as models for the others in this area.

It is important to have business-oriented skills in order to assess the validity, value, and marketability of a technology,[2] particularly when technology transfer is demand-centered and market-oriented, as noted. Whose role is it to perform market assessments or to determine the commercial value of the technologies? The unclear responsibilities and lack of attention in this area point to a need for position descriptions and performance expectations that reflect these roles appropriately. If individuals in the laboratories do not have or acquire these skills, another option is to contract out these activities which makes it all the more important to understand the associated time and costs. Some researchers gave the impression of being "intuitively aware" of their technologies' commercial prospects. This type of information could supplement the formal analyses, but it would require good communications between the researchers and technology transfer staff. However, as noted, in the post-legislation cases there was a tendency for the research and technology transfer function to be even less connected as manifested by the effort to "objectivize."

In terms of the researchers' roles, the findings also indicated an undercurrent of caution towards technology transfer in both the pre-legislation and post-legislation cases. A 1985 FLC study[3] determined that less than half of the *technology transfer* respondents would be willing to visit the site of a potential outside user. If the same survey were administered today to *researchers*, the results might not be that different. According to the post-legislation cases, the laboratory researchers were still somewhat reluctant to pursue technology transfer which is likely the result of a lack of awareness on legislative intent. The cases in both time frames showed that it took a long time for laboratory researchers to learn about legislative provisions encouraging technology transfer (eg., CRADAs or consulting opportunities related to inventions).[4] The agencies could focus more attention on elevating awareness of the legislation and ensuring that regulations are consistent with the legislation. For example, the cases showed that certain laboratories still have policies or rules prohibiting entrepreneurial leaves of absence (while others are inconsistent in their approaches).[5] Researchers may become entrepreneurial if their laboratories adopt explicit policies and programs in this area. Also, government technology transfer and commercialization training programs may help to address this awareness problem. Most existing training programs are focused on technology transfer personnel. The Navy is one of the few agencies focusing this type of training on scientific and technical personnel; perhaps this is an option to be considered by other agencies. In addition to the laws, the cases suggested certain other topics for "awareness-raising" such as protecting intellectual property while marketing technologies, yet sharing enough information so that the user has a need to license the technology.[6] Another potential training topic suggested by the cases is the repercussions of inadvertently disclosing proprietary and sensitive company information accessible through public-private partnerships.

Throughout most of the cases in both time frames, a tension was "sensed" between the technology transfer function and the research function, although this was not necessarily explicitly identifiable. Technology transfer inherently brings together different professions (and cultures) who have a responsibility to work together closely. In such an environment, difficulties or points of irritation would naturally become more apparent, whereas they may be hidden in other scenarios. Thus, success presents the challenge of addressing heightened sensitivities and perhaps misunderstandings about technology transfer. The examples cited were oriented toward maintaining some objectivity which indicates that more of an attempt is being made to be specific about roles and responsibilities in response to this highly sensitive working environment.

To encourage more trusting relationships among those who are part of the technology transfer process, perhaps compensation packages can be designed so that a laboratory team (the technology transfer staff along with the research team) is judged on how it works together toward successfully implementing technology transfer mechanisms and agreements. The technology transfer staff, alone, cannot cover all the necessary contacts with the partnering firms (the corporate research team, marketing staff, etc.). And, as noted, the laboratory researchers were discouraged from conducting outreach past a certain point in time. Yet, coordinating as a laboratory team *could* accomplish this objective. Encouraging teamwork is probably best accomplished through through laboratory compensation packages rewarding teamwork rather than through legislative amendments.

As a closing thought on the topic of roles, one laboratory technology transfer officer recently noted that the truest metric of technology transfer success is the barometer of the laboratory researchers and their research managers; if they don't feel the technology transfer function is supporting them, he said, then there is a problem.[7] So, the question of roles hinges on relationships between the researchers and technology transfer officers. Supporting that relationship and managing the expectations related to it is of paramount importance.

Funding, Financing

The technology transfer legislation did not affect other aspects of laboratory activities, but nevertheless raised policy implications based upon the findings. In terms of government funding, the post-legislation cases showed an increase in small-firm involvement, and several cases involved Small Business Innovation Research (SBIR) recipients, but no recipients of the Small Business Technology Transfer (STTR) program which was established specifically for small firms to partner with government laboratories or universities. This may reflect the fact that the STTR program is relatively new[8] and the fact that the amount of government funding for SBIR is more than that for STTR. On the other hand, the STTR program may need to be promoted to various audiences such as FLC awardees.

Intellectual Property

The findings also showed that laboratory (and company) patenting increased in the post-legislation cases. The enhanced interest in patenting presumably was the result of the improved incentives for both researchers and laboratories provided by the legislation. However, increased patenting on the part of the laboratories may yield to more selective patenting in the long run due to the cost of patenting and the potential for increased royalty revenues based upon market analyses and technology evaluations. Many high technology firms strategically manage their intellectual property, and the case of the thermoplastic material case provided an example of licensing a technology to "round out" the firm's intellectual property portfolio in a technology area.

Technology Transfer Mechanisms

Findings related to technology transfer mechanisms showed a lack of differences between the pre-legislation and post-legislation cases. In both time frames, the laboratories emphasized CRADAs and licenses to transfer technologies. They used other technology transfer mechanisms to a lesser degree, and the post-legislation cases exhibited slightly less variety in mechanisms than the pre-legislation cases. Neither the pre-legislation nor the post-legislation cases had many examples of mechanisms like scientific user facilities and reimbursable work as they relate to technology transfer, and there were no examples of personnel exchanges.

Is it effective for government laboratories to rely on one or two standard mechanisms such as CRADAs and licensing? In their favor, licensing-related issues for both time frames declined over time. And CRADAs have certain advantages over older mechanisms like procurement contracts or memorandums of understanding (such as allowing joint development work without creating conflicts of interest). Nevertheless, in both time frames, both researchers and private partners expressed some negative views toward CRADAs, indicating the laboratories should not rely exclusively on this mechanism.[9] However, the implications are that both government and industry may need incentives to use other mechanisms. With personnel exchanges, for example, possible incentives could include permitting tax deductions on the part of industry or earmarking funds for government personnel to exchange scientific and technical personnel.

The researchers' comments about license royalty-sharing indicated their strong feelings in this area, and the strong feelings increased in the post-legislation cases. This shows that royalty-sharing as an incentive is working but, as a result, certain implementation issues related to royalty-sharing may need to be examined. As suggested by the cases, for example, the "greed factor" may come into play when multiple inventors are involved. This will necessitate dispute mechanisms such as at the level of individual laboratories, or at the agency level, or government-wide. Also, the scientists may not be getting the level of business support needed from their laboratories in the form of regular statements to track their royalty income (such as through the laboratory finance office). Further, there may be a need for guidelines regarding the disposition of royalties or other income from unpatented technology licensing[10] or commissions from revenues resulting from other technology transfer mechanisms such as user fees. Lastly, references to the lack of uniformity among agency incentives raised the issue as to whether the minimum royalty-sharing provisions of the 1995 Technology Transfer and Advancement Act apply to contractor-operated laboratories. (DOE, with its numerous contractor-operated laboratories, is taking the position that the act applies only to government-owned and -operated laboratories.)

User Groups

In terms of process, technology transfer was practiced more efficiently and effectively in the post-legislation cases than in the pre-legislation cases. The post-legislation cases involved more small firms, indicating that fairness-of-access prevailed for the small-firm partners. They also exhibited more user-initiated contacts, indicating more market pull than technology push. In addition, the users narrowed from broad groups to targeted markets in the post-legislation cases. Further, the post-legislation cases exhibited improved systemic effects of the legislation such as more "institutionalized" university relationships.

Barriers to Commercialization

The barriers to commercialization did not appear to be "in-house" problems such as lack of management support, in either the pre-legislation or post-legislation cases.[11] The laboratories' private partners experienced problems in their commercialization efforts, particularly in the post-legislation cases. The private partners had problems in scaling-up, difficulties in marketing highly-technical products, and partnership problems. However, it is questionable as to whether public solutions are appropriate. How far should government laboratories go in assisting the private sector with commercialization roadblocks? Economic conditions and prevailing politics dictate that *typical* private sector problems are not amenable to, or appropriate for, public policy solutions and, therefore, don't warrant government involvement.

In any case, the overall benefits of the technology transfer legislation to the private sector partners were clear. It is hoped that ultimately realizing such benefits on an individual basis would help to ameliorate any difficulties experienced during the commercialization process. The laboratory/company relationships in the cases appeared to be viable, productive and mutually-beneficial relationships, even though the laboratories and companies each had different objectives, with the laboratories being focused on their missions and the companies on profitability.

User Benefits/Economic Impact/Outcomes

The findings and analysis of the series of cases indicated that the technology transfer legislation had positive effects in terms of user benefits, economic impact, and outcomes. The indicators showed that, in comparing the pre- and post-legislation cases, new products (generated as a result of technology transfer) increased from sixteen in the pre-legislation cases to 114 in the post-legislation cases. Similarly, sales revenues increased from "a few thousand dollars" to $7.7 million. New companies increased from zero to six, and new jobs increased from zero to 348 net jobs generated. Also, the post-legislation cases exhibited fewer technology transfer and commercialization failures (inactive licenses, etc.) than the pre-legislation cases.

In addition, the cases showed that technology transfer contributions to dual use and similar types of government gains increased in the post-legislation time frame. The Pentagon has enhanced its dual use initiatives in the 1990s, but the findings in the cases resulted from efforts to transfer technologies rather than from those programs, indicating government technology transfer is helping to diversify the military-industrial base.

Furthermore, in addition to greater economic impact and government gains, the findings showed that government laboratories and their partnering firms became faster in getting products to market. Perhaps as a result of their close working relationships with private-sector partners, the laboratories are becoming more sensitive to the issue of timing in moving a technology from the laboratory into the commercial sector.

International Activity

In both the pre-legislation and post-legislation cases, government laboratory technology transfer was the domain of domestic companies, but these firms were doing quite a bit of international commercial activity. Government policy circles have worried for years about whether the laboratories should give preference to domestically-owned firms in carrying out their technology transfer activities. The cases indicated this is not necessary now, but it may be worth monitoring in the future.

The second implication in the international area relates to the international commercial activity becoming more prominent in the post-legislation period. Certain companies in the cases sought patents overseas, yet the U.S. laboratories did not file for foreign patent rights. This is surprising with the increased emphasis on global technology sectors and recent changes in NAFTA, GATT, and U.S. Department of Commerce regulations. The lack of interest in foreign patenting probably comes from a combination of the lack of good market data plus the large costs in foreign patenting, but this cannot be determined strictly from the cases. The agencies may need criteria for determining the need for foreign filing so that their technologies are protected, marketed and used overseas if appropriate.

Economic Development, Technical Assistance

In both the pre-legislation and post-legislation cases, the findings indicated a lack of interaction between the government laboratories and state and local governments as *users*. In both time frames, they also indicated a lack of interaction between high-technology companies and state and local governments as *service providers*.

The small amount of work being undertaken by the laboratories *for* state or local government users was surprising. One of the two cases that involved state and local government work was the Navy penetrometer case at Stennis Space Center, which was not a surprise since the government entities located at Stennis are traditionally known for their outreach to local communities.[12] Why aren't other laboratories making more of an effort to reach out to neighboring jurisdictions and assist their school systems, public works departments, economic development groups, etc.? Given that the incentives for companies to be involved in technology transfer are rooted in commercial returns, perhaps the commercial incentives provided by the legislation are similarly causing the laboratories to focus on the more lucrative commercial aspects and to ignore state and local government needs for technical assistance.[13]

The total lack of use of state- and locally-provided economic development and management/ technical assistance services by the partnering companies was surprising. Either those services are doing an inadequate job of outreach to potentially needy high-tech firms, or the services aren't needed, or the services may be targeted toward a different audience than the partners in the cases in this study. In any event, the interview question addressing this topic did not directly relate to the overall research question regarding whether recent technology transfer legislation has made a difference. Rather, it was asked as a matter of peripherally-related interest. Perhaps the interview question should have addressed the use of public (or private) technology venture funds rather than economic development broadly.

Lessons Learned – Technology Transfer Evaluation

The literature review chapter of this dissertation attempted to show that the measurement and evaluation of technology transfer is a major issue in the technology transfer community for a variety of cited reasons. There is a relative lack of experience in technology transfer evaluation, as well as a lack of measurement models. There are a handful of individuals who are successfully making progress in this area as it relates to government laboratories, including Chapman, Bozeman, Roessner, Link, Papadakis, and others. This list expands to include Tornatzky, Feller, and even others from the closely-related state and university communities. Each of these researchers has attempted to explain some of the contradictions and complexities in "metrics" for the technology transfer field (and how it differs from traditional evaluation), and their observations were highlighted in that chapter and/or referenced in footnotes. Based upon the early findings of this study's literature review, the methodology for this study was carefully designed to try to overcome the expected problems and pitfalls. In the tradition of the experts noted, the lessons learned about technology transfer metrics through this study are shared below.

In this study, in both the pre-legislation and post-legislation cases, the researchers and partners described their successes many ways. Therefore, documenting technology transfer results in a consistent fashion for comparison or benchmarking[14] purposes was problematic in both time frames. The findings from this study suggest that the agencies could consider some approaches to addressing this and other problems if they are not already doing so. Various recommendations are discussed in more detail below. In short, some minimum requirements for technology transfer metrics are: acceptance of a wide variety of indicators; complex intangible measures; particularly long measurement periods; mandatory feedback (with no enforcement capabilities!); ironclad data protection efforts; interagency coordination to eliminate double counting; and documentation of failures (with no incentives to do so!).

Experiment With a Greater Variety of Indicators

The interviewees in this study were not required to answer the question about economic impact or outcomes in any particular way. As a result, the findings both validated the typically-used indicators for measuring technology transfer and revealed a variety of additional indicators such as the: amount of commercial space occupied, percentage of market share, level of teamwork, new market opening, taxes paid by the private partners, gain of a unique patent position, and speed-to-market advantage. With a little experimentation on the part of the agencies, some of these indicators could become used consistently enough to be comparable. Many agencies already track or document other indicators used in the cases (license royalty revenues, number of publications and literature citations, and number of awards), and they may want to think about using these statistics more publicly if they are interested in furthering technology transfer. The variety of indicators shows that technology transfer involves several sets or types of measures: those related to commercial activity, those related to scientific milestones, and those related to laboratory missions. The overall recommendation here is for the agencies to continue concentrating on the most important and most useful impact indicators, but also be flexible in adding others to the repertoire which may, with repeated use over time, yield unexpected results and useful patterns.

Legitimize Intangible Measures for Less-Common Mechanisms

From the cases, it appears that some of the less common technology transfer mechanisms cannot be associated with typical indicators. Several cases, particularly in the pre-legislation time frame, did not involve *commercial* successes although the technologies contributed to public objectives:

- The not-fully-commercialized penetrometer was used to help neighboring state and local jurisdictions, providing productivity gains in that arena.
- The commercially-unsuccessful laser-based method was used in cancer research in hospitals and other settings; also, it contributed to improving the laboratory's image.
- The commercially irrelevant radiation therapy quality assurance program provided medical benefits, contributing to improved quality of life.
- The tracer technology was used for testing purposes, making indirect contributions to commercial success (such as when a builder improves air circulation in its structures).

Rather than CRADAs and licenses, these cases involved technology transfer mechanisms such as standards-setting, scientific user facilities, technical assistance, and consulting. With these mechanisms, there appeared to be few accepted means to measure their contribution because of the distinctively intangible nature of the resulting successes. As long as the laboratories continue to use these albeit lesser-used mechanisms, the agencies need to acknowledge their inherently less-tangible measurement approaches as legitimate. Otherwise, the means for categorizing the successes is lost. Thus, the agencies could consider providing incentives for using (and systematically documenting) the lesser-used mechanisms, particularly technical assistance to state and local governments.

Accept Longer Measurement Periods

Although the findings showed that the time to market has decreased, they also confirmed the long-term nature of technology transfer relative to political administrations. This verifies that longer time frames are necessary to properly assess legislative impact in this area. This also has significant implications for the political tendency to change direction with each new administration.

Team With Partners to Measure, and Document in Agreements

As technology transfer is increasingly practiced with partners, its measurement and evaluation also needs to be a teaming effort. Laboratories and agencies cannot determine ultimate economic or societal impacts without input from their private partners. For example, bartering products for services or intellectual property, as was the situation in a couple of the cases, makes economic impact measurement difficult and complicated, and only the private partner can help to sort out the value of such trades. To assist this measurement process, the partnering agreement (eg., CRADA) could specify the precise type of information (ie., which measures will require quantification) to be disclosed by the partner, the time period, and how the information would be protected and reported (in the aggregate, individually, anonymously, etc.).

Take Care with Confidential Company Information

The study also suggests the laboratories should take care with confidential company financial and business information. A systematic series of cases is useful for research purposes such as this study, but the experience of implementing this study suggests that a retrospective case study approach would be an expensive means for the agencies to collect basic economic impact statistics in the aggregate.[15] In addition to the expense, another problem inherent to an individual case study format is that privately-held companies in extremely competitive markets are not willing to divulge sales revenues, certain financial data[16] and other proprietary information.[17] The point is that, with more CRADAs and similar public-private partnering, senstitive company information is often available to the in-house laboratory personnel working with the company. Laboratory personnel must be careful when handling this type of information, because if improperly revealed to one of the company's competitors, it could undermine the company's competitive position or result in the loss of intellectual property rights.

Avoid Double-Counting

The cases exhibited examples of both incremental and revolutionary applications of technology. Revolutionary applications are easier to track than incremental ones because they can be directly attributed to a particular source.[18] However, since incremental applications do occur, in tabulating government-wide technology transfer metrics, it is important to avoid the double counting that may result when agencies tally up the technology transfer results of their entire laboratory system. This points out the importance of an interagency coordination effort.

Document Failures

The fact that several of the spinoff businesses failed but were replaced by successful ones shows that it is important for the agencies and laboratories to document both successes and failures, thereby revealing the net result or reason for failure.[19] For example, the original spinoff company the gravity meter case employed twenty persons before going out of business; however, the two "replacement" spinoff companies employed ten people altogether. As another example, among the failed products and inactive technology licenses in the pre-legislation cases were those of the thermoplastic material case. The inventor thought one or two of the licenses had been rescinded (perhaps under march-in rights), but was not certain (and neither the technology transfer office nor the company knew). Since the government has rarely exercised march-in rights, this is another example of a failure from which we could learn, if the information were only available. Naturally, agencies will be reluctant to document failures because this may reveal program weaknesses. However, it is important to keep in mind that metrics are not just for program justification, but also for program improvement.

Ultimately, the call for standards in measuring technology transfer will become louder and may lead to the creation of a forum to sort out and build consensus upon the measurement and evaluation methods. This will allow comparisons among common programs and/or laboratories, contributing to a more rational budgeting process and improved communication. Over time, this would promote more efficient allocation of scarce budget dollars.

Notes

1 For example, the microwave oven scientist was motivated and enlivened by his interactions with the younger scientists at the partnering company whom he mentored. Similarly, in the artificial heart case there was a good deal of interconnectivity and mutual respect between the laboratory researchers and their CRADA partners.

2 For example, one researcher mentioned that he filled the laboratory's technology transfer position on a part-time basis which is not uncommon in smaller laboratories. In such situations, the individuals are more likely to have scientific skills than market analysis skills.

3 *Interagency Study of ORTA Organization and Operation and Lessons Learned Case Studies in Technology Transfer*, Federal Laboratory–Industry Interaction Working Group of the Federal Laboratory Consortium, DOE/METC-85/6019, May 1985.

4 In the past, for example, even if an agency granted authority to its laboratories to allow laboratory employees to do consulting, the individual laboratories did not necessarily permit it to happen. This was not only a disincentive to the researchers, but also a hindrance to the partnering companies that needed help.

5 Although it is probably less of a necessity for those policies to be consistent throughout the laboratory system.

6 The 1980s alginate herbicide/ pesticide case shows that laboratory researchers may need time to garner some experience in handling potential partners. The lead scientist in the 1990s microwave oven case addressed the potential disclosure problems involved in serving as marketing agents. Further, increased international activity and use of the internet means that scientists must be careful in the timing of publishing articles because this could mean the loss of *both* domestic and foreign rights.

7 Christopher D. McKinney, Oak Ridge National Laboratory, Presentation on Panel Session, "Why Can't We All Get Along?", Association of University Technology Managers 1998 Annual Meeting, February 27, 1998.

8 Enacted in 1992, initiated fiscal year 1994.

9 For example, in one case, it was implied that the guarantee of exclusive licenses offered to CRADA partners by the 1995 act hinders the laboratory's negotiating position in terms of agreed-upon royalty rates, which is counter-productive to the royalty incentives for laboratory researchers. The researcher in another case had some negative comments about CRADAs and the technology transfer

function, but asked not to have this documented on the record.

10 Although less common than *patent* licensing.

11 As noted earlier, the military and weapons laboratories are often cited in this area, yet the cases did not exemplify this problem. For example, in the post-legislation cases, the following laboratories and researchers aggressively marketed their technologies: the Air Force team funding the speech coder; the Navy light-stick researchers; and the DOE Oak Ridge weapons complex.

12 The other case involving state and local government work included one of the lightstick companies working with local high schools.

13 Not to be confused with the state and local government "market" that purchases high-tech equipment such as computers and police cars, albeit a disaggregated one with widely varying requirements.

14 Benchmarking can be referred to as the gathering of standardized measures across programs. It helps to identify best practices and program improvement opportunities.

15 Individual cases can also be useful for other purposes such as communication with politicians, technology marketing, or public relations, but for those uses the cases would not necessarily need common elements or formats.

16 For example, the capital investments and other leveraged outside investments that were discussed under the topic of funding *could* be used as additional indicators of success; however, the firms were not willing to share enough detail for this indicator to be useful.

17 Information on industrial activities, such as launching new products or agreements, is also sensitive company information. In two cases, the laboratory researchers said major business deals were "in the works" but couldn't be discussed until after official announcements. In both cases (the Oatrim case and one of the herbicide/ pesticide cases), the details were soon revealed in newsletter articles.

18 For example, the alginate-based herbicide/ pesticide case shows that, as incremental applications and variations on the base technology gradually unfold in other geographical areas and institutions, the original inventor begins to lose credit. This is unfortunate, and it is obvious why the inventor would be interested in tracking these variations and may have difficulties doing so.

19 Currently, there is a debate within the small business policy community about the accuracy of small firm statistics because it is charged that they only count firms starting up or in existence during a given year and do not include firms that cease to exist.

Appendix A
Interview Questionnaire

1. *Role of Laboratory Researchers and Other Personnel:* Please comment on the contribution and background of the involved laboratory personnel in terms of the technology transfer that occurred. Besides the person/team who received the award, were there other laboratory personnel involved? Is this a large group, in terms of the overall size of the laboratory? In terms of support to the transfer of the technology, could you compare your (research) activities with those of other parts of the laboratory (management support, help from the lab's Office of Research and Technology Applications, etc.)? In terms of the chief scientist and laboratory researchers involved in the transfer of the technology, what was the level of professional technology transfer experience before this case?
2. *Technology and Applications:* Please describe the technology and its applications.
3. *University Involvement:* If there was a university partner involved, are you aware of any university benefits such as student or faculty benefits due to university partnering?
4. *Funding, Financing:* Did the funding come solely from the laboratory or also from cost-sharing with a partner or another source? Was other outside funding or financing involved in the transfer of the technology? This could include, for example, SBIR funds or other government technology funding programs (Technology Reinvestment Project, Advanced Technology Program, etc.), corporate allocations or matching private capital, venture capital, public offerings, loans, or other sources.
5. *Intellectual Property:* What were the results of the formal invention disclosure in the laboratory? Did it lead to patents being granted (or applied for), the technology being licensed, and/or articles being published? If the technology was licensed, what were the terms (e.g., exclusive, non-exclusive, etc.)? What is the annual income in fees or royalties?
6. *Technology Transfer Mechanisms:* What mechanism(s) was used to transfer the technology? Possibilities include: publications or presentations at professional or trade association meetings, laboratory-sponsored

conferences, cooperative research/ CRADAs or other strategic alliances, licensing, work for others (reimbursable), use of facilities, technical assistance, informal collaborations, education and training, researcher exchange, etc.

7. *User Group:* What was the technology's intended primary and/or secondary user or user group? Was the technology applied commercially, or was it used to solve a state or local government problem or some other more specific application area?

8. *Barriers to Commercialization:* Were there any barriers to transfer that were particularly evident or troublesome? Also, are you aware of why such barriers existed?

9. *User Benefits/ Economic Impact/ Outcomes:* Are you aware of any benefits to the users resulting from the technology being developed and transferred? For example, an increase in any private company's market share in the way of product sales or jobs generated. Are you aware of any costs being cut or intangible gains (competitive information gained, unique patent/intellectual property position secured, etc.)?

10. *International Activity:* Was there any international activity on the technology? For example, overseas patents or sales.

11. *Government Gains:* Were there any unanticipated government gains (other than the directly-intended R&D results). For example, technological spinbacks to the laboratory or agency, R&D costs saved or avoided, resources shared, etc.

12. *Economic Development, Technical Assistance:* Did the transfer involve outside assistance (management, legal, etc.) such as from a state Small Business Development Center, NASA regional technology transfer center, or other university-based technical assistance center?

13. *Elapsed Time:* What was the time frame involved in transferring the technology? This can be stated in terms of the stage of development, in terms of specific years, etc. What was the technology's stage of development as technology marketing efforts were initiated—conceptual, technical feasibility, development, prototype, production, support, etc.?

14. *Other Factors:* Were there any other factors not discussed that would be important to understanding this transfer of technology?

Appendix B
All 1985-86 and 1992-93
Award Winners

Agricultural Research Service (Beltsville, Maryland) – Leadership in the development and introduction of a unique method of biological pest control which saved Agriculture $13 million in 1982 and at least that amount in each subsequent year.

Air Force Engineering and Services Center (Tyndall Air Force Base, Florida) – Success in aggregating the critical resources needed to significantly advance and transfer fire fighting agents, protective clothing and equipment for fire fighters in the United States.

Argonne National Laboratory (Argonne, Illinois) – Leadership and initiative in transferring to the private sector a new chemical extractant which has application in nuclear industry and in separations research.

Argonne National Laboratory – Development and transfer to private industry of the Synthesis of the Synthetic Metal Precursor.

Brookhaven National Laboratory (Upton, New York) – Developing and assisting in the commercialization of polymer concrete materials used by industry for producing construction materials; used by States and municipalities for bridge and highway repair; and used by the United States Air Force for runway repair under damp and cold conditions.

Brookhaven National Laboratory – Efforts in developing polymeric tape insulation for power cables as an American-produced alternative to paper/polymer laminated insulation (the currently dominant material) produced exclusively by the Japanese.

Forest Products Laboratory (Madison, Wisconsin) – Design and patenting of the Truss-Framed System for residential construction, and an aggressive technology transfer effort resulting in national acceptance by the public and private sectors.

Lawrence Livermore National Laboratory (Livermore, California) – Efforts in

285

facilitating collaboration with U.S. industry in the development of the Nova Laser System.

NASA Lewis Research Center (Cleveland, Ohio) – Development of PMR-15 polyimide matrix resins and transfer of the technology to advanced aerospace composite applications.

NASA Marshall Space Flight Center (Huntsville, Alabama) – Significant biomedical engineering contributions in the development of a precise Ocular Screening System used to detect eye abnormalities in children.

*National Institute of Standards and Technology (Gaithersburg, Maryland) – Outstanding achievement in creating and operating microcomputer-based electronic bulletin board systems that support technology transfer to individuals in industry and government.

*National Institute of Standards and Technology – Key efforts in establishing a national network of radiation therapy measurement calibration laboratories in cooperation with AAPM, government laboratories, and industry.

Naval Civil Engineering Laboratory (Port Hueneme, California) – Transfer of the large amount of data generated on sprayed polyurethane foam roof systems to state and local government and into the private sector using a variety of technology transfer procedures.

Naval Command Control Ocean Surveillance Center (San Diego, California) – Technical innovation in the printed circuit technology, and leadership in transferring this technology to industry.

*Naval Oceanographic Office (Stennis Space Center, Mississippi) – Conceiving and coordinating development by Louisiana State University of a prototype Expendable Bottom Penetrometer System for automatic sea bed classification and measurement of sediment shear strength.

Naval Underwater Systems Center (New London, Connecticut) – Tireless and creative efforts to assist state and local government in Rhode Island to implement cable television and other telecommunications systems.

Oak Ridge Institute for Science and Education (Oak Ridge, Tennessee) – Developing new technologies, disseminating information, and providing radiation dose assistance to government agencies, industry, laboratories and physicians.

Pacific Northwest Laboratory (Richland, Washington) – Transferring advanced decontamination techniques successfully to the commercial nuclear industry where they are in widespread use.

Pacific Northwest Laboratory – Transferring power plant dry cooling technology to the commercial power industry.

Sandia National Laboratories (Albuquerque, New Mexico) – Development of a process to increase the efficiency of solar cells using ion-beam hydrogenation, and transfer to the manufacturing industry.

Sandia National Laboratories – Development of a stabilized aqueous foam material, and transfer to the security system and fire fighting industries.

1986

*Agricultural Research Service (Beltsville, Maryland) – Effective dissemination of the alginate process technology which incorporated chemical and biological pesticides in granular formulations to industry and research organizations.

Agricultural Research Service – Transfer of science and technology resulting in the commercial production of lactose-reduced milk and milk products for lactose intolerant individuals.

Air Force Wright Laboratories (Wright Patterson Air Force Base, Ohio) – Leadership in developing and transferring state-of-the-art computer programs with very special application.

*Air Force Wright Laboratories – Establishing, demonstrating and transferring a major break-through in computer system integration manufacturing technology.

Ames Laboratory [DOE], Iowa State University (Ames, Iowa) – Development and commercialization of a new photoacoustic cell for Fourier transform infrared spectroscopy.

Argonne National Laboratory (Argonne, Illinois) – Developing a proliferation-resistant nuclear fuel for research reactors and transferring the fabrication technology to commercial fuel element producers.

Army Construction Engineering Research Laboratory (Champaign, Illinois) – Development and implementation of a comprehensive technology transfer plan for the U.S. Army Corps of Engineers' Repair, Evaluation, Maintenance, and Rehabilitation research program.

Army Construction Engineering Research Laboratory – Design and development of a powerful state-of-the-art numerical modeling system for predicting flow and sedimentation and for leadership in making the system available to practicing engineers.

Army Natick RD&E Center (Natick, Massachusetts) – Development and application of a unique system for the sensory and objective description and classification of edible fish species for use by government laboratories and the American Fishing Industry.

Army Natick Research, Development and Engineering Center – Pioneering efforts to establish and foster the ASTM Retort Pouch Subcommittee F-2.4 and for providing singular technical support to contractors on all aspects of packaging and packing for the Meal, Ready-to-Eat combat ration.

Army Close Combat Armaments Center (Dover, New Jersey) – Leadership in the technology transfer of new fracture analysis methods.

*Brookhaven National Laboratory (Upton, New York) – Initiative in developing and commercializing an air infiltration measurement system that will result in more comfortable homes and energy efficient buildings.

Energy Technology Engineering Center (Canoga Park, California) – Promoting technology transfer between academic and Federal laboratory sectors by suggesting graduate thesis topics of mutual interest to the Department of Energy and university research departments.

Federal Highway Administration (McLean, Virginia) – Outstanding effort in the transfer of traffic engineering technology to the public and private sectors at the national and international level.

Lawrence Livermore National Laboratory (Livermore, California) – Transfer of the LLNL-developed Cray Time Sharing System to firms in private industry seeking to provide less costly supercomputer environments.

Lawrence Livermore National Laboratory – Success in using cooperative agreements to transfer underground coal gasification technology to industry and foreign countries.

Los Alamos National Laboratory (Los Alamos, New Mexico) – Transfer of an optical keratoplasty device to Radtech, Inc.

*NASA Langley Research Center (Hampton, Virginia) – Development of Polyimidesulfone (PISO2), a thermoplastic, high modulus polymer, and transfer of production technology to the chemical, electronics and aerospace industries.

National Institute of Standards and Technology (Gaithersburg, Maryland) – Leadership and innovation in the development and transfer of computer codes and data to allow the rapid implementation of large scale enhanced oil recovery and supercritical solvent separation projects in the United States.

National Institute of Standards and Technology – Development of the Automated Manufacturing Research Facility and unusual creativity and innovation in transferring federally developed technology to industry, universities, and the general public.

Naval Civil Engineering Laboratory (Port Hueneme, California) – Creativity in the transfer of important technology on maintenance painting to state and local governments and the private sector.

Naval Command Control Ocean Surveillance Center (San Diego, California) – Outstanding performance while providing initiative, management and leadership in the area of non-metallic materials technology transfer.

Naval Research Laboratory (Washington, D.C.) – Establishment of the Liquid Encapsulated Czoohralski (LEC) gallium arsenide growth and wafer finishing technology.

Naval Surface Warfare Center; Carderock Division (Bethesda, Maryland) – Distinguished leadership in the development of chemically controlled release technology and the subsequent transfer and adoption of this technology worldwide.

Oak Ridge Associated Universities (Oak Ridge, Tennessee) – Developing new treatment techniques, providing technical assistance, and disseminating information to government agencies, industry, and hospitals to improve the handling of radiation accidents.

Oak Ridge National Laboratory (Oak Ridge, Tennessee) – Contributions toward the commercialization of the fiber optics luminoscope resulting in a licensing agreement with Environmental Systems Corporation of Knoxville, Tennessee.

Oak Ridge National Laboratory – Noteworthy contributions toward the commercialization of aluminide alloys and, specifically, toward the achievement of a licensing agreement with Cummins Engine Company, Inc.

*Pacific Northwest Laboratory (Richland, Washington) – Initiative and uncommon creativity in transferring a biobarrier technology into products that prevent unwanted root growth and vegetation from roadways, septic tanks, sewer lines, buried gas lines, and irrigation heads.

Pacific Northwest Laboratory – Initiative and uncommon creativity in transferring the "SAFT-UT Signal Processing for High Resolution Imaging" to private industry for its application at electric utilities.

1992

*Agricultural Research Service, National Center for Agricultural Utilization Research (Peoria, Illinois) – Developing and successfully transferring Oatrim technology thereby providing a new fat substitute which improves health by lowering blood cholesterol.

Air Force Armstrong Laboratory (Dayton, Ohio) – Transferring technology from DOD to the private sector involving local industries. A joint program involving DOD, industry, and a university was developed.

Air Force Armstrong Laboratory (Brooks Air Force Base, Texas) – Demonstrating uncommon creativity and initiative in the transfer of technology and winning national recognition of an unprecedented technology in a very short time.

Air Force Phillips Laboratory (Kirtland Air Force Base, New Mexico) – Contribution to new forms of advanced electronics packaging for applications inside and outside of the military.

Argonne National Laboratory (Argonne, Illinois) – Transfer of neutron diffraction techniques which have been developed to determine strains and stresses in engineering composites for validation of analytical models and optimization of fabrication procedures.

Army Construction Engineering Research Laboratory (Champaign, Illinois) – Completing the transfer of ceramic coated anode technology through publication of the design guidance contained in the engineering technical letter.

*Forest Products Laboratory (Madison, Wisconsin) – Outstanding efforts in technology transfer and paper industry implementation of the paper restraint for paper testing.

Forest Service, Northeast State and Private Forestry (Radnor, Pennsylvania) – Identifying the ready-to-assemble furniture market for wood products, developing a business plan, and transferring technology leading to increased employment and economic development.

*Idaho National Engineering Laboratory (Idaho Falls, Idaho) – Outstanding achievement in the development and successful implementation of cooperative research and development agreements for transfer of rapid solidification technology.

Jet Propulsion Laboratory (Pasadena, California) – Excellence in technology transfer using the unique and innovative process designed into the JPL/NASA technology affiliates program.

*Lawrence Berkeley Laboratory (Berkeley, California) – Development and transfer of a highly sensitive device, the squid magnetometer, using the new high temperature superconducting materials for medical and geophysical applications.

Lawrence Berkeley Laboratory – Transferring the LBL Phosnox process for combined removal of SO_2 and NOX from flue gas.

Lawrence Berkeley Laboratory – Synchrotron-radiation research depends upon optics formed into complicated shapes with unprecedented precision, and the laboratory researcher and industry collaborators made several breakthroughs in manufacturing and characterizing these devices.

Lawrence Livermore National Laboratory (Livermore, California) – Transfer of electrochemical plannerization and electropolishing technology to a computer manufacturer.

*Lawrence Livermore National Laboratory – Outstanding development and transfer of state-of-the-art solid state laser technology to Hampshire Instruments for use in the HI model 3500 X-ray lithography system.

Los Alamos National Laboratory (Los Alamos, New Mexico) – Transferring the rapid DNA sequencing technology to industry, resulting in a patent issued, a patent applied for, and a CRADA with Life Technologies, Inc.

Los Alamos National Laboratory – Transferring the technology of their optical high-acidity sensor to industry which may result in the transfer being completed in one year's time.

*National Institute of Standards and Technology; Boulder Laboratories (Boulder, Colorado) – Transfer of optical fiber current sensor technology from NIST to the 3M Company.

*National Institute of Standards and Technology; Boulder Laboratories – Transferring the fundamental optical, electronic, and mechanical technology in the JILA absolute gravity measurement device to Axis Instruments Company.

National Institutes of Health; National Heart, Lung and Blood Institute (Bethesda, Maryland) – Pioneering research and development that has brought NIH technology from the theoretical realms of the laboratory to clinical applications.

National Oceanic and Atmospheric Administration; Environmental Research Laboratory (Boulder, Colorado) – Initiative and creativity in transferring new static pressure probe technology to the private and public sectors, making a significant contribution to aircraft safety.

Naval Command Control Ocean Surveillance Center (San Diego, California) – Persistence, dedication and marketing effort to develop and transition the thin film silicon on sapphire (TFSOS) technology which could impact the fabrication of advanced microelectronics products of the late 1990s.

*Naval Research Laboratory (Washington, D.C.) – Exceptional creativity and initiative in the effective technology transfer of phthalonitrile monomer/prepolymer technology for a broad spectrum of applications.

Oak Ridge National Laboratory (Oak Ridge, Tennessee) – Significant contributions to the invention, development, licensing, and commercialization of the Cl₂EAN OUT™ process for dechlorination of waste streams.

Pacific Northwest Laboratory (Richland, Washington) – Determination in transferring the portable blood irradiator technology to an international center for treatment of leukemia and blood diseases, where it can help save lives.

Pacific Northwest Laboratory – Vision and persistence in transferring the electro-optic liquid soil sensor, a simple, inexpensive device with environmental and agricultural applications.

Pacific Northwest Laboratory – Dedicated and innovative technology transfer of the tempest software for three-dimensional transient hydrothermal analysis.

*Pittsburgh Energy Technology Center (Pittsburgh, Pennsylvania) – Outstanding efforts in transferring an advanced flow diagnostic technique developed for fossil fuels to assist the medical community in improving artificial heart pumps.

*Sandia National Laboratories (Albuquerque, New Mexico) – Combining the vision, industry needs, and laboratory resources into a microelectronics quality/reliability center to transfer quality technologies to the IC industry.

1993

Agricultural Research Service (Beltsville, Maryland) – Transfer of technology in development of the first commercial in virtobaculovirus biological pesticide system.

Agricultural Research Service, Southern Regional Research Center (New Orleans, Louisiana) – Effective international dissemination and commercial licensing of a new core-spinning technology for producing unique composite yarns of predominantly cotton content and almost 100% cotton surface.

Air Force Rome Laboratory (Griffiss Air Force Base, New York) – Creativity and initiative in the transfer of Lanthanum Hexaboride thin film coating technology for use in improving efficiency of fluorescent lamps and X-ray medical equipment.

*Air Force Rome Laboratory (Hanscom Air Force Base, Massachusetts) – Initiative in the transfer of robust speech compression technique designed to provide reliable, high quality speech communications while increasing the capacity of narrowband channels.

Air Force Wright Laboratory (Wright-Patterson Air Force Base, Ohio) – Initiative in transferring "smart dipstick" technology for measuring remaining useful lubricant life to industry and implementing broader applications in transportation and food processing industries.

*Ames Laboratory [DOE], Iowa State University (Ames, Iowa) – Unusual devotion and effort in transferring a new laser-based method for indirect fluorescence of biological samples.

*Army Construction Engineering Research Laboratory (Champaign, Illinois) – Initiative in the development of support structures and mechanisms to transfer Geographic Resources Analysis Support System (GRASS) technologies to government, private sector, and educational institutions throughout the U.S. and around the world.

Army Research Institute for the Behavioral and Social Sciences (Alexandria, Virginia) – Exemplary activities in transferring the Job Skills Education Program. A commercial version of this Army-developed program could help millions of American workers improve their job performance.

Forest Service, Forest Products Laboratory (Madison, Wisconsin) – Development and rapid adoption of economical and innovative techniques to improve the quality and reliability of structural wood products.

Forest Service, Pacific Northwest Experiment Station (Portland, Oregon) – Transfer of the IMPLAN system (economic impact assessment technology) to federal, state and local government agencies, academic institutions and private businesses.

Forest Service (Ft. Collins, Colorado) – Transfer of the IMPLAN system (economic impact assessment technology) to federal, state and local government agencies, academic institutions and private businesses.

Lawrence Berkeley Laboratory (Berkeley, California) – Development and transfer to the U.S. building industry of the technology base for "superwindows," windows with better thermal performance than insulating walls.

Lawrence Berkeley Laboratory – Development of a new polymeric material which can significantly extend the active lifetime of enzymes and allow their use in harsh industrial environments.

Los Alamos National Laboratory (Los Alamos, New Mexico) – Transfer of the KIVA software, which has lead to the widespread use of the technology by U.S. engine manufacturers such as General Motors, Ford, Chrysler and Cummins Engine Company.

*Los Alamos National Laboratory – Transfer of the resonant ultrasound inspection technology, which resulted in a license agreement between Los

Alamos National Laboratory and Quatro Corporation, and in a product that is now being marketed.

NASA Lewis Research Center (Cleveland, Ohio) – Transfer to the aerospace and construction industries the first comprehensive Fastener Design Manual directed toward the design engineer.

National Institute for Occupational Safety and Health (Cincinnati, Ohio) – Reducing lead exposure in radiator repair shops through effective technology transfer.

National Institutes of Health, National Institute on Aging (Bethesda, Maryland) – Pioneering research and development and an unsurpassed commitment to transferring NIH/NIA technology to benefit mankind.

National Oceanic and Atmospheric Administration, Environmental Research Laboratories (Boulder, Colorado) – Creativity transferring the 915 MHz profiler technology to the private sector through use of a CRADA.

*Naval Air Warfare Center, Weapons Division (China Lake, California) – Efforts in transferring technology from DOD to the private sector resulting in two license agreements.

*Naval Civil Engineering Laboratory (Port Hueneme, California) – Outstanding dedication and initiative in transferring paints and coatings expertise to the private sector via direct technical consultation and assistance.

*Oak Ridge National Laboratory (Oak Ridge, Tennessee) – Significant contributions to the invention, development, licensing and commercialization of the Variable Frequency Microwave Furnace.

Pacific Northwest Laboratory (Richland, Washington) – Use of the CRADA mechanism to revitalize the licensee's interest in commercializing the glycine nitrate process.

Pacific Northwest Laboratory – Transferring MEPAS© environmental assessment software to Mesa State College, thereby improving Mesa's curriculum, training workers for DOE, and enhancing MEPAS' marketability.

Pacific Northwest Laboratory – Personal effort and innovation in forming a new company through an alliance with an existing business to transfer and commercialize Waste Acid Recovery Systems.

Pacific Northwest Laboratory – Insight, initiative and persistence in trailblazing the rapid technology transfer of ReOpt software, the first scientific approach to identifying technologies for waste cleanup.

Pittsburgh Energy Technology Center (Pittsburgh, Pennsylvania) – Leadership, creativity, and initiative in effecting the first transfer of a process patented by a DOE Energy Center to private industry.

Sandia National Laboratories (Albuquerque, New Mexico) – Initiative in the transfer of technology from Sandia's Mesh Generation Consortium which is providing U.S. industry with engineering software to reduce the time required for design iterations using advanced mesh generation algorithms and adaptive analysis techniques.

*Documented in *Winners in Technology Transfer: Success Stories from the Federal Laboratory Consortium*, Washington, DC: Federal Laboratory Consortium, Special Reports Series No. 2, ISSN: 1075-9492, August 1994.

Appendix C
Interviewees

Penetrometer for Seabed Classification/Measurement

Mr. Carey Ingram, Supervisory Oceanographer
Special Support Division
Naval Oceanographic Office
Stennis Space Center, Mississippi

Dr. Joseph N. Suhayda, Professor
Civil Engineering Department
Louisiana State University
Baton Rouge, Louisiana

Advanced Thermoplastic Polymer Material

Dr. Terry St. Clair
NASA Langley Research Center
Hampton, Virginia

Ms. Rosa Webster
Technology Transfer Office
NASA Langley Research Center
Hampton, Virginia

Mr. Milton Evans, President
High Technology Systems, Inc.
Clifton Park, New York

Substance Tracer Technology

Dr. Russell N. Dietz, Head
Tracer Technology Center
Brookhaven National Laboratory
Upton, New York

Slow-Release, Alginate-Based Herbicide/ Pesticide

Mr. William J. Connick, Jr.,
Research Chemist
Southern Regional Research Center
Agricultural Research Service
New Orleans, Louisiana

Dr. Ramon Georgis
Biosys, Inc.
Columbia, Maryland

Mr. James Walter, Director
Research and Development
Thermo Trilogy Corporation
Columbia, Maryland

Controlled-Release, Chemically-Imbedded Herbicide/ Pesticide Material

Dr. Peter Van Voris
Pacific Northwest National
Laboratory
Richland, Washington

Mr. Harry E. Barnes, Biobarrier
Manager
Reemay, Inc.
Old Hickory, Tennessee

Mr. Rodney Ruskin, CEO
Geoflow, Inc.
Sausalito, California

Radiation Therapy Quality Assurance

Dr. Robert Loevinger
Radiation Physics
National Institute of Standards
and Technology
Gaithersburg, Maryland

Dr. Geoffrey S. Ibbott, Asst.
Professor and Director
Department of Radiation
Medicine
University of Kentucky
Lexington, Kentucky

Laser-Based Method to Light Up Biological Samples

Dr. Edward S. Yeung, Program
Director
U.S. Department of Energy –
Ames Laboratory
Iowa State University
Ames, Iowa

Mr. Craig Ranger, President
Lachat Instruments, Inc.
Milwaukee, Wisconsin

Voice Coder for Telecommunications

Mr. Luigi Spagnuolo, Acting
Branch Chief
INFOSEC Technology Office
Electromagnetics and Reliability
Branch
Rome Laboratory
Hanscom Air Force Base,
Massachusetts

Dr. John C. Hardwick
Digital Voice Systems, Inc.
Burlington, Massachusetts

Paper Quality Tester

Mr. Theodore L. Laufenberg
Forest Products Laboratory
U.S. Department of Agriculture
Madison, Wisconsin

Mr. Peter Davis
Isthmus Engineering and
Manufacturing Co-op
Madison, Wisconsin

Variable-Frequency Microwave Oven

Dr. Robert J. Lauf, Senior
Development Staff Member
Metals and Ceramics Division
Oak Ridge National Laboratory
Oak Ridge, Tennessee

Mr. Don W. Bible, Development
Staff Member
Instrumentation and Controls
Division
Oak Ridge National Laboratory
Oak Ridge, Tennessee

Mr. Richard S. Gerard, President
and CEO
Lambda Technologies, Inc.
Raleigh, North Carolina

Gravity Meter

Dr. James Faller
Joint Institute for Laboratory
Astrophysics
University of Colorado/National
Institute of Standards and
Technology
Boulder, Colorado

Dr. Steve ONeil, Director of
Technology Transfer & Industry
Outreach
University of Colorado
Joint Institute for Laboratory
Astrophysics – Industry Liaison
Boulder, Colorado

Dr. Tim Niebauer, President
Micro-g Solutions
Erie, Colorado

Dr. Mike Winters, President
Winters Electro-Optics
Longmont, Colorado

"Oatrim" Fat Substitute

Dr. George E. Inglett, Biopolymer
Materials Specialist
National Center for Agricultural
Utilization Research
Agricultural Research Service
Peoria, Illinois

Mr. Stephen B. Grisamore,
General Manager
Mountain Lake Specialty
Ingredients Company
Omaha, Nebraska

Mr. Mark Freeland, Director of
Textural Technologies
Rhone-Poulenc, Inc.
Cranbury, New Jersey

Mr. Lanny Babbitt
Quaker Oats Company
Chicago, Illinois

Chemiluminescent "Light Sticks"

Dr. Herbert Richter, Supervisory
Research Chemist
Naval Warfare Center – Weapons
Division
China Lake, California

Ms. Martha Harrington
Technology Transfer Office
Naval Warfare Center – Weapons
Division
China Lake, California

Mr. Fred Kaplan, President and
CEO
Omniglow Corporation
Novato, California

Artificial Heart Flow Diagnostics

Mr. Franklin D. Shaffer,
Mechanical Engineer
Pittsburgh Energy Technology
Center
Pittsburgh, Pennsylvania

Dr. Harvey Borovetz, Professor of
Surgery & Director of Biomedical
Engineering
Presbyterian–University Hospital

Department of Surgery/School of
Medicine
University of Pittsburgh
Pittsburgh, Pennsylvania

Appendix D
Pre- and Post-legislation Summaries

Pre-legislation Findings Summary

This section of the appendix summarizes the findings for the selected pre-legislation cases in Chapter 3 organized according to the questions addressed in the interviews. The cases appear in the following order:
a. Penetrometer
b. Thermoplastic Polymer
c. Substance Tracer
d. Alginate Herbicide
e. Root-Control Barrier
f. Radiation Measurement Standards.

Role of Laboratory Researchers and Other Personnel

a. For the river bottom penetrometer, Mr. Ingram worked with state and local government officials on a county problem, serving as lead scientist for a team of intergovernmental personnel surveying the river bottom. He developed the concept for a new surveying tool and consulted with a nearby university as they developed the prototype.

b. Dr. St. Clair, who invented the thermoplastic material, developed both the material and a process for producing it. Then he proactively sought industry partners to further develop and manufacture it by speaking at a number of workshops and conferences co-sponsored by industry trade associations, particularly aerospace-related ones.

c. Dr. Dietz at Brookhaven demonstrated the tracer technology to a wide range of potential user communities both in this country and overseas, and even produced a video showing the technology's capabilities. Dr. Dietz has worked with all types of user groups; in fact, his tracer technology center performs tracer services for some fifteen to twenty user groups each year. He has also conducted experiments with other federal agencies, such as EPA and NASA, and with the Commission of European Communities. In addition, he has written scientific papers and spoken at conferences.

d.   Mr. Connick, inventor of the alginate process, was sought out by others, rather than undertaking proactive technology transfer approaches. He published a number of scientific papers and made presentations over the years that "caught on" in the scientific community. His work has been cited by other scientists nationally and internationally.

e.   Upon coming up with their idea serendipitously at a social bridge game, the team of PNNL researchers working on the root-control technology tested a variety of chemicals and materials for the applications they had in mind. For one of the applications, they designed and constructed the prototype themselves; for two other applications, they worked with private companies involved in those product lines. They wrote a number of scientific papers and teamed with private sector partners for a number of these papers.

f.   For the radiation therapy quality assurance case, Dr. Loevinger at NIST proposed to the American Association of Physicists in Medicine (AAPM) that national methods were needed in the area of ensuring radiation therapy dosage measurements. He also wrote scientific papers in this area.

Technologies and Applications

a.   The river bottom penetrometer was a sophisticated instrument developed by combining parts from different off-the-shelf instruments with a computer and printer to create a new river surveying instrument.

b.   The PISO2 thermoplastic material is advanced in its thermal properties. It would be relatively inexpensive to mass-produce and would have a variety of potential uses in industry and space programs if commercialized.

c.   Part of the interest of the perfluorocarbon tracer technology is that it has so many applications. The technology's commercial possibilities include: building ventilation analysis, underground cable leak detection, utility applications, underground storage tank leak identification, explosives detection, petroleum reservoir analysis, pre-fire warning, environmental monitoring, and disaster emergency management. The instrumentation involved in tracer systems could also be commercialized.

d.   The alginate-based herbicide/ pesticide is unique because it can incorporate either chemical or biological control agents that attack only their intended targets and do not affect their surrounding environments. Also, they work slowly over a period of time, rather than having to be re-applied. The alginate technology can be applied to a variety of "undesirables": plant diseases like root rot, unwanted water and soil-based weeds, young insects and other pests, and fungi that attack grains and crops. It is being used in bioremediation of toxic chemicals underground.

e. The PNNL chemical slow-release technology fights plant roots extending down into radioactive waste sites. In the commercial arena, the same technology can be applied to control plant roots ruining underground watering systems, tree roots "uprooting" sidewalks, weeds overtaking gardens and landscaped areas, or roots intruding into sewer pipes. Longer-term applications being developed include insect and rodent control and protecting decaying telephone poles, railroad ties, and buried power and gas lines.

f. In the case of NIST establishing a radiation therapy quality assurance program, the technology that was transferred was the traceability of x-ray dosimetry (measurement) systems to national and international standards. Before this was transferred to other sources in this country, these measurements were performed in a "vague, uncontrolled, and amateurish" way[1] in many institutions. Although NIST performed calibrations for some institutions, there was no systematic process in place for other institutions to be calibrated. With this technology, these other institutions can be calibrated by other organizations besides NIST. The calibrations have traceability to NIST for ensuring their accuracy.

The Laboratories

a. The laboratory for the penetrometer was the Naval Oceanographic Office located with other agencies at the Stennis Space Center. Stennis is known for its outreach to local governments in the state of Mississippi. Stennis was one of the earliest FLC members to initiate a formal technical assistance project with the counties in the state. With this type of reputation, it is not surprising that the substance of the case had to do with helping a local economic development organization.

b. The thermoplastic material was developed at the NASA Langley field center which had an interest in this material for aircraft structures.

c. The tracer technology was developed at Brookhaven National Laboratory (BNL). Although the technology has not materialized as a commercial venture in any of its application areas, it is consistent with BNL's culture to provide this service to outside users on a fee per service basis. BNL's culture is based on a reputation for providing access to its unique facilities for both proprietary work as well as basic research.

d. The alginate-based herbicide/ pesticide started at the USDA Crop Protection Research Laboratory in New Orleans, but eventually it branched to a variety of other USDA and university laboratories. The federal laboratories included the USDA Southern Weed Science Laboratory in Mississippi, the USDA Aquatic Weed Research Laboratory in Florida, and the Agricultural Research Service headquarters in Beltsville, Maryland's Biological Control of Plant Diseases group. Ultimately, the technology involved the USDA Subtropical Agricultural Research Laboratory in Texas and joint work with some university laboratories.

    e.   The technology involving chemically-imbedded herbicide/ pesticide materials was invented at Pacific Northwest National Laboratory (PNNL). Although the technology was tested in Colorado, it was intended to help solve the problem posed by the laboratory's next door neighbor, the huge DOE Hanford Site. The technology was needed to control underground plant roots threatening major underground storage tanks full of nuclear wastes.

    f.   The radiation therapy quality assurance technology was conceived by a scientist at NIST. NIST was the perfect agency for initiating such an activity. Its missions include assisting associations and groups to establish standards and funding research related to standards. NIST also provides calibration services linking customers' equipment to national standards.

University Involvement

    a.   For the penetrometer technology, the Naval Oceanographic Office contracted with a nearby university to develop a prototype and test the new instrument. The university, in turn, worked with a company intending to commercialize the technology.

    b.   In the case of the thermoplastic material, the small firm obtaining the NASA SBIR contract conducted the R&D jointly with a professor at a nearby university.

    c.   Universities laboratories that do environmental testing are occasionally customers for BNL's tracer technology-based service.

    d.   Universities have utilized the alginate-based process for a great deal of laboratory testing work because it is suited for obtaining accurate results in that type of work. In addition to independent university research, Mr. Connick has been conducting joint research with the University of Arkansas which has resulted in a joint patent and publications. Also, USDA's company partners in this area have used a number of university researchers as consultants to test their products.

    e.   There was no university involvement in the root control technology.

    f.   For the NIST radiation therapy quality assurance program, some of the hospitals in the national quality assurance system were university hospitals.

Funding, Financing

a. Naval Oceanographic Office funds supported the conceptualization, prototype/testing contract, and procurement contract for the prototype copies.

b. The work at NASA Langley on the thermoplastic material was funded through NASA. There was little interaction between the NASA inventor and the private-sector licensees. Laboratory funds were not involved with commercialization with one exception: the small firm still working on further developing the technology received NASA SBIR funding and state R&D funding. However, the firm does not have any outside funding for its work on the technology.

c. The tracer technology demonstrations, tests, and experiments have been, for the most part, jointly funded by the organizations involved in them. BNL covered Dr. Dietz' time and that of other BNL researchers involved. These projects are viewed as a way to prove the feasibility of the technologies resulting from the R&D being performed at the laboratory.

d. All of the work on the alginate-based technology at the USDA centers and laboratories was covered by USDA funds. In addition, the USDA Beltsville center provided cooperative research funding to the Grace-Sierra Crop Protection Company for its work in this area.

e. Initial funding for the chemically imbedded herbicide/ pesticide material was provided by PNNL, Rockwell International (the contracting operator for the nearby Hanford Site), and the Office of Nuclear Energy at DOE headquarters. The PNNL research team currently devotes only five percent of its time to this project, which is funded by companies and military services interested in the pest control applications.

f. In the radiation therapy quality assurance case, Dr. Loevinger's work with the AAPM committee was considered part of his NIST job responsibilities, and therefore was covered by his NIST salary. NIST's charge for its primary level, national standards calibrations for radiation therapy instruments averages $500. Each of the five certified secondary level regional calibration laboratories are voluntary and not federally subsidized. They remain self-sufficient or near break-even by charging for the calibration services they provide to tertiary-level institutions involved in radiation therapy. (There are other benefits of being certified as a regional calibration laboratory, but they are more intangible.)

Intellectual Property

a. The Navy chose not to apply for a patent for the two versions of the penetrometer.

b. Two patents were jointly filed by NASA and MIT for the thermoplastic material, with the inventors being the NASA scientist and an MIT graduate student.

c. The only application area that was patented for the tracer technology was the pre-fire warning system. Dr. Dietz commented that, until recently, the laboratory researchers tended not to patent.

d. The alginate technology has been patented for many applications: two patents for use with chemical herbicides, two for biological control of plant diseases, two for weed control, and one for bioremediation. USDA has applied for a patent for its use with nematodes. In the meantime, Biosys has developed proprietary knowledge on its formulation and production. The company has a patent application in process in the United States and patent applications filed in four other countries.

e. For the chemically imbedded material, there have been no less than seven invention disclosures at the laboratory and seven patent applications filed by DOE. In addition, the partnering companies have filed patents.

f. This topic was not applicable to the radiation calibration services.

Technology Transfer Mechanisms

a. The transfer mechanism with the Navy's penetrometer was the contract with the university for developing the prototype.

b. NASA licensed its thermoplastic material to two companies. The small company with the SBIR contracts also has ownership rights to the technology through the Bayh-Dole Act.

c. A variety of mechanisms have been involved in the transfer the tracer technology from BNL to outside users. These mechanisms have included such diverse means as:

- BNL sale of equipment with an exclusive license to a trade association,
- A CRADA with a small ventilation company,
- Successful early demonstrations and subsequent transfer of a proprietary knowledge-based system to a public/private electric power consortia which offers the service commercially,
- Unsuccessful joint tests between BNL and a utility,
- Successful joint tests between BNL and a commercial laboratory,

- The loaning of equipment to an instrument company,
- A collaborative BNL/company procurement project,
- Fee-for-service provision to the petroleum industry, and
- Traditional marketing of a patent license for a pre-fire warning system.

d. For the alginate-based herbicide/ pesticide case, in the late 1980s the cooperative research funding USDA provided to Grace-Sierra evolved into one or more CRADAs, since the company was interested in protecting its intellectual property rights. Because there were company changes, it is difficult to sort out the point when the various CRADAs started and ended, and how they corresponded to the company takeovers. Biosys, Inc. did not have a formal collaborative arrangement with Mr. Connick's center, but eventually signed a CRADA with one of the USDA laboratories in Texas. Mycogen Corporation signed an exclusive license with USDA for all the alginate-based patent applications.

e. For the chemically imbedded material, Agrifim Irrigation International obtained an exclusive worldwide license from PNNL's contracting operator, Battelle. Agrifim sub-licensed the technology to Geoflow™ Subsurface Irrigation, an Agrifim division, for the production of underground watering systems. In addition, Agrifim sub-licensed the technology to Torro for use in termite control products. Reemay, Inc. signed an exclusive license to manufacture a geotextile fabric containing herbicide pellets. Mantaline Corporation obtained a license to manufacture sewer line gaskets with the technology.

f. For the radiation therapy quality assurance case, as a result of the NIST scientist's recommendation to the AAPM a national system was established to ensure more accurate radiation therapy dosage measurements, including a task force and a permanent subcommittee. The subcommittee chose a system of three, and later five, regional calibration laboratories. These five laboratories have their calibration equipment calibrated directly by NIST and provide a secondary standards level calibration to other organizations.

User Groups

a. The penetrometer was unusual in that the user group was strictly state and local government officials. Had the penetrometer instrument become commercialized, there would have been other user groups.

b. The user groups for the thermoplastic material are companies from aerospace, electronics, and other sectors.

c. The users of the tracer technology include power companies, hospitals, trade associations, university and commercial testing laboratories, instrumentation companies, the petroleum industry, railroads, other federal agencies, and similar groups overseas.

d. The users of the alginate-based herbicide/ pesticide include those involved in plant and crop diseases, weed control, or insect infestation (e.g., farmers, professional greenhouses, nurseries, landscape firms, even homeowners).

e. The users of the chemically imbedded herbicide/ pesticide material overlap the users of the alginate-based herbicide/ pesticide. They include farmers, municipalities, facilities maintenance companies and, again, homeowners.

f. Users of the NIST radiation therapy quality assurance system would be those institutions offering radiation therapy for diseases such as cancer. The end users of the equipment are the patients undergoing such treatments, about 600,000 in this country.

Barriers to Commercialization

a. At the time of the penetrometer development, Mr. Ingram implied[2] that Navy researchers did not receive much in the way of incentives for transferring technologies so it was not worth pushing through the system. Mr. Ingram did note that in recent years, however, CRADAs have made it easier to transfer technologies.

b. Part of the barriers to commercializing NASA's thermoplastic material have revolved around corporate changes and re-directions in the two original licensees. The third company with current rights to the technology is a small firm with a lack of corporate resources for commercialization and the wherewithal to compete against products manufactured by General Electric, DuPont, and other larger players in the materials markets.

c. The level of demand for tracer technology in the application areas noted above apparently is not large enough to support an entire business. A commercial service based upon the technology appears to succeed only as a sideline business for smaller companies. There are mixed reviews about the success of its being offered through associations or consortia of companies.

d. The major barrier to commercializing the alginate-based herbicide/ pesticide involves the costs associated with scaling up from laboratory and market testing to full-scale manufacturing levels. The raw materials, including the alginate/clay mixture, are expensive and, at higher levels of production, the product is more labor-intensive to produce and requires more quality control.

e. In terms of the chemically imbedded herbicide/ pesticide material, there are some inherent marketing problems being faced by the partnering companies. First, there seems to be a bias against products that work long-term. Commercial ventures prefer throw-away products that only last for a short period of time, so it is difficult to market a product that lasts for two years or longer. Second, traditional chemical pesticide treatments involve larger quantities because they must be re-applied often. On the other hand, slow-release or controlled-release technologies are sold in smaller quantities. This requires a shift in thinking for consumers and distributors alike. Third, when a new product is introduced that doesn't replace an old product, it is difficult to create product visibility or to create a market. Also, the raw materials going into the end product are expensive, so that value must be sold to the ultimate consumer.

f. There were no commercialization barriers in the radiation standards case.

User Benefits/Economic Impact/Outcomes

a. The penetrometer prototype was used to perform the originally intended river survey work; in fact, this was how the prototype was field tested. Only six prototype clones were produced by Sippican Corporation so there was little overall impact on the economy with this technology.

b. Both of the original NASA licenses for the thermoplastic material are dead because of corporate re-structuring and new business strategies. In NASA's current database of available technologies, the material is stated as accessible through the NASA/SBIR firm that is still doing development work. The product has not been popular, however.

c. The National Association of Home Builders' tracer technology service did not succeed in the long run and was discontinued. It is now being offered (along with other related services) by a small environmental service company using BNL for the analysis portion of the work. Provision of on-line tracer services by the private ventilation service company did not succeed, and the company now offers it on an as-needed basis along with other services. It is not known what the monetary benefits to the Electric Power Research Institute have been for its provision of the underground leak detection service to its member electric power companies; however, the service is still being offered. The demonstrations show that this leak detection method is less costly than traditional methods. The other miscellaneous utility testing work was not as successful as anticipated and was not pursued. Tests on the underground storage tanks were very successful, but the user company decided not to adopt this line of business. The explosives detection instrumentation development has been held up by the necessity for a Nuclear Regulatory Commission license, although it is still being pursued. Another company's joint procurement collaboration with BNL became inactive once the project was completed; although the company still markets certain tracer technology instruments and/or services along with its other lines of business. The petroleum reservoir analysis service is being provided by BNL, only.

   d.   For the alginate-based herbicide/ pesticide, Grace-Sierra test marketed GlioGard™, the first bio-fungicide on the market, for about two years. During this phase, the company sold thousands of pounds and received favorable feedback; however, they ran into problems when scaling up, so they changed the formulation and the product's name to SoilGard™. About that time the Grace Company sold several products (including SoilGard) to Thermo Trilogy Corporation. Thermo trilogy is re-negotiating ownership rights with USDA. Biosys, Inc. manufactured a product called BioSafe[R] that was marketed for about six years by Ortho. Ortho eventually sold its retail line to Monsanto, which subsequently cancelled most of its products (including BioSafe). Meanwhile, Biosys signed a related CRADA with a USDA laboratory in Texas that evolved into a new line of three products for the company: Vector[R], Lesco™ Vector[R], and BioVector[R]. The company's market share increased from 27 to 84 percent based upon two of the products, and the third product is reportedly doing well in its first year of introduction. EcoScience Corporation temporarily had a product called Aqua-Fyte™, which it field tested through an EPA Experimental Use Permit. But the company has been undergoing financial difficulties and doesn't produce it any more. After a number of years of keeping up its license, Mycogen Corporation terminated its license in 1993 without having commercialized any products.

   e.   Regarding the PNNL chemically imbedded herbicide/ pesticide material, Agrifim's division Geoflow installed its Rootguard[R] products on at least fifty agricultural sites, 28 landscape sites, and fifteen turfgrass sites. In addition, Geoflow's Wasteflow™ systems have been installed in at least five sites. Reemay, Inc. manufactures two products: Biobarrier[R] for root control, and Biobarrier[R] II for weed control. Although the company assumed the main market for its root control product would be DOE nuclear waste sites, it found that these DOE sites are very independent. A license from one site does not imply an inside track with other sites. However, now Reemay is actively selling its product to municipalities for public works applications and is finding the market receptive. The company cites at least five examples of cities and counties that have used the product. The Mantaline Corporation license is now dead for unexplained reasons.

f. The NIST radiation therapy quality assurance system does not involve commercialized products; instead, an important medical technology was transferred. There was one early study of the new system that determined it was working well. Anecdotal evidence indicates that the system was sorely needed at the time it was established. The system helps to ensure the public health and safety of about 600,000 patients in the United States undergoing radiation therapy for cancer each year. It also helps to reduce the probability of lawsuits related to negligence and improper calibrations and dosages, thereby holding down health costs.

International Activity

a. There was no international activity on the penetrometer.

b. Although there have been expressions of interest in the thermoplastic material, there is no international activity on that technology either.

c. Dr. Dietz of BNL used the tracer technology to conduct successful experiments for the Commission of European Communities. He simulated global and regional pollution from sources such as nuclear and/or chemical disasters.

d. An international consortium is using the alginate-based technology in its research on agriculture-related chemicals pesticides and herbicides. Also, Mr. Connick's work has spread to other overseas scientists, as well, who have similarly cited him in their research.

e. Both companies involved in producing the chemically imbedded herbicide/ pesticide material, Geoflow and Reemay, are selling, testing, or marketing worldwide.

f. In addition to linking its customers' radiation therapy equipment to national standards, NIST's calibration standards ultimately link the measurements of precision equipment to international standards.

Government Gains

a. The Navy was able to realize some unknown (possibly classified) benefit from the six penetrometers produced by Sippican, however, additional penetrometers were not under contract after that.

b. Although the NASA thermoplastic material was originally invented as an aerospace application, it is not being used in the space program.

    c.   As it turns out, the BNL's tracer technology can be applied to the space station and can be used by EPA to do environmental monitoring—two originally unanticipated uses of the technology.

    d.   The alginate-based application related to bioremediation could be used to clean up military sites, government explosive sites, or chemical dumps. It could also be used to help with the Superfund's cleanup activities.

    e.   Ironically, PNNL's chemically imbedded herbicide/ pesticide material was originally developed for DOE use. It has not caught on with the DOE nuclear waste sites, but it is currently being considered for applications at military bases for the long-term control of insects. It may also be incorporated into military uniforms that repel bugs.

    f.   There has been more than one spinoff of Dr. Loevinger's radiation therapy quality assurance program. Another program establishes certified laboratories among the federal laboratories for other types of calibrations. And another such program was established for x-ray protection instruments used by radiation workers in both the public and private sectors.

Economic Development, Technical Assistance

The only company that received any outside economic development services was the minority-owned company, High Technology Services, with the NASA-funded thermoplastic material. The company has been helped by both NASA and state programs aimed at assisting small firms.

Elapsed Time

    a.   The river bottom technical assistance work began in 1982, and the first penetrometer patent was filed in 1983. The prototype was developed and tested and "written up" in 1984. Sippican Corporation's holotypes were made in the mid- to late-1980s. So the technology development took roughly five years from conceptualization to pre-commercialization, but the instrument is not moving.

    b.   The thermoplastic material was developed in the early 1980s, patented in 1983 and 1984, and first licensed in 1985 (ending in failure). The SBIR contract with the small firm was from 1990 to 1992. So, it was over fifteen years from invention to pre-commercialization stages. The material is still not being mass-produced.

    c.   The tracer technology has been under development at BNL for over two decades. The building ventilation work reached it peak in the late 1980s and early 1990s. The successful underground utility work is still going on through EPRI. The other services (underground storage tank leak detection, petroleum reserve analyses, etc.) continue to be offered by BNL. Some tracer instruments and services are sold by private sector firms, as well. What will result from efforts to commercialize the explosive detector and pre-fire system remains to be seen.

    d. Mr. Connick first developed the alginate-based chemical pesticide almost two decades ago. He continued his work and teamed with other USDA sites during the 1980s, with patents and articles resulting during that time frame. The work with Grace-Sierra began in the mid-1980s before CRADAs were possible. The company's first product hit the market in the early 1990s, having been under development for about seven years. Biosys also began working with USDA in the mid- to late-1980s, but did not begin its CRADA work in Texas until later. The three products developed under the CRADA proceeded from basic research to commercialization and market introduction in a record time of three to four years. The scale-up phase only took six months, and the introductory market promotion lasted eight months. EcoScience Corporation's experiments took place in the 1992-1993 time frame.

    e. The chemically imbedded herbicide/ pesticide material was conceived about two decades ago, with the research gearing up in the late-1970s. DOE conducted field tests at the Colorado sites in the early 1980s. Patents and publications appeared beginning in 1982 and 1983. Battelle negotiated licenses from 1983 to 1986, at a time when DOE did not have all the technology transfer procedures in place for the agency. Reemay's Biobarrier root control product was commercialized in eight years and has been on the market for four to five years.

    f. In the radiation therapy quality assurance case, Dr. Loevinger began working with the AAPM subcommittee over twenty-five years ago, after joining NIST in the late 1960s. The first set of regional calibration laboratories took form in the early 1970s. After a NIST-funded (but independently-implemented) study of the system in 1976, two additional regional laboratories were added. Other than that change, the system continues successfully to the present day.

Post-legislation Findings Summary

This section of the appendix summarizes the findings for the selected post-legislation cases in Chapter 4 organized according to the questions addressed

in the interviews. The cases appear in the following order:
a. Laser Method to Light Samples
b. Voice Coder
c. Paper Quality Tester
d. Variable-Frequency Microwave Oven
e. Gravity Meter
f. Oatrim Fat Substitute
g. Chemiluminescent Light Stick
h. Artificial Heart Blood Flow Analysis.

Roles of Laboratory Researchers and Other Personnel

   a. Dr. Yeung at Ames Laboratory developed both the laser-based method of lighting biological samples and an instrument for accomplishing the illumination method. To get independent evaluations of the technology, he reached out to a medical laboratory instrument company, a university medical chemistry department, a hospital, the research arms of two drug companies, and a private laboratory overseas. He provided a variety of services to these organizations. For example, for the instrument company, he assembled a prototype instrument using various parts and components already manufactured by the company in hopes that the company would choose to mass-produce the instrument necessary to perform his fluorescent method. He also made presentations and published widely in scientific journals, both in this country and others. Even after business arrangements with the outside organizations were in place, he continued his work testing and further developing the technology, making significant improvements.

   b. The Rome Laboratory team sought Air Force support for the voice coding technology against all odds, since it was competing for funding with existing industry telecommunications standards. Also, there were no valid in-house Air Force requirements for the technology. However, early-on they recognized the advantages of this technology over existing ones and persisted in their fight to convince others of the technology's superiority.

   c. After developing a laboratory prototype, the inventor of the paper quality testing technology, Mr. Gunderson, worked with others in his research unit to develop a technology transfer plan and market assessment. They contacted a hundred potential users, and solicited competitive CRADA proposals from equipment manufacturers they had identified. They also worked with the eventual partner in the equipment design and development. In addition, they published a number of scientific papers in scholarly journals.

Some of these papers were co-authored with others from partnering organizations.

d. For the variable-frequency microwave oven, the Oak Ridge team combined their areas of expertise in different areas to come up with a new technology. They shopped for an off-the-shelf commercially available component, the traveling wave tube, and eventually discovered a small defense contractor in another state. The company donated the needed component to ORNL, and ORNL bought the power supply necessary for the ORNL researchers to develop a large prototype system, while the company developed a bench-top version. The researchers presented their findings at various conferences and in scientific papers. The ORNL researchers' relationship with the company partner has flourished over the years with a variety of joint patents and papers being produced.

e. Dr. Faller, who chairs NIST/JILA, developed the mechanical, electronic, and optical technology incorporated into the original prototype gravity meter. He was the principal technical advisor for the agency's contract to procure several gravity meters. Dr. Faller also headed the effort to test and evaluate the commercial gravity meters when they were later delivered to NIST.

f. Dr. Inglett invented the technology to manufacture the Oatrim fat substitute. He wrote scientific articles for journals and provided USDA information to interested companies. Eventually, he announced Oatrim's development at a national meeting of a professional society and subsequently received thousands of inquiries from companies. Consequently, he arranged a technology transfer conference at the laboratory, which was attended by seventy industry representatives. Once the technology was licensed, Dr. Inglett assembled an information packet on the licensees and hosted corporate visits to the laboratory. He did much "hand-holding" with the partnering companies during their product development work, visiting their pilot plants and inviting them to his laboratory to view processing techniques. In addition, he initiated "human studies" of the fat substitute, which were very successful. After all this, Dr. Inglett continued his research to improve processing and address production problems involved in scaling up from pilot plant levels.

g. In the light stick case, Dr. Richter and his team of scientists at China Lake researched and experimented with chemiluminescent technologies for a long period of time. Over the years, they performed a series of tests on a variety of chemical compounds and then continued testing to determine the most sustained, temperature-resistant, and non-flammable combinations. They performed both laboratory tests in glass vials and field tests in a variety of environmental conditions. Dr. Richter worked closely with the China Lake technology transfer officers. He produced a sample kit containing potential commercial products and performed demonstrations for manufacturing companies and all levels of government agency users. The group wrote a number of scientific papers and presented at conferences. The early private sector contacts resulted from this exposure.

h. Mr. Shaffer at FETC invented and developed the artificial heart blood flow analysis system. Mr. Shaffer and another scientist in the laboratory's Fundamental Combustion Group helped the partnering medical researchers to set up the FETC-like system at Baxter Healthcare and trained them to use it. Also, the FETC scientists, along with various combinations of these partnering researchers, co-wrote a number of scientific papers.

Technologies and Applications

a. The laser-based method to light up biological samples helped researchers monitor those processes in detail, even for small volumes of samples. Previous detectors to perform these functions were based upon conventional light sources. This is applicable to the field of medical research.

b. The voice coding technology involved a revolutionary approach for compressing voice patterns digitally so that they could be transmitted long distances and then reassembled at the receiving end. Traditional coding techniques involved "linear prediction" programming rather than this new "sinewave-based" programming. Speech/voice compression technology is used in land mobile radios and mobile satellite telephones (in place of cellular service where that is not available). This technology is also used in digitally based voice answering machines and in desktop computer video conferencing.

c. FPL developed a paper quality testing device that allows accurate measurements of paper product deterioration as it is stored over time (such as rolls of paper in warehouses) and exposed to humidity. Better quality control assessments allow paper manufacturers to reduce the use of expensive coatings and wood fiber, which help preserve paper quality in the long run. The quality testing technique has also been found to work well in testing the plastic-type materials used in computer circuit boards, which degrade over time due to exposure to heat and other conditions.

d. The variable-frequency microwave furnace had advantages over conventional fixed-frequency ovens in that it could vary frequency to heat dead spots in a sample for more even heating. It is usable for uniform computer circuit board etching, application of synthetic films to industrial equipment, ceramic heating, and resin curing.

e. NIST/JILA's gravity meter was a highly accurate device based upon absolute rather than conventional relative measurements of gravity. It is used for oil prospecting, measuring volcanic seismographic activity, and other purposes.

f. The fat substitute was called Oatrim by USDA because it is derived from oats. Upon incorporation as an ingredient in prepared foods, such as dairy products, dressings and sauces, meats, cereals, etc., it tastes like fat yet it has less than one-ninth the calories of fat. Also, it lowers blood cholesterol levels because it decreases bad cholesterol and increases good cholesterol.

g. Chemiluminescent light sticks are now known by the general public as the novelty items that glow in the dark and that are fun for kids. They are sold at parades, festivals, carnivals, etc. They were created by mixing certain chemical compounds and dyes together. They were originally developed for military purposes such as marking targets and locating downed pilots.

h. FETC's blood flow analysis technology used a laser to make a fluid fluorescent. The fluid was viewed using multiple exposures of digital photography. The velocity and other properties were measured using software which accompanies the digital photography. When this technology was applied to artificial hearts, it was the first time blood flow was visualized and measured on the internal surface of an artificial heart. This information was used to counter the tendency in artificial hearts for blood to clot, an often fatal complication.

The Laboratories

a. The laser-based method to light up biological samples was developed at Ames Laboratory. DOE's Ames Laboratory has research programs in biochemistry and environmental sciences, yet the laboratory is known for its accomplishments in fields like materials science, metallurgy, and superconductivity. Dr. Yeung's work in this area, his international visibility, and contacts with outside organizations improved the laboratory's industrial and scientific standing.

b. The laboratory involved with the voice coding technology was the Air Force's Rome Laboratory which comprises seventy laboratories around the country involved primarily in communications technology and coordinated from headquarters in Rome, New York. Rome Laboratory's Electromagnetics and Reliability branch is located at Hanscom Air Force Base in the Boston technology corridor. This part of the laboratory focuses on telecommunications equipment, antennas, microelectronics, and related areas.

c. The paper and plastic quality testing technology was developed at the USDA Forest Products Laboratory, one of the eight Forest Service laboratories that are part of the USDA. Of the 350 employees at the laboratory, only about a hundred are scientists and engineers because the laboratory has a contingent of economists who conduct market research for the laboratory's products.

d. The variable-frequency microwave furnace was developed at DOE's Oak Ridge National Laboratory, a DOE multi-program laboratory. However, some of the funding for the technology development was obtained from the DOE Y-12 Plant next door to ORNL, which is one of DOE's defense-oriented laboratories.

e. The gravity meter was developed at the NIST's Joint Institute of Laboratory Astrophysics (JILA) located on the University of Colorado campus at Boulder, Colorado. The NIST employees at JILA report to the NIST Physics Laboratory located at the Gaithersburg, Maryland headquarters. The NIST/JILA structure is an interesting one not replicated in too many other federal laboratory/university locations around the country.

f. The fat substitute was invented at the USDA's National Center for Agricultural Utilization Research in Peoria, Illinois.

g. Chemiluminescent light sticks for marking military targets were developed at the China Lake, California, site of the Naval Warfare Center's Weapons Division.

h. The artificial heart blood flow diagnostics were developed at the Pittsburgh Energy Technology Center, which applied the same techniques normally used to analyze the flow of fuel through pipes to artificial heart pumps. This case "represents technology transfer in its finest sense, because it embodies the application of technology from one discipline—fundamental engineering in fossil fuels—to a quite different one—medical technology."[3]

University Involvement

a. In the laser-based biological samples case, Dr. Yeung served as a long-term consultant to Northeastern University's research program in High-Performance Capillary Electrophoresis so that they could become familiar with his method.

b. The voice coding technology was developed at the Massachusetts Institute of Technology through Air Force (and other) funds. The MIT research team published a number of papers on the technology in the early years of its development.

c. Universities were part of the network established by FPL for evaluating its paper quality testing technology, and also part of the panel screening the competitive CRADA proposals.

d, e, f, g. There was no university involvement in the microwave oven, gravity meter, Oatrim, or light stick technologies.

h. The University of Pittsburgh's Presbyterian-University Hospital is serving as one of the FDA test sites for the artificial heart pump; in fact, doctors at the university's medical school initiated the flow diagnostics work with FETC because they felt they did not have the proper testing techniques they needed. The school's director of biomedical engineering approached FETC to request assistance with measuring and analyzing blood flow.

Funding, Financing

a. Ames Laboratory and DOE headquarters' programs funded Dr. Yeung's work on the laser-based method. Dr. Yeung's work with the outside organizations was covered by those organizations through either independent consulting or contracts with Ames Laboratory. Iowa State University's Research Foundation funded the patent application process since the patent was issued to Iowa State, the laboratory's managing organization.

b. The voice coding technology was funded by university funds and grants and contracts from federal agencies including the Air Force and some intelligence community funding. The resulting spinoff company has not received any federal funding since its founding.

c. In the paper quality tester case, at the same time that FPL was receiving funds from Isthmus Engineering for the CRADA work, the commercial product was provided at cost to the laboratory in return for the laboratory's technical assistance. For the same technology, the laboratory put into place ten cooperative research agreements totaling almost $1 million to public and private organizations so they could independently evaluate the testing technology.

d. In the variable-frequency microwave oven case, the Oak Ridge National Laboratory researchers obtained funding from three DOE sources to allow them to continue work on the microwave oven, including two DOE headquarters programs, the Office of Industrial Technologies and the Laboratory Technology Transfer Program. The third source was the Advanced Manufacturing Program at the defense-oriented Y-12 Plant next door.

e. Funding for the R&D on the gravity meter was provided by NIST and the Defense Mapping Agency. NOAA provided the funds for the initial procurement and provided partial funding for the NIST/JILA lead scientist's time. Among the spinoff companies, at least two of the original Axis principals invested almost $1 million of their own money toward starting up the company and developing the gravity meter. Micro-g received NOAA SBIR funds for further development work. The other spinoff has relied on small "angel"-type investments.

    f.   The Agricultural Research Service funded Dr. Inglett's time to research the Oatrim fat substitute. Both the ConAgra and the Rhone-Poulenc/Quaker partnerships invested "millions of dollars" into their Oatrim products. This involved both further development of the technology, as well as development of the production process. Their high expectations were dashed when they found they had licensed a laboratory process which was "worlds apart" from pilot plant product level and then to full-scale mass production.

    g.   Various Navy and Army offices sponsored the chemiluminescent light stick research. The Marine Corps Exploratory Development Program provided funds for the patent process.

    h.   The CRADA involving the artificial heart flow analysis technique was one of the Pittsburgh Energy Technology Center's (PETC) first CRADAs. Therefore, the agency was not sure how to handle the receipt of private industry funds when Baxter Healthcare offered it. Therefore, this particular CRADA and even the more recent follow-on CRADA involved equipment from Presbyterian-University Hospital and Baxter valued at $500,000.

Intellectual Property

    a.   The original patent and later extensions on the laser-based method were issued to Iowa State University, DOE/Ames Laboratory's contracting operator.

    b.   The spinoff company that developed the voice coding technology received at least four patents with several pending, both foreign and domestic.

    c.   Two patents were issued to USDA for different applications of the paper quality testing technology; certain aspects of the technology were not patentable.

    d.   Oak Ridge National Laboratory was issued two patents for the variable-frequency microwave furnace system with a variety of laboratory, company, and non-profit scientists listed as inventors. There were three additional patents pending related to the CRADA work.

    e.   The gravity meter was not patented; however, it was a proprietary instrument for Axis Instruments, which produced the first devices.

f. There were at least two USDA patents on the Oatrim fat substitute with Dr. Inglett registered as the inventor.

g. The chemiluminescent light stick technology resulted in a number of patents over the years. There are two Navy patents that are still current with three China Lake scientists listed as the inventors.

h. A patent was issued to DOE for the artificial heart blood flow analysis technique with Mr. Shaffer listed as the inventor. Also, Baxter Healthcare received a patent on its artificial heart pump device.

Technology Transfer Mechanisms

a. In the laser-illuminated biological samples case, Ames Laboratory granted a license to Lachat Instruments, a relatively young but stable high-tech company with established product markets, for further developing and eventually selling the required instrument for accomplishing the laser-based method. The technology was also transferred to at least three other research facilities through consulting contracts to Dr. Yeung.

b. The transfer strategy used by the Rome Laboratory team to push the voice coding technology involved establishing a strong presence at standards meetings and presenting supportive arguments for their technology before standards committees. Consequently, the technology was entered into a number of federal, state, local and commercial competitions and independent evaluations where it performed well and was highly rated on a technical basis. Eventually, this new technology was accepted as the new national standard in a number of telecommunications areas.

c. The paper quality testing technology was transferred through a one-year CRADA and an exclusive license on both patents to a small cooperatively organized testing equipment manufacturing firm. The time frame for the license is equivalent to the life of the patents and its field of use is paper products. In addition to these arrangements, the laboratory issued a variety of cooperative research agreements for testing purposes.

d. For the microwave oven, Oak Ridge National Laboratory granted a non-exclusive license to Microwave Laboratories, Inc. and eventually signed a CRADA with the company.

e. NIST implemented a competitive public solicitation to transfer the gravity meter technology and signed a five-year procurement contract with Axis Instruments Company, a small high-tech start-up in Boulder. Axis was to manufacture at least two gravity meters built to specifications being designed and developed by NIST/JILA. The first two instruments were turned over to

NOAA from NIST. Axis also obtained rights, in exchange for building a gravity meter, to a new type of iodine laser from the International Standards Bureau in Paris, and agreed to pay them royalties.

   f.   For the Oatrim fat substitute, USDA granted three non-exclusive licenses to: ConAgra Specialty Grain Products Company (a $25 billion company and the second largest food manufacturer in the United States); Rhone-Poulenc, Inc. (a $16 billion French company); and Quaker Oats (a $6 billion company). A CRADA was later signed on a related technology.

   g.   For the light sticks, the Navy signed non-exclusive licenses with American Cyanamid Corporation and Chemical Devices Corporation, now called Omniglow, Inc., extending to 1993. Before the licenses expired, Omniglow brought a lawsuit against American Cyanamid for filing a patent excluding the government (and Omniglow, as a licensee) from rights to this technology which the government actually owned. In spite of the government not joining in the case, Omniglow won the case and was granted American Cyanamid's light stick technology and business by the court. Omniglow subsequently canceled both licenses, saying they weren't necessary. The company now holds the lion's share of the chemiluminescent patents worldwide.

   h.   The artificial heart flow technology involved a multi-partner CRADA to cooperatively perform the FDA-required testing of the heart pump. The partners are PETC and the University of Pittsburgh's Presbyterian-University Hospital and its schools of both Medicine and Engineering.

User Groups

   a.   For the laser-illuminated method, any type of laboratory (clinical, pharmaceutical, industrial, university, etc.) could benefit from its use. For example, pharmaceutical laboratories can test drugs on a cell-by-cell basis, whether blood cells or cancer cells.

   b.   Telecommunications equipment manufacturers are starting to make use of the voice coder. They are finding it to be a superior technology over existing standard technologies in this area.

   c.   The users of paper quality testing machines are paper product manufacturers. The technology allows quality testing before the products have left the manufacturing facilities. As a result, manufacturers will be able to experiment with less expensive combinations of coatings, fibers, and recycled ingredients so that products are not over designed and over processed to compensate for deterioration. Also, there will be less discarded scrap paper.

d. As with other technologies, initial users of the variable-frequency microwave oven in the short term include laboratories in all sectors: university research laboratories, commercial testing laboratories, etc. Ultimately larger user groups will include semiconductor manufacturers and other types of industrial equipment manufacturers and companies that perform ceramic heating or resin curing, in addition to the military.

e. Government agencies use gravity meters to measure global climate change and warming or to make earthquake and volcano predictions based upon, for example, ocean water levels. They can also contribute to oil prospecting and exploration.

f. The immediate users of the Oatrim fat substitute are food companies. The ultimate end users are their adult customers who are concerned about weight and cholesterol problems as they age.

g. The primary users of the manufactured light sticks are U.S. military personnel and policemen, but also many other countries' military, law enforcement, and public safety agencies are using this technology. Children, as well, are "users" of light stick products sold in stores and amusement parks, as are doctors and medical clinics (for the chemiluminescent biomedical applications). In addition, other manufacturing companies use the technology for industrial safety applications, and fishermen use light sticks as bait and lures.

h. The ultimate beneficiaries of the artificial heart flow diagnostic technology are critically ill patients with heart diseases. There are thirty to fifty thousand people waiting for heart transplants every year. The intermediate users are the health care organizations that make use of artificial hearts in their practice.

Barriers to Commercialization

a. In the laser-based case, Dr. Yeung's major barriers were having to make a special effort to convince skeptics that his technically proficient method was also easy to operate. Also, the licensee did not want to expend any effort on their own towards commercialization. They wanted a market-ready product.

b. The voice coding technology is used in the telecommunications industry, which maintains elaborate standards in all areas of technology in order for equipment to be functionally compatible. The problem in promoting the new speech coding technology was that it was fundamentally different from all of the existing standards for this type of equipment. This required novel approaches to transferring the technology, because it first needed to be accepted within the industry.

   c.   When the paper quality testing device was first created, there was no market demand for the technology because it was a first-of-its kind. This made promoting its use difficult. Consequently, the laboratory used cooperative research agreements, among other mechanisms, to help promote the technology. But this strategy ran into problems when each organization required different machine configurations and data analysis methodologies, which made comparisons and overall technology assessment difficult.

   d.   Regarding the NIST procurement contract to obtain the initial gravity meters, NIST was still in the process of design work and developing product specifications while they were also in the process of negotiating the contract, which made the contract implementation difficult. In addition, years later, the two spinoff companies have to deal with the prospect of market saturation, given that their products are highly technical and specialized. As a result of this level of sophistication, most of their customers or competitors are publicly funded agencies, difficult to compete against as a small firm.

   e.   The Oatrim fat substitute experienced challenges that were surmountable. First, it had to gain U.S. Food and Drug Administration approval but being a natural product, the process was not as lengthy as for some products. Secondly, like other new products, market impact was initially slow because consumers need to receive samples and become educated. Also, although the partnerships offered advantages to the companies entering into them, they also brought certain inherent management, cultural, and communication challenges.

   f.   When China Lake began to seriously transfer the light stick technology, neither the laboratory nor the industry partners had extensive licensing experience, so the licensing process took at least a year.

User Benefits/Economic Impact/Outcomes

   a.   For the laser-based method, Lachat Instruments intended to market a spinoff product for monitoring contamination in drinking water and wastewater. However, the company expected the prototype to be readier for the market and was not willing to put effort into developing it further, and the license is now dead. The other outside users are still using the technology for research purposes, not for commercial gain.

   b.   The outcome of the Air Force-funded voice coding technology at MIT was a spinoff company, Digital Voice Systems, Inc. (DVSI). DVSI sells an Improved Multi-Band Excitation or IMBE™ Speech Compression System.

The available accompanying hardware is either an IMBE™ VC-20 or an IMBE™ VC-100 Voice Codec Module. The company has recently introduced a new Advanced Multi-Band Excitation AMBE^R system implemented by an AMBE-1000™ Coder. The AMBE coder hardware costs anywhere from $38 (or less) up to $99, depending upon the size of the order. (Orders for 100,000 or more units can be negotiated.) DVSI has averaged ten employees since its founding. Being a privately-held company, the principals are not willing to divulge sales revenues, but they add that the company has been doing well.

    c. The commercial version of the FPL's paper quality testing device was developed by Isthmus Engineering and Manufacturing Co-op, which averages $10 million in yearly sales and is known for the high-quality of its products and product servicing. Within eight months of the signing of the one-year CRADA, Isthmus delivered a commercial version of the testing device called a Vacuum Compression Apparatus. The company subsequently sold three additional machines, for a total of $415,000 in sales. The fourth machine is called a Thin Film Analyzer because it tests the effects of humidity and other conditions on the materials in computer circuits. As with most of Isthmus' business, the latter machine sale was the result of its word-of-mouth referral rather than through marketing. In fact, it is this latter application that may open up new markets for the technology, because in the paper industry new technologies have supplanted the technology developed by the FPL that are less expensive, simpler, and easier to implement. The original technology is still being used by those doing specialty research and requiring more sophisticated approaches. At its inception as a leading-edge product, it created a market need for the testing technology.

    d. Oak Ridge National Laboratory's original partner on the microwave oven went out of business when its defense contracts were cut back. Subsequently, Lambda Technologies was spun out of a portion of the defunct business with only one product line and a smaller more focused group of employees. The new company acquired the mother company's inventory, patents, and licenses, although the legal agreements needed to renegotiated with ORNL. The new company delivered a prototype to DOE under terms of its CRADA agreement and introduced its first product, the Vari-Wave, in 1996. The company is producing two to three units a month in response to orders, and it is anticipated this will possibly soon grow to thirty to fifty thousand orders. Initially, the product is appealing to university research laboratories, but is being marketed as a multi-functional oven that can also serve as an analytical or measurement tool. Three models are available with varying wattages, frequencies, and sizes, in addition to a single-function basic model with programmable options. The company is gearing up to respond to orders for specialized manufacturing equipment (such as for the gluing of athletic shoe soles). The company has almost twenty employees. By 2000, the company hopes to do $50 million a year or more in sales. The original carry-over CRADA has ended, and a new one has been signed.

e.   Axis Instruments called its gravity meter an "FG5 Gravimeter." In addition to its NIST contract to build the two NOAA gravity meters, Axis Instruments obtained contracts with another eight agencies overseas, building up to about $3 million in revenues and twenty employees. However, the company went out of business. Subsequently, the Axis chief scientist bought the rights to the gravity meter and the Axis inventory, agreeing to certain royalties, and formed Micro-g Solutions. This new spinoff was conservative in its spending and managed to grow to seven or eight employees. Micro-g sold eight gravity meters by mid-1996 for approximately $2.4 million in sales. The company has a CRADA with NIST to continue development work and is in the process of developing a smaller instrument. Revenues for the new product are projected to be $10 million per year. Between Axis and Micro-g, the government has recovered a total of about $1.8 in taxes from the two companies over five years. Another Axis principal, the company's physicist, obtained rights to the Paris iodine laser and formed a separate spinoff company, Winters Electro-Optics, Inc., which now pays the BIPM royalty payments, and also has a two-year CRADA with NIST. Winters agreed to handle the former Axis laser warrantees, and found it was easier to replace the old lasers with the new Parisien version which essentially amounted to the company's start-up costs. Winters has sold roughly forty lasers representing $1.2 million in sales over three years, providing a little over $425,000 in gross revenues.

f.   In the fat substitute case, ConAgra is a publicly traded company, but since the company is in a joint venture with another company to produce Oatrim, they need to get permission from their partner to release information specifically related to Oatrim production. However, some observations can be deduced. Since Oatrim is an ingredient in ConAgra companies' Healthy Choice product line, a $1.2 billion brand overall for the parent company, it must be used extensively. Also, ConAgra's annual report stated that its gross margin increased in 1996 due to margin improvements in, among others, specialty food ingredients which would include Oatrim. Specialty ingredients also contributed to the company's pre-tax earnings increase. ConAgra's Mountain Lake Partnership and ARS-Peoria formed a CRADA resulting in a new fat substitute called Z-Trim that combined oats, corn and soybeans. The patent application was filed in November 1995, and Dr. Inglett announced the new technology at a conference in August 1996. So, ConAgra as the CRADA partner, had first right of refusal to an exclusive license up until a year after patent filing, even though, ironically, the patent may not be issued within that year-long deadline. Without a patent or some indication from the Patent and Trademark Office that it is likely, it would not necessarily be in the company's interests to risk negotiating a license without the intellectual property rights being firmly in hand. The Rhone-Poulenc/Quaker™ partnership started mass-producing an improved food ingredient product called Beta-Trim in September 1996. Both of the contacts for this partnership indicated disappointment in their sales and business growth, noting that, although they are selling their product to fifty or sixty companies at this point, significant sales should translate into $10 to 100 million (which they have not reached) for companies of their sizes. However, say they have major technical changes in progress that they hope will soon result in some breakthroughs.

g. Largely as a result of acquiring the light stick technology from China Lake, and also as a result of acquiring American Cyanamid's light stick business and license, Omniglow Corporation has grown from being a small start-up business with three employees and one product, to three hundred employees and over a hundred products. During recent regional conflicts, Omniglow sold over fifteen million light stick units to DOD, amounting to at least $150 in annual product sales in this area. Although the military, law enforcement, and public safety agencies are still the company's largest customer group, Omniglow also sells products made for a variety of other customers, including retail toy and novelty stores, a number of major amusement parks like Disneyland, both the commercial and recreational fishing industries, industrial safety users, and now the biomedical market. Overall, since the early 1990s, the company has produced over 250 million light sticks.

h. The FETC blood flow diagnostics case involved a CRADA established to help with FDA clinical testing of Baxter's artificial heart, since earlier testing had indicated that blood clotting was a serious problem. It was quickly apparent that the application of FETC's fuel flow analysis system to artificial hearts was successful. The FETC system for performing flow diagnostics ultimately contributed to redesign of the heart pump. It also helped with transplant operation techniques and patient management. After eighteen months of animal testing, the new design was approved for clinical trials in 1995 and reapproved for investigational testing again in 1996. Statistics on the number of patients implanted or supported over time by the artificial heart device have been steadily improving. In the meantime, Baxter Healthcare cannot realize revenues in the United States until the device has passed all the required regulatory hurdles and is available on the open market. The development stages for products that require FDA approval are costly. Profits are difficult to measure because of the tremendous development involved (costing "multiple millions of dollars"). Each heart pump costs about $80,000. The immediate success of the heart pump work caused the first CRADA to be amended early-on to include artificial lung applications. Testing for this application proceeded from animal testing to clinical trials in 1993, and the flow diagnostics have been successful.

International Activity

a. Other than some foreign exposure and contact with overseas laboratories, there are no international business arrangements on the laser-based method such as foreign patents.

b. In addition to federal, state, local and commercial standards competitions, the voice coding technology was entered into several international competitions for the inherently global satellite market. Again, the technology performed in a superior fashion to existing standards and resulted in its adoption as the new standard.

c. In the paper quality tester case, FPL had a cooperative research agreement with the Swedish Pulp and Paper Research Institute and, as a result of this relationship, jointly sponsored a technical conference with the institute that had previously not been an international conference.

d. For the variable-frequency microwave oven, at least two foreign patents have been filed in more than one application area.

e. Many of the gravity meters have been sold to overseas customers, some of it involving a great deal of export-related government paperwork. However, the individuals involved pointed out that all of this has helped the United States' balance of payments.

f. The USDA waived its right to file foreign patents on the Oatrim fat substitute so Dr. Inglett, himself, patented Oatrim outside of the United States. He subsequently granted ConAgra an exclusive worldwide license, and the Rhone-Poulenc/Quaker partnership negotiated a sub-license from ConAgra for fourteen countries. More recently, ConAgra dropped its license in eight countries so each company is now selling in six foreign countries.

g. Omniglow, which is selling the light stick technology, has overseas sales offices in Canada, England, and Japan. It sells military/law enforcement products in twenty-five other countries and novelty products in thirty other countries. The company also worked with the Japan Tuna Association to further develop, test, and market the technology for use in the fishing industry.

Omniglow has joint venture manufacturing facilities under construction (or consideration) in China, Indonesia, and Eastern Europe.

h. The original user of FETC's blood flow analysis technique, Baxter Healthcare, is now using its artificial heart device in Europe. Their widest use is in England, France, and Germany. In addition, most health care companies developing artificial organs now use FETC's flow analysis technique, and each have set up facilities duplicating the original FETC facility at their laboratories in Europe, Australia, and Korea.

Government Gains

b. A manufacturing base and skill base is being established for the new voice coder which can provide DOD with commercially available equipment that is state of the art. Also, state and local government emergency management personnel are upgrading from analog to digital radios, which require a speech coder.

c. In the paper quality tester case, the U.S. Treasury Department's Bureau of Engraving and Printing signed an interagency agreement with FPL to test its technology. Over time, the laboratory received over $600,000 from the Bureau. The Bureau also purchased one of the original machines from Isthmus Engineering.

d. In the microwave oven case, traveling microwave tubes, a technology developed for military use, was adapted for civilian use. The adapted variable-frequency microwave furnace is contributing to the larger economy.

e. In addition to the original two NOAA gravity meters, the U.S. military purchased two gravity meters. Furthermore, the absolute gravity meter makes it possible for a new mass standard for kilogram replacement to be adopted, so the NIST Gaithersburg office has already purchased a gravity meter in anticipation of this turn of events in the next few years. Development of a less expensive, less precise but still highly accurate absolute gravity measuring device will allow broader audiences to appreciate a new level of accuracy in this area.

f. It was pointed out that the Oatrim fat substitute, being derived from a raw agricultural farm commodity and converted into a value-added product, has helped the United States' balance of trade, which helps the government indirectly.

g. Although the chemiluminescent light sticks were originally developed for fairly specific military purposes (primarily marking bomb targets and secondarily identifying downed aircraft pilots), once they were put into use, they found additional applications. There was an entire spectrum of military uses for the light stick technology that became apparent serendipitously during combat situations. The additional uses mostly involved alternate lighting to flashlights and identification in enemy territories.

h.   In the blood flow diagnostics case, the original FETC application is benefitting from the input of researchers in the medical field. Plus, the original application is being spread into other FETC research activities.

Economic Development, Technical Assistance

None of the companies highlighted in this grouping has made use of state, regional or local economic development or technical assistance services of any type.

Elapsed Time

a.   Dr. Yeung began promoting his laser-based technology through outside contacts in late 1989. He worked with Northeastern University from 1990 to 1992. The patent was issued in 1991. At that time he began working with Lachat Instruments in hopes that this relationship would ultimately result in instrument production and sales.

b.   The MIT work on the voice coding technology began in the early 1980s, with related scientific papers being published during the decade. Digital Voice Systems, Inc. was founded in 1988, although the technology was not really ready for full-scale production until the early 1990s. Certain important standards competitions announced their selection of this technology in 1992 and also in 1993 when the first patent was issued.

c.   The FPL's paper quality testing device was patented in 1982 and 1984. The laboratory solicited industry partners in 1988. CRADA and license agreements were signed in late 1990, and the CRADA was completed in early 1992. Cooperative research agreements were implemented from 1991 through 1993. The international conference was co-sponsored in 1994. Additional CRADA activity focused on different applications continued through 1995.

d.   For the variable-frequency microwave oven, the patent application was filed in 1991. The initial CRADA and license with the original company were executed two months later in early 1992. Jointly written scientific papers were published in 1992 and 1993. The original company closed its doors, and the spinoff company was started in 1994. Agreements between the laboratory and the new company were re-executed after that time.

e. In the gravity meter case, the NIST/JILA absolute gravity measurement research has been going on since the early 1960s. In the mid-1980s, the team began developing a portable device. In 1989, NOAA requested the two gravity meters which set in motion the procurement. NIST transferred the device in 1990. The first two instruments were delivered in 1992, then went through a six-month evaluation phase. So the technology development lasted a short eighteen months. Axis Instruments was in business from 1990 to 1993, at which time Micro-g and Winters Electro-Optics were spun off.

f. Dr. Inglett recognized the Oatrim fat substitute's potential in 1988. The patent application was filed in 1990. He announced Oatrim at a conference in 1990 while the patent was still pending. He held the technology transfer conference at the laboratory later that year. The first license was signed even later the same year with ConAgra. Their first product reached the market a year later. Widespread product marketing began in 1991, while the pilot plant was being completed. So overall scale-up from bench-top to commercial production was accomplished in record time with the ConAgra venture. The time from bench-top to commercialization, eleven months, was exceptionally fast, a process that would normally take about two years. By the end of 1991, the technology was licensed to the two other companies. These last two licensees soon formed a partnership. Comments received on the Rhone-Poulenc/Quaker partnership time frame were not consistent, but this may be a matter of semantics regarding the phases of product development. The Quaker Oats representative said[4] they were hoping for a one-year commercialization period. The Rhone-Poulenc representative said[5] that their average time for a product going from "ground zero" to being a commercial success was five to seven years. In mid-1996, they had been "at it" for four years. They are still within their average six-year product time line, although somewhat behind.

g. The light stick technology was under development at China Lake for more than twenty-five years before it was transferred in 1989. More specifically, the first invention disclosure was filed in 1973 and patented shortly after that. A number of demonstrations of the technology began in the 1970s. The two patents currently in effect were issued in 1986 (about the time Omniglow was founded) and 1987, and efforts to formally transfer the technology began in 1988. The two licenses were signed in 1989. In 1990, Omniglow branched out to markets beyond the military market, both in the United States and abroad.

h. For the PETC blood flow analysis system, the early jointly authored scientific papers were published from 1989 to 1992. The original five-year CRADA was initiated in 1991. Since this CRADA recently expired, it was followed up by a new CRADA to extend the analysis technique to identify cancer cells using fiber optics.

Notes

1   Interviews with Dr. Loevinger, August 27, 1996, November 12, 1996, and December 9, 1997.
2   Interview with Mr. Ingram, September 4, 1996.
3   FLC Award for Excellence in Technology Transfer nomination form, December 3, 1991.
4   Interview with Mr. Babbitt, October 4, 1996.
5   Interview with Mr. Freeland, October 11, 1996.

References

Technology Transfer, General

Federal Laboratory Consortium. *Winners in Technology Transfer: Success Stories from the Federal Laboratory Consortium.* Special Report Series No. 2. Washington, D.C. ISSN 1075-9492. August 1994.

Roberson, B. F. and R. O. Weijo. "Using Market Research to Convert Federal Technology to Marketable Products." *Journal of Technology Transfer* (Fall): 27-33. 1988.

Rood, Sally A. "Legislative-Policy Initiatives as a Problem-Solving Process: The Case of Technology Transfer." *Journal of Technology Transfer* 14 (1/Winter): 14-25. 1989.

Rood, Sally A. and Larkin S. Dudley. "Technology Commercialization: Combining Public and Private." *Policy Studies Journal* 18 (1): 188-202. Fall 1989.

Rood, Sally A. and Annice Brown. "Technology Transfer: Bringing R&D to the Marketplace." *Acquisition Issues* 1 (November): 1-12. 1991.

Rood, Sally, A. editor. *Technology Transfer Metrics Summit Proceedings.* Chicago, Illinois: Technology Transfer Society. June 1997.

Defense Conversion

Aerospace Industries Association of America, Inc. *Key Technologies for the 1990s, An Overview.* Washington, D.C.: November 1987.

Alic, John, Lewis Branscomb, Harvey Brooks, Ashton Carter and Gerald Epstein. *Beyond Spinoff: Military and Commercial Technologies in a Changing World.* Cambridge, Massachusetts: Harvard Business School Press. 1992.

Carnegie Commission on Science, Technology, and Government. *A Science and Technology Agenda for the Nation: Recommendations for the President and Congress.* 1992.

Carnegie Commission on Science, Technology, and Government. *Science, Technology, and Government for a Changing World.* Concluding Report of the Carnegie Commission. ISBN 1-881054-11-X. April 1993.

Congressional Budget Office. *Using R&D Consortia for Commercial Innovation: SEMATECH, X-Ray Lithography, and High Resolution Systems.* July 1990.

Congressional Research Service. *Critical Technologies: Legislative and Executive Branch Activities.* Washington, DC: Library of Congress. 93-734 SPR. 1993.

Council on Competitiveness. *Gaining New Ground: Technology Priorities for America's Future*. Washington, D.C. 1991.

Department of Commerce. Economic Development Administration. *From War to Peace: A History of Past Conversions*. 1993.

Department of Commerce. Technology Administration. *Emerging Technologies: A Survey of Technical and Economic Opportunities*. Spring 1990.

Department of Defense. *Critical Technologies Plan for the Committees on Armed Services, United States Congress*. March 15, 1990.

National Academy of Sciences. Committee on Science, Engineering, and Public Policy. *Science, Technology and the Federal Government: National Goals for a New Era*. Washington, D.C.: National Academy Press. 1993.

National Academy of Sciences. Panel on the Government Role in Civilian Technology. *The Government Role in Civilian Technology: Building a New Alliance*. Washington, D.C.: National Academy Press. ISBN 0-309-04630-0. 1992.

Office of Science and Technology Policy, National Science and Technology Council, National Critical Technologies Review Group. *National Critical Technologies Report (1995)*. March 1996.

Office of Science and Technology Policy. Report of the National Critical Technologies Panel. March 1991.

Office of Technology Assessment. *Building Future Security: Strategies for Restructing the Defense Technology and Industrial Base*. Washington, D.C.: U.S. Government Printing Office. 1992.

Office of Technology Assessment. *Redesigning Defense: Planning the Transition to the Future U.S. Defense Industrial Base*. Washington, D.C.: U.S. Government Printing Office. 1991.

Shapley, Deborah and Rustom Roy. *Lost at the Frontier: U.S. Science and Technology Policy Adrift*. Philadelphia, Pennsylvania: Institute for Scientific Information Press. 1985.

Van Opstal, Debra. *Integrating Commercial and Military Technologies for National Strength*. Report of the CSIS Steering Committee on Security and Technology. Washington, D.C.: Center for Strategic and International Studies. March 1991.

White, Richard H., James P. Bell, J. Scott Hauger, Michael S. Nash, Merle Roberson, An-Jen Tai and Caroline F. Ziemke. *A Survey of Dual-Use Issues*. Alexandria, Virginia: Institute for Defense Analyses. IDA Paper P-3176. March 1996.

International Competitiveness

Carnegie Commission on Science, Technology and Government. *Technology and Economic Performance: Organizing the Executive Branch for a Stronger National Technology Base*. September 1991.

Center for Strategic and International Studies. *Global Innovation/ National Competitiveness*. Washington, D.C. 1996.

Competitiveness Policy Council. *A Competitiveness Strategy for America*. Second Report to the President and Congress. March 1993.

Competitiveness Policy Council. *Building a Competitive America*. Report to the President and Congress. 1992.

Competitiveness Policy Council. *Enhancing American Competitiveness*. Progress Report. October 1993.

Competitiveness Policy Council. *Implementing Technology Policy for a Competitive America*. Report of the Critical Technologies Subcouncil. August 1993.

Competitveness Policy Council. *Promoting Long-Term Prosperity*. Third Report to the President and Congress. May 1994.

Competitiveness Policy Council. *Pursuing a New Technology Policy*. Report of the Critical Technologies Subcouncil. Erich Bloch, Chairman. May 1994.

Competitiveness Policy Council. *Saving More and Investing Better: A Strategy for Securing Prosperity*. Fourth Report to the President and Congress. September 1995.

Competitiveness Policy Council. *Technology Policy for a Competitive America*. Report of the Critical Technologies Subcouncil. March 1993.

Congressional Budget Office. *Using Federal R&D to Promote Commercial Innovation*. A Special Study Prepared at the Request of the Senate Budget Committee. April 1988.

Congressional Research Service. *Analysis of 10 Selected Science and Technology Policy Studies*. CRS Report to Congress. Prepared by William C. Boesman, Science Policy Research Division. 97-836 SPR. Updated October 24, 1997.

Council on Competitiveness. *Competitiveness Index 1996: A Ten-Year Strategic Assessment*. Washington, D.C. ISBN 1-889866-18-0. 1996.

Council on Competitiveness. *Gaining New Ground: Technology Priorities for America's Future*. Washington, D.C. 1991.

Council on Competitiveness. *Technology Policy Implementation Assessment 1993*. Washington, D.C. 1993.

Cyert, Richard M. and David C. Mowery, editors. *Technology and Employment: Innovation and Growth in the U.S. Economy*. National Academy of Sciences, Committee on Science, Engineering and Public Policy, Panel on Technology and Employment. Washington, D.C.: National Academy Press. 1987.

Department of Commerce. *Commerce ACTS: Advanced Civilian Technology Strategy*. Draft for Public Comment. November 1993.

Department of Commerce. Economics and Statistics Administration. Office of the Chief Economist. *Technology, Economic Growth and Employment*. 1994.

Executive Office of the President. Office of Science and Technology Policy. National Science and Technology Council. *Science and Technology Shaping the Twenty-First Century*. 1997.

Executive Office of the President. Office of Science and Technology Policy. *U.S.*

Technology Policy. September 26, 1990.

General Accounting Office. *Competitiveness Issues: The Business Environment in the United States, Japan, and Germany.* Report to Congressional Requesters. GAO/GGD-93-124. August 1993.

Guile, Bruce R. and Harvey Brooks, editors. *Technology and Global Industry: Companies and Nations in the World Economy.* National Academy of Engineering Series on Technology and Social Priorities. Washington, D.C.: National Academy Press. 1987.

House of Representatives. Committee on Science, Space and Technology. *Technology Policy and its Effect on the National Economy: Report Prepared by the Technology Policy Task Force.* Washington, DC: U.S. Government Printing Office. 1988.

McLaughlin, Glenn and Richard E. Rowberg. "Linkages Between Federal Research and Development Funding and Economic Growth." Congressional Research Service series, *Economic Policymaking in Congress: Trends and Prospects.* February 21, 1992.

Mowery, David and Nathan Rosenberg. *Technology and the Pursuit of Economic Growth.* Cambridge: Cambridge University Press. 1989.

National Academy of Engineering. *Technology and Economics.* Washington, D.C.: National Academy Press. ISBN 0-309-04397-2. 1991.

National Academy of Sciences. *The Positive Sum Strategy: Harnessing Technology for Economic Growth.* Washington, D.C.: National Academy Press. 1986.

National Governors Association and the Conference Board. *The Role of Science and Technology in Economic Competitiveness.* Final Report Prepared for the National Science Foundation. September 1987.

National Institute of Standards and Technology. *Technology and Economic Growth: Implications for Federal Policy.* Prepared by Gregory Tassey, Senior Economist. October 1995.

National Research Council. Board on Science, Technology and Economic Policy. *International Friction and Cooperation in High-Technology Development and Trade: Papers and Proceedings.* Charles W. Wessner, editor. Washington, D.C.: National Academy Press. 1997.

National Research Council, with Hamburg Institute for Economic Research and Kiel Institute for World Economics. *Conflict and Cooperation in National Competition for High-Technology Industry.* Washington, D.C.: National Academy Press. 1996.

Nelson, Richard R. *High-Technology Policies: A Five-Nation Comparison.* New York: Columbia University. 1988.

Office of Science and Technology Policy. National Science and Technology Council. Committee on Civilian Industrial Technology. *Technology in the National Interest.* July 1996.

Office of Science and Technology Policy. National Science and Technology Council. Committee on Fundamental Science. *Science in the National Interest.*

August 1994.

Porter, Michael E. *The Competitive Advantage of Nations*. Cambridge, Massachusetts: Harvard University Press. 1990.

President William J. Clinton and Vice President Albert Gore, Jr. *Technology for America's Economic Growth, A New Direction to Build Economic Strength*. February 22, 1993.

Rosenberg, Nathan, Ralph Landau and David C. Mowery, editors. *Technology and the Wealth of Nations*. 1992.

Rushing, Francis W. and Carole Ganz Brown, editors. *National Policies for Developing High Technology Industries: International Comparisons*. Westview Special Studies in Science, Technology, and Public Policy. Boulder, Colorado: Westview Press. 1986.

Schact, Wendy H. and Glenn J. McLoughlin. *Technology and Trade: Indicators of U.S. Industrial Innovation*. Congressional Research Service Review. October 1986.

Saxenian, Annalee. *Regional Advantage: Culture and Competition in Silicon Valley and Route 128*. Cambridge, Massachusetts: Harvard University Press. 1994.

Smilor, Raymond and George Kozmetsky, editors. *Creating the Technopolis: Linking Technology Commercialization and Economic Growth*. Cambridge, Massachusetts: Ballinger Publishing Company. 1988.

Stowsky, Jay and Richard H. White. *Anchoring U.S. Competitiveness: Revisiting the Economic Rationale for Technology Policy*. Alexandria, Virginia: Institute for Defense Analyses. IDA Document D-1777. September 1995.

Tassey, Gregory. *The Economics of R&D Policy*. Westport, Connecticut: Quorum Books. ISBN 1-56720-093-1. 1997.

White House. *Technology for Economic Growth: President's Progress Report*. November 1993.

Science and Technology – Budget Issues, Government Role

Branscomb, Lewis M. "From Technology Politics to Technology Policy." *Issues in Science and Technology* (Spring): 41-48. 1997.

Bush, Vannevar. *Science: The Endless Frontier*. Report to the President. First Edition. 1945.

Congressional Research Service. *Industrial Competitiveness and Technological Advancement: Debate Over Government Policy*. CRS Issue Brief prepared by Wendy H. Schacht, Science Policy Research Division. IB91132. Updated December 5, 1997.

Congressional Research Service. *R&D Partnerships: Government-Industry Collaboration*. CRS Report for Congress prepared by Wendy H. Schacht. 95-499 SPR. Updated January 12, 1998.

Congressional Research Service. *Research and Development Funding: Fiscal Year*

1998. CRS Issue Brief prepared by Michael E. Davey, Science Policy Research Division. IB07023. Updated December 17, 1997.

Congressional Research Service. *Research and Development Funding in a Constrained Budget Environment: Alternative Support Sources and Streamlined Funding Mechanisms.* 1996.

Congressional Research Service. *Research and Development: Priority Setting and Consolidatoin in Science Budgeting.* CRS Issue Brief prepared by Genevieve J. Knezo, Science Policy Research Division. IB94009. Update January 15, 1998.

Congressional Research Service. Science Policy Research Division. *The Federal Role in Technology Development.* CRS Report for Congress prepared by Wendy H. Schacht. 95-50 SPR. Updated January 12, 1998.

Council on Competitiveness. *Endless Frontier, Limited Resources: U.S. R&D Policy for Competitiveness.* Washington, D.C. April 1996.

Harvard University. Center for Science and International Affairs. *Investing in Innovation: Toward a Consensus Strategy for Federal Technology Policy.* Project on Technology Policy Assessment. Sponsored by the Competitiveness Policy Council. April 24, 1997.

Kash, Don E. "Technology Policy Requires Picking Winners." Commentary in *Economic Development Quarterly* 6 (3/August): 227-240. 1992.

National Academy of Sciences. Committee on Criteria for Federal Support of Research and Development. *Allocating Federal Funds for Science and Technology.* Frank Press, Committee Chair. Washington, D.C.: National Academy Press. 1995.

"When the State Picks Winners." Editorial, *The Economist* (January 9): 13-14. 1993.

White House. Council of Economic Advisors. *Supporting Research and Development to Promote Economic Growth: The Federal Government's Role.* October 1995.

White, Robert M. *U.S. Technology Policy: The Federal Government's Role.* Paper Commissioned by the Competitiveness Policy Council. September 1995.

Government Laboratory Policy

Congressional Research Service. *DOE Laboratories: Capabilities and Missions.* Washington, D.C.: Library of Congress. 93-752 SPR. 1993.

Congressional Research Service. *Restructuring DOE and Its Laboratories: Issues in the 105th Congress.* CRS Issue Brief prepared by William C. Boesman, Science, Technology and Medicine Division. IB7012. Updated January 9, 1998.

Department of Energy. *Alternative Futures for the Department of Energy National Laboratories.* Secretary of Energy Advisory Board Office, Task Force on Alternative Futures. February 1, 1995.

Department of Energy. Secretary of Energy Advisory Board (SEAB). *Report to the Secretary on the DOE National Laboratories.* Prepared by the SEAB Task

Force on the DOE National Laboratories. July 1992.

General Accounting Office. *Department of Energy: A Framework for Restructiring DOE and Its Missions*. Report to the Congress. GAO/RCED-95-197. August 1995.

General Accounting Office. *Department of Energy: National Laboratories Need Clearer Missions and Better Management*. Report to the Secretary of Energy. GAO/RCED-95-10. January 1995.

General Accounting Office. *DOE's National Laboratories: Adopting New Missions and Managing Effectively Pose Significant Challenges*. Testimony before the Subcommittee on Energy and Power. Committee on Energy and Commerce, House of Representatives. GAO/T-RCED-94-113. February 3, 1994.

General Accounting Office. *Energy Research: Opportunities Exist to Recover Federal Investment in Technology Development Projects*. Report to the Chairman, Subcommittee on Energy and Environment, Committee on Science, House of Representatives. GAO/RCED-96-141. June 1996.

General Accounting Office. *National Laboratories: Are Their R&D Activities Related to Commercial Product Development?* Report to Congressional Requesters. GAO/PEMD-95-2. November 1994.

Office of Technology Assessment. *After the Cold War: Living With Lower Defense Spending*. Washington, D.C.: U.S. Government Printing Office. OTA-ITE-524. February 1992.

Office of Technology Assessment. *Defense Conversion: Redirecting R&D* and *Summary*. Washington, D.C.: U.S. Government Printing Office. 1993.

Technology Transfer Background, Policy, Practice

Atlantic Council. *Transfer of Technology to Industry from U.S. Department of Energy Defense Programs Laboratories*. 1992.

Battelle. *Interactions of Science and Technology in the Innovation Process: Some Case Studies*. Final Report prepared for the National Science Foundation. Contract NSF-C 667. Columbus, Ohio: Battelle Columbus Laboratories. March 19, 1973.

Birch, David. *Job Creation in America: How Our Smallest Companies Put the Most People to Work*. New York: Free Press. 1987.

Brett, Alistair M. "Federal Laboratory Spin-Off Companies: Development of Case Studies for Training in Effective Domestic Technology Transfer." Virginia Polytechnic Institute and State University. August 9, 1989. Unpublished.

Brett, Alistair, David V. Gibson and Raymond W. Smilor, editors. *University Spin-off Companies: Economic Development, Faculty Entrepreneurs, and Technology Transfer*. Lanham, Maryland: Rowman & Littlefield Publishers, Inc. 1991.

Burton, Daniel F. *Industry as a Customer of the Federal Laboratories*. Washington,

D.C.: Council on Competitiveness. 1992.

Congressional Research Service. *Cooperative R&D: Federal Efforts to Promote Industrial Competitiveness*. CRS Issue Brief prepared by Wendy H. Schacht, Science Policy Research Division. IB89056. Updated December 5, 1997.

Congressional Research Service. *Cooperative Research and Development Agreements (CRADAs)*. CRS Report for Congress prepared by Wendy H. Schacht, Science, Technology and Medicine Division. 95-150 SPR. Updated January 12, 1998.

Congressional Research Service. *Technology Transfer: Use of Federally Funded Research and Development*. CRS Issue Brief prepared by Wendy Schacht, Science Policy Research Division. Issue Brief IB 85031. Updated December 5, 1997.

Department of Commerce. Technology Administration. Office of Technology Policy. *Listening to Industry: Business Views on Technology Policy*. Draft for Public Comment. June 1994.

Federal Laboratory Consortium. *Technology Transfer in a Time of Transition: A Guide to Defense Conversion*. 1994.

Federal Laboratory Consortium for Technology Transfer. Federal Laboratory-Industry Interaction Working Group. *Interagency Study of ORTA Organization and Operation and Lessons Learned Case Studies in Technology Transfer*. DOE/METC-85/6019. May 1985.

Federal Laboratory Consortium for Technology Transfer. *FLC Performance Report to Congress and the Federal Agencies: Fiscal Years 1995-1996*. Activities and Accomplishments of the FLC pursuant to the Federal Technology Transfer Act of 1986. November 1997.

Gibson, David V. and Everett M. Rogers. *R&D Collaboration on Trial*. Cambridge, Massachusetts: Harvard Business Review Press. 1994.

Grissom, Fred E., Jr. and Richard L. Chapman. *Mining the Nation's Brain Trust: How to Put Federally-Funded Research to Work for You*. Reading, Massachusetts: Addison-Wesley Publishing Company, Inc. ISBN 0-201-55015-6. 1992.

Gutterman, Alan S. and Jacob N. Erlich. *Technology Development and Transfer: The Transactional and Legal Environment*. Westport, Connecticut: Quroum Books. ISBN 1-56720-021-4. 1997.

Lepkowski, Wil. "R&D Policy: Cooperation is the Current Byword." AAAS Science and

Technology Policy Yearbook 1998. Albert H. Teich et al, editors. American Association for the Advancement of Science. ISBN 0-87168-611-2. p. 223-236. 1997.

Link, Albert N. and Gregory Tassey, editors. *Cooperative Research and Development: The Industry-University-Government Relationship*. Norwell, Massachusetts: Kluwer Academic Publishers. 1989.

McKenney, Bruce A. *National Benefits from National Labs: Meeting Tomorrow's*

National Technology Needs. Final Report of the CSIS National Benefits from National Laboratories Project.Washington, D.C.: Center for Strategic and International Studies. ISBN 0-89206-224-X. 1993.

Meyer, Christopher. *Relentless Growth: How Silicon Valley Innovation Strategies Can Work in Your Business.* New York: Free Press. 1997.

Office of Technology Assessment. *Innovation and Commercialization of Emerging Technology.* Washington, D.C.: U.S. Government Printing Office. OTA-BP-ITC-165. September 1995.

Preston, John T. "Success Factors in Technology Transfer." *Preparing the Way: Technology Transfer in the 21st Century, Technology Transfer Society 16th Annual Meeting Proceedings, June 9-11, 1991, Denver, Colorado.* 1991. (Revised September 26, 1992 and entitled "Success Factors in Technology Development.")

Rogers, Everett M. *Diffusion of Innovations.* New York: Free Press. ISBN 0-02874074-2. First edition, 1962. Third edition, 1983. Fourth edition, 1995.

Rogers, Everett M. with the assistance of F. F. Shoemaker. *Communication of Innovations: A Cross-Cultural Approach.* New York: Free Press. 1971.

Rood, Sally and Diane Palmintera. *Tapping Federal Laboratories and Universities to Improve Local Economies: The Role of the Mayor and City Government.* Washington, D.C.: U.S. Conference of Mayors. October 1988.

Tarter, C. Bruce. "National Laboratory Partnerships: What Works and What Doesn't." AAAS Science and Technology Policy Yearbook 1998. Albert H. Teich et al, editors. American Association for the Advancement of Science. ISBN 0-87168-611-2. p. 265-278. 1997.

Tornatzky, Louis G., J. D. Eveland, et al. *The Process of Technological Innovation: Reviewing the Literature.* National Science Foundation, Division of Industrial Science and Technological Innovation, Productivity Improvement Research Section. May 1983.

Tornatzky, Louis G., Mitchell Fleischer, et al. *The Processes of Technological Innovation.* Lexington, Massachusetts: Lexington Books. ISBN 0-669-20348-3. 1990.

Wigand, Rolf T., Slawomir J. Marcinkowski and Igor Plonisch. "Transferring Technology on the Information Highway." *Technology Commercialization and Economic Growth, Technology Transfer Society Proceedings, 20th Annual Meeting, July 16-19, 1995, Washington, D.C.*: 267-276. 1995.

Williams, Frederick and David V. Gibson, editors. *Technology Transfer: A Communication Perspective.* Newberry Park, California: Sage Publications. ISBN 0-8039-3741-5. 1990.

Yin, Robert K., et al. *A Review of Case Studies of Technological Innovations in State and Local Services.* Santa Monica, California: RAND Corporation. R-1970-NSF. February 1976.

Fundamental Science, Basic Research – Program Evaluation

Cozzens, Susan E. *Methods for Evaluating Fundamental Science.* Critical Technologies Institute/ RAND. Draft Paper. DRU-875/2-CTI. October 1994.

Cozzens, Susan E. "Strategic Evaluation and the Keystone Model of Basic Research." *AAAS Science and Technology Policy Yearbook.* Albert H. Teich, S. D. Nelson and C. McEnaney, editors. Washington, D.C.: American Association for the Advancement of Science. Chapter 21: 281-291. 1994.

Cozzens, Susan E. "U.S. Research Assessment: Recent Developments." *Scientometrics* 34 (3): 351-362. 1995.

David, Paul, David Mowery and W. Edward Steinmueller. "Assessing the Economic Payoffs from Basic Research." *Economics of Innovation and New Technology* 2: 73-90. 1992.

Gunderson, Norman E. and Elizabeth Rodriquez. "The Government Performance and Results Act of 1993: How it Will Affect Federal Scientific Programs." *AAAS Science and Technology Policy Yearbook 1994.* Albert H. Teich, S. D. Nelson and C. McEnaney, editors. Washington, D.C.: American Association for the Advancement of Science. 1994.

Kostoff, Ronald N. "Assessing Research Impact: Federal Peer Review Practices." *Evaluation Review* 18 (1/February): 31-40. 1994.

Kostoff, Ronald N. "Assessing Research Impact: Semiquantitative Methods." *Evaluation Review* 18 (1/February): 11-19. 1994.

Kostoff, Ronald N. *Handbook of Research Impact Assessment.* Seventh edition. DTIC Report ADA296021. Summer 1997.

National Science Foundation. *Performance Assessment at the National Science Foundation: Proposals for NSF's Response to the Government Performance and Results Act.* Discussion Paper. November 13, 1995.

Office of Science and Technology Policy. *Evaluation of Fundamental Research Programs: A Review of the Issues.* Report on Discussions in the Practitioners' Working Group on Research Evaluation. Susan E. Cozzens, Convenor and Rapporteur. August 15, 1994.

Office of Science and Technology Policy. National Science and Technology Council. Committee on Fundamental Science. Subcommittee on Research. *Assessing Fundamental Science.* July 1996.

Office of Technology Assessment. *Research Funding as an Investment: Can We Measure the Returns? A Technical Memorandum.* Washington, D.C.: U.S. Congress. OTA-TM-SET-36. April 1986.

Popper, Steven W. *Economic Approaches to Measuring the Performance and Benefits of Fundamental Science.* Santa Monica, California: RAND. 1995.

Wagner, Caroline and Ann Flanagan. *Workshop on the Metrics of Fundamental Science: A Summary.* Critical Technologies Institute/ RAND. February 1995.

Research & Development – Program Evaluation

Bernstein, Jeffrey and M. Ishaq Nadiri. "Interindustry Spillovers, Rates of Return, and Production in High-Tech Industries." *American Economic Review Papers and Proceedings* 78: 429-434. 1988.

Bernstein, Jeffrey and M. Ishaq Nadiri. "Productivity Demand, Cost of Production, Spillovers, and the Social Rate of Return to R&D." National Bureau of Economic Research (NBER) Working Paper Series. Cambridge, Massachusetts: NBER. Working Paper no. 3625. 1991.

Congressional Budget Office. "A Review of Edwin Mansfield's Estimate of the Rate of Return from Academic Research and its Relevance to the Federal Budget Process." CBO Staff Memorandum. April 1993.

Foster Associates, Inc. *A Survey of Net Rates of Return on Innovation.* Three Volumes. National Science Foundation. May 1978.

General Accounting Office. *Managing for Results: Key Steps and Challenges in Implementing GPRA in Science Agencies.* GAO/T-GGD/RCED-96-214. July 10, 1996.

General Accounting Office. *Measuring Performance: Strengths and Limitations of Research Indicators.* Report to Congressional Requesters. GAO/RCED-97-91. March 1997.

General Accounting Office. *Measuring Performance: Challenges in Evaluating Research and Development.* Testimony Before the Subcommittee on Technology, House Committee on Science. GAO/T-RCED-97-130. April 10, 1997.

Griliches, Zvi. "Issues in Assessing the Contribution of Research and Development to Productivity Growth." *The Bell Journal of Economics* 10 (Spring). 1979.

Griliches, Zvi. "Patent Statistics as Economic Indicators: A Survey." *Journal of Economic Literature* 28 (December): 1661-1707. 1990.

Griliches, Zvi, editor. *R&D, Patents, and Productivity.* Chicago, Illinois: University of Chicago Press. 1984.

Illinois Institute of Technology Research Institute. *Technology in Retrospect and Critical Events in Science (TRACES).* Prepared for the National Science Foundation. Contract NSF-C535. December 15, 1968.

Logsdon, John M. and Clair B. Rubin. "Research Evaluation Activities of Ten Federal Agencies." *Evaluation and Program Planning* 11: 1-11. 1988.

Mansfield, Edwin. "Rates of Return from Industrial R&D." *American Economic Review Papers and Proceedings* 55 (2): 310-322. May 1965.

Mansfield, Edwin. "Academic Research and Industrial Innovation." *Research Policy* 20: 1-12. 1991.

Mansfield, Edwin. "How Economists See R&D." *Harvard Business Review* 59 (6/November-December): 98-106. 1981.

Mansfield, Edwin. *Industrial Research and Technological Innovation: An*

Econometric Analysis. Cowles Foundation for Research in Economics. New York, New York: W. W. Norton Books. 1968.

Mansfield, Edwin. "Social Returns from R&D: Findings, Methods and Limitations." *Research/ Technology Management* (November-December). 1991.

Mansfield, Edwin, Anthony Romeo, M. Schwartz, D. Teece, S. Wagner and P. Brach, *Technology Transfer, Productivity and Economic Policy*. New York, New York: W. W. Norton Books. 1982.

Mansfield, Edwin, John Rapoport and Anthony Romeo. *Social and Private Rates of Return from Industrial Innovations, Volume 2 – Detailed Descriptions of 17 Case Studies*. Philadelphia, Pennsylvania: University of Pennsylvania. 1975.

Mansfield, Edwin, John Rapoport, Anthony Romeo, Samuel Wagner, and George Beardsley. "Social and Private Rates of Return from Industrial Innovation." *Quarterly Journal of Economics* 41: 221-40. 1977.

Nadari, M. Ishaq. "Innovations and Technological Spillovers." National Bureau of Economic Research (NBER) Working Paper Series. Cambridge, Massachusetts: NBER. Working Paper no. 4423. August 1993.

Narin, Francis, Dominic Olivastro and Kimberly A. Stevens. "Bibliometrics/ Theory, Practice, and Problems." *Evaluation Review* 18 (1/February): 65-76. 1994.

Narin, Francis, Kimberly S. Hamilton, and Dominic Olivastro. "The Increasing Linkage Between U.S. Technology and Public Science." *AAAS Science and Technology Policy Yearbook 1998*. Albert H. Teich et al, editors. American Association for the Advancement of Science. ISBN 0-87168-611-2. P. 101-121. 1997.

National Science Board. *Science and Engineering Indicators 1993*. Washington, D.C.: U.S. Government Printing Office. NSB 93-1. 1993.

Project HINDSIGHT. U.S. Department of Defense. Office of the Director of Defense Research and Engineering. Final Report. AD495905. October 1969.

Robert R. Nathan Associates, Inc. *Net Rates of Return on Innovation*. Three Volumes. National Science Foundation. October 1978.

Solow, Robert M. "Technical Change and the Aggregated Production Function." *Review of Economics and Statistics* 39: 312-320. 1957.

Terleckyj, N. "Effects of R&D on the Productivity Growth of Industries: An Exploratory Study." Washington, D.C.: National Planning Association. 1974.

Federal Technology Funding Programs and Consortia – Program Evaluation

Braid, Robert B., Marilyn A. Brown, C. Robert Wilson, Charlotte A. Franchuk and Colleen G. Rizy. *The Energy-Related Inventions Program: Continuing Benefits to the Inventor Community*. Prepared for Department of Energy by Oak Ridge National Laboratory, Energy Division. ORNL/CON-429. October 1996.

Brown, Marilyn A. "The Energy-Related Inventions Program: Evaluation

Challenges and Solutions." *Technology Transfer Metrics Summit Proceedings.* Sally A. Rood, editor. Chicago, Illinois: Technology Transfer Society. 171-185. June 1997.

Brown, Marilyn A. and C. R. Wilson. "Government Promotion of Energy Innovations: An Evaluation of the Energy-Related Inventions Program." *Policy Studies Journal* 20 (1): 87-101. 1992.

Brown, Marilyn, T. Randall Curlee and Steven R. Elliott. "Evaluating Technology Innovation Programs: The Use of Comparison Groups to Identify Impacts." *Research Policy* 24 (4): 669-684. 1995.

Department of Commerce. Technology Administration. *The Advanced Technology Program: A Progress Report on the Impacts of an Industry-Government Technology Partnership.* April 1996.

Finan, W. F. and A. N. Link. *Evaluation of the Value of the Semiconductor Research Corporation to its Corporate Members.* Washington, D.C.:Technecon Analytic Research. SRC Technical Report T94177. November 1994.

General Accounting Office. *Federal Research: Assessment of Small Business Innovation Research Programs.* GAO/RCED-89-39. January 1989.

General Accounting Office. *Federal Research: Interim Assessment of the Small Business Innovation Research and Technology Transfer Programs.* GAO/T-RCED-96-93. March 6, 1996.

General Accounting Office. *Federal Research: Interim Report on the Small Business Innovation Research Program.* GAO/RCED-95-59. March 1995.

General Accounting Office. *Federal Research: Small Business Innovation Research Shows Success But Can Be Strengthened.* GAO/RCED-92-37. 1992.

General Accounting Office. *Measuring Performance: The Advanced Technology and Private-Sector Funding.* Report to the Ranking Minority Member, Committee on Science, House of Representatives. GAO/T-RCED-96-47. January 1996.

General Accounting Office. *Performance Measurement: Efforts to Evaluate the Advanced Technology Program.* Report to the Ranking Minority Member, Committee on Science, House of Representatives. GAO/RCED-95-68. May 1995.

Gibson, David V. and Raymond W. Smilor, editors. *Technology Transfer in Consortia and Strategic Alliances.* Lanham, Maryland: Rowman & Littlefield Publishers, Inc. ISBN 0-8476-7717-6. 1992.

Link, Albert N. *Advanced Technology Programs Case Study: Early Stage Impacts of the Printed Wiring Board Research Joint Venture, Assessed at Project End.* Prepared for National Institute of Standards and Technology. NIST GCR 97-722. November 1997.

National Institute of Standards and Technology. Advanced Technology Program. Office of Economic Assessment. *Acceleration of Technology Development by the Advanced Technology Program: The Experience of 28 Projects Funded in 1991.* Frances Jean Laidlaw, Industry Consultant. NISTIR-6047. September

1997.

Powell, Jeanne W. *Advanced Technology Program Development, Commercialization, and Diffusion of Enabling Technologies: Progress Report for Projects Funded 1993-1995*. U.S. Department of Commerce. NISTIR 6098. December 1997.

Rorke, Marcia L. and Harold C. Livesay. *A Longitudinal Examination of the Energy-Related Inventions Program*. Rockville, Maryland: Mohawk Research Corporation. 1986.

Silber, Bohne. *Survey of Advanced Technology Program 1990-1992 Awardees: Company Opinion About the ATP and its Early Effects*. Clarkesville, Maryland: Silber & Associates. January 30, 1996.

Small Business Administration. *Results of Three-Year Commercialization Study of the SBIR Program*. Small Business Innovation Research Program. Document #90-00.147. No date.

Small Business Administration, Office of Innovation, Research and Technology. *Small Business Innovation Development Act: Tenth Annual Report*. 1993.

Solomon Associates. *The Advanced Technology Program: An Assessment of Short Term Impacts, First Competition*. Submitted to the Advanced Technology Program, National Institute of Standards and Technology. February 1993.

Wallsten, Scott. *Can Government-Industry R&D Programs Increase Private R&D? The Case of the Small Business Innovation Research Program*. California: Stanford University. November 1997. [See also Testimony presented at Hearing on the Small Business Technology Transfer Program before the House Committee on Science, Subcommittee on Technology, September 4, 1997.]

White, Richard H. and An-Jen Tai. *The Economics of Commercial-Military Integration and Dual-Use Technology Investments*. Institute for Defense Analyses. IDA Paper P-2995. June 1995.

White, Richard H., Jay Stowsky and Scott Hauger, editors. *Assessing the Economic and National Security Benefits from Publicly Funded Technology Investments: An IDA Round Table*. Alexandria, Virginia: Institute for Defense Analyses. IDA Paper P-3138. September 1995.

State Technology Funding Programs – Program Evaluation

Alaska Science and Technology Foundation. *Review of State Technology-Based Economic Development Programs: The Lessons Learned*. February 14, 1995.

Bartsch, Charles. *Enhancing Competitiveness: Selected State Technology Transfer Initiatives*. Washington, D.C.: Northeast-Midwest Institute. January 1994.

Battelle. *The Edison Technology Center: An Economic Impact Study*. December 1996.

Bergland, Dan and Christopher Coburn. *Partnerships: A Compendium of State and Federal Cooperative Technology Programs*. Christopher Coburn, editor.

Columbus, Ohio: Battelle Press. 1995.

Bozeman, Barry and Julia Melkers, editors. *Evaluating R&D Impacts: Methods and Practice*. Boston, Massachusetts: Kluwer Academic Publishers. 1993.

Collaborative Economics. *Index of the Massachusetts Innovation Economy*. Westborough, Massachusetts: Massachusetts Technology Collaborative. 1997.

Connecticut Academy of Science and Engineering. *Science and Technology Policy: Lessons From Six American States*. Report to the Carnegie Commission on Science, Technology and Government. October 1, 1994.

Corporation for Enterprise Development. *1997 Development Report Card for the States*. Washington, D.C. 1997.

Feller, Irwin and Gary Anderson. "A Benefit-Cost Approach to the Evaluation of State Technology Development Programs." *Economic Development Quarterly* 8 (2/May): 127-140. 1994.

Melkers, Julia and Susan Cozzens. "Developing and Transferring Technology in State S&T Programs: Assessing Performance." *Journal of Technology Transfer* 22 (2/Summer): 27-32. 1997.

Melkers, Julia, Daniel Bugler and L. A Wilson. *Evaluation of the Alaska Science and Technology Foundation, Phase I: Final Report*. University of Alaska Southeast. May 3, 1994.

Riggle, James D. et al. *Summary Report: Virginia's Center for Innovative Technology Economic Impact and Customer Assessment Study Fiscal Year 1977*. George Mason University, The Institute of Public Policy, Center for Regional Analysis. October 1997.

SRI International. *New York State Centers for Advanced Technology Programs: Evaluating Past Performance and Preparing for the Future*. New York State Science and Technology Foundation. April 1992.

Manufacturing Extension – Program Evaluation

Dziczek, Kristin, Daniel Luria and Edith Wiarda. "Assessing the Impact of a Manufacturing Extension Center." *Technology Transfer Metrics Summit Proceedings*. Sally A. Rood, editor. Chicago, Illinois: Technology Transfer Society, 186-197. June 1997.

General Accounting Office. *Technology Transfer: Federal Efforts to Enhance the Competitiveness of Small Manufacturers*. Report to the Ranking Minority Member, Committee on Small Business, U.S. Senate. GAO/RCED-92-30. November 1991.

Haines, Ruth. *Project Reporting and Evaluation: NIST Manufacturing Extension Partnership*. December 1993.

Jarmin, Ronald S. *Measuring the Impact of Manufacturing Extension*. Washington, D.C.: U.S. Bureau of the Census, Center for Economic Studies. August 1996, revised January 1997.

Luria, Dan. *A Framework for Evaluating the NIST/MTC's: A Summary Based on the Recommendations of the NIST/MTC Evaluation Working Group.* Ann Arbor, Michigan: Industrial Technology Institute and Midwest Manufacturing Technology Center. 1993.

Mt. Auburn Associates. *Technology Transfer to Small Manufacturers: A Literature Review.* Final Report. Submitted to U.S. Small Business Administration. Submitted by Mt. Auburn Associates, Inc. with Regional Technology Strategies, Inc. August 1995.

Nexus Associates, Inc. *Evaluation of the New York Manufacturing Extension Partnership.* Final Report. Prepared for the New York State Science and Technology Foundation/ Empire State Development. Gen#95037. March 18, 1996.

Roessner, J. David. "Evaluating Government Innovation Programs: Lessons from the U.S. Experience." *Research Policy* 18: 343-359. 1989.

Shapira, Philip. *Best Practices for Industrial Modernization.* Prepared for the National Institute of Standards and Technology, U.S. Department of Commerce. Contract# 43NANB212963. Atlanta, Georgia: School of Public Policy, Georgia Institute of Technology. December 1993.

Shapira, Philip and Jan Youtie. *Assessing GMEA's Economic Impacts: Towards a Benefit-Cost Methodology.* GMEA Evaluation Working Paper E9502. Atlanta, Georgia: Georgia Tech Economic Development Institute. 1995.

Shapira, Philip, Jan Youtie and J. D. Roessner, "Current Practices in the Evaluation of U.S. Industrial Modernization Programs," *Research Policy* 25: 185-214. 1996.

Technology Transfer Evaluation

National Institute of Standards and Technology

Department of Commerce. National Institute of Standards and Technology. *Setting Priorities and Measuring Results at the National Institute of Standards and Technology.* January 1994.

Link, Albert N. *Economic Impact Assessments: Guidelines for Conducting and Interpreting Assessment Studies.* National Institute of Standards and Technology, Program Office. Planning Report 96-1. May 1996.

Link, Albert N. *Evaluating Public Sector Research and Development.* Westport, Connecticut: Praeger Publishers. ISBN 0-275-95368-8. 1996.

National Institute of Standards and Technology. *NIST Industrial Impacts: A Sampling of Successful Partnerships.* NIST Special Publication 872. First printing September 1994; revised February 1996.

Tassey, Gregory. *Rates of Return from Investments in Technology Infrastructure.* National Institute of Standards and Technology, Program Office. Planning

Report 96-3. June 1996.

Tassey, Gregory. *Technology Infrastructure and Competitive Position.* Norwell, Massachusetts: Kluwer Academic Publishers. 1992.

NASA

Anderson, Robert J. et al. *A Cost-Benefit Analysis of Selected Technology Utilization Office Programs.* Princeton, New Jersey: Mathtech. 1977.

Bush, Lance B. *An Analysis of Technology Transfer at NASA.* NASA Technical Memorandum 110270. Hampton, Virginia: Langley Research Center. July 1996.

Chapman, Richard L. "Alternative Methods to Evaluate Technology Transfer." *Technology Commercialization and Economic Growth: Technology Transfer Society 20th Annual Meeting Proceedings, July 16-19, 1995, Washington, D.C.:* 1-9. 1995.

Chapman, Richard L. "An Exploration of the 'Spinback' Phenomenon." *Journal of Technology Transfer* 19 (3-4/December): 78-86. 1994.

Chapman, Richard L. "Case Studies in the Tracking and Measuring of Technology Transfer." In *Technology Transfer Partnerships: Technology Transfer Society 19th Annual Meeting Proceedings, June 22-24, 1994, Huntsville, Alabama.* Kenneth E. Harwell, Kathy Wagner and Carl Ziemke, editors: 164-171. 1994.

Chapman, Richard L. "Measuring Technology Transfer Success: Overcoming the 'If You Can't Count It, It Doesn't Count' Syndrome." *Technology Transfer Tools: Technology Transfer Society 18th Annual Meeting Proceedings, June 26-29, 1993, Ann Arbor, Michigan*: 13-19. 1993.

Chapman, Richard, Loretta C. Lohman and Marilyn J. Chapman. *An Exploration of Benefits from NASA Spinoff.* Littleton, Colorado: Chapman Research Group, Inc. Contract 88-01 with NERAC, Inc. June 1989.

Craft, Harry, W. Sheehan and A. Johnson. "NASA's Southeastern Regional Initiative in Technology Transfer and Commercialization." *46th International Astronautical Congress, October 2-6, 1995, Oslo, Norway.* American Institute of Aeronautics and Astronautics, Inc. IAA-95-IAA.1.2.08. 1995.

Evans, Michael K. *The Economic Impact of NASA R&D Spending.* Bala Cynwyd, Pennsylvania: Chase Econometrics Associates, Inc. April 1976.

Johnston, F. Douglas and Martin Kokus. *NASA Technology Utilization Program: A Summary of Cost-Benefit Studies.* Prepared for Office of Technology Utilization, National Aeronautics and Space Administration. Denver, Colorado: Denver Research Institute, Industrial Economic Division. Contract NASW-3021. December 1977.

Johnston, F. Douglas, with Martin Kokus, Jana Henthorn and Stephen Quist. *NASA Technology Utilization Program: A Cost-Benefit Evaluation.* Prepared for Office of Technology Utilization, National Aeronautics and Space Administration. Denver, Colorado: Denver Research Institute. Contract NASW-

3021. December 1979.

Lohman, Loretta C. and Richard L. Chapman. *"Lessons Learned" about the Collection of Spinoff Benefits Data*. Littleton, Colorado: Chapman Research Group, Inc. NERAC Contract #87-01. March 1989.

Mathematica, Inc. Mathtech Division. *Quantifying the Benefits to the National Economy from Secondary Applications of NASA Technology*. Washington, D.C.: National Aeronautics and Space Administration. NASA Contract Report CR-2673/CR-2674. June 1975, revised March 1976.

Midwest Research Institute. *Economic Impact and Technological Progress of NASA Research and Development Expenditures*, Three Volumes. Kansas City, Missouri: Midwest Research Institute. NASA Contract Report CR-195946. September 1988.

Midwest Research Institute. *Economic Impact of Stimulated Technological Activity*. Three Volumes. Kansas City, Missouri: Midwest Research Institute. October 1971.

University of Tennessee Space Institute and the Tennessee Valley Aerospace Region. *Technology Transfer Research Project: Identification and Analysis of the Factors Present in Successful Technology Transfer Cases*. Prepared by Brett Pichon and Bobbie Woodard. Sponsored by the Tennessee Valley Authority. June 17, 1993.

Department of Energy

Chapman, Richard and Dana Moran. "Measuring the Results of Partnerships for Technology Transfer: Lessons Learned at the National Renewable Energy Laboratory." *Technology Transfer Models for Growth and Revitalization: Technology Transfer Society Proceedings, 21st Annual Meeting, July 21-23, 1996, Cleveland, Ohio*. William Grimberg, Sally Kickel and Lydia Skapura, editors: 145-154. 1996.

Department of Energy. *Our Commitment to Change: A Year of Innovation in Technology Partnerships*. September 1994.

Department of Energy. *Setting Priorities and Measuring Results*. Oak Ridge Centers for Manufacturing Technology. 1995.

Department of Energy. *Success Stories: The Energy Mission in the Market Place*. 1995.

Department of Energy. *The Transfer and Commercial Impact of the U.S. Department of Energy's Award-Winning Technologies*. Prepared for Office of the Deputy Under Secretary for Technology Partnerships, U.S. Department of Energy. Prepared by Oak Ridge Institute for Science and Education, Training and Management Systems Division. February 1995.

General Accounting Office. *DOE's Success Stories Report*. GAO/RCED-120R. April 15, 1996.

General Accounting Office. *Energy R&D: Observations on DOE's Success Stories*

Report. Testimony before the Subcommittee on Energy and Environment, Committee on Science, House of Representatives. GAO/T-RCED-96-133. April 17, 1996.

Ham, Rose Marie, David Mowery and Hank Chesbrough. *Managing and Evaluating Single-Firm CRADAs: An Assessment of Five Recent Cases at Lawrence Livermore National Laboratory*. Berkeley, California: Center for Research Management. Consortium on Competitiveness and Cooperation Working Paper No. 95-7. September 1995.

Los Alamos National Laboratory. *New Mexico Regional Impact Report*. May 1997.

Shea, Moira M. "Technology Partnerships: Measuring Performance, The Integrated Technology Transfer System." *Technology Commercialization and Economic Growth: Technology Transfer Society 20th Annual Meeting Proceedings, July 16-19, 1995, Washington, DC*: 35-39. 1995.

Sheahen, Thomas P., Robert E. Rosenthal, Robert A. Hawsey, Stephen W. Freiman and James G. Daley. "Evaluation of Technology Transfer by Peer Review." *Journal of Technology Transfer* 19 (3/4 December): 100-109. 1994.

Department of Defense

Department of Defense. Director of Defense Research and Engineering. *Survey of Laboratories and Implementation of the Federal Defense Laboratory Diversification Program*. February 1994.

Guilfoos, Stephen J. "Measuring Transfer Effectiveness or Why Don Quixote Tilts at Windmills." *Technology Transfer Partnerships: Technology Transfer Society 19th Annual Meeting Proceedings, June 22-24, 1994, Huntsville, Alabama*. Kenneth E. Harwell, Kathy Wagner and Carl Ziemke, Editors: 172-176. 1994.

Lesko, John and Michael Irish. *Technology Exchange: A Guide to Successful Cooperative R&D Partnerships*. Battelle and Economic Strategy Institute. Columbus, Ohio: Battelle Press. ISBN 1-57477-037-3. 1995.

Lesko, John, Phillip Nicolai and Michael Steve. *Technology Exchange in the Information Age: A Guide to Successful Cooperative R&D Partnerships*. Second Edition. Columbus, Ohio: Battelle Press. 1998.

Department of Agriculture

Chapman, Richard L. and Marilyn J. Chapman. *An Exploration of Benefits From ARS and Cooperative Research*. Littleton, Colorado. Chapman Research Group, Inc. 1992.

Multi-Agency – Congress and GAO

General Accounting Office. *Constraints Perceived by Federal Laboratory and Agency Officials*. Briefing Report to the Chairman, Committee on Science,

Space and Technology, House of Representatives. GAO/RCED-88-116BR. March 1988.

General Accounting Office. *Copyright Law Contraints on the Transfer of Certain Federal Computer Software With Commercial Aplications*. Statement of John M. Ols, Jr., Director in the Resources, Community, and Economic Development Division, Before the Committee on Commerce, Science and Transportation, United States Senate. GAO/T-RCED-91-91. September 13, 1991.

General Accounting Office. *Diffusing Innovations: Implementing the Technology Transfer Act of 1986*. Report to the Chairman, Committee on Science, Space and Technology, House of Representatives. GAO/PEMD-91-23. May 1991.

General Accounting Office. *Federal Agencies' Actions to Implement Section 11 of the Stevenson-Wydler Technology Innovation act of 1980*. GAO/RCED-84-60. August 24, 1984.

General Accounting Office. *Technology Transfer: Barriers Limit Royalty Sharing's Effectiveness*. Report to Congressional Committees. GAO/RCED-93-6. December 1992.

General Accounting Office. *Technology Transfer: Federal Agencies' Patent Licensing Activities*. Report to Congressional Requesters. GAO/RCED-91-80. April 1991.

General Accounting Office. *Technology Transfer: Implementation of CRADAs at NIST, Army, and DOE*. Testimony before the Subcommittee on Energy, Committee on Science and Technology, U.S. House of Representatives. GAO/T-RCED-93-53. June 10, 1993.

General Accounting Office. *Technology Transfer: Implementation Status of the Federal Technology Transfer Act of 1986*. Report to Congressional Requesters. GAO/RCED-89-154. May 1989.

General Accounting Office. *Technology Transfer: Improving Incentives for Technology Transfer at Federal Laboratories*. Testimony before the Subcommittee on Science, Technology and Space, Committee on Commerce, Science and Transportation, U.S. Senate. GAO/T-RCED-94-42. October 26, 1993.

General Accounting Office. *Technology Transfer: Improving the Use of Cooperative R&D Agreements at DOE's Contractor-Operated Laboratories*. Report to Congressional Requesters. GAO/RCED-94-91. April 1994.

General Accounting Office. *Technology Transfers: Benefits of Cooperative R&D Agreements*. Report to the Vice Chairman, Joint Economic Committee, U.S. Congress. GAO/RCED-95-52. December 1994.

House of Representatives. Committee on Small Business. Subcommittee on Regulation, Business Opportunities and Energy. *Technology Transfer Obstacles in Federal Laboratories: Key Agencies Respond to Subcommittee Survey*. Washington, D.C.: U.S. Government Printing Office. Committee Print 101-3. March 1990.

Multi-Agency - DOC, Interagency Committee, FLC

Chapman Research Group, Inc. *Managing the Successful Transfer of Technology from Federal Facilities: A Survey of Selected Laboratories and Facilities in the Mid-Continent Region of the Federal Laboratory Consortium*. Federal Laboratory Consortium. 1997.

Department of Commerce. *Technology Transfer Under the Stevenson-Wydler Technology Innovation Act: The Second Biennial Report*. Report to the President and the Congress from the Secretary of Commerce. January 1993.

Department of Commerce. *The Federal Technology Transfer Act of 1986: The First Two Years*. Report to the President and the Congress from the Secretary of Commerce. July 1989.

Department of Commerce, Office of Technology Policy. *Effective Partnering: A Report to Congress on Federal Technology Partnerships*. Richard J. Brody, Project Director. April 1996.

Federal Laboratory Consortium. *Remaining Issues in Federal Technology Transfer: An Update*. For distribution at the 1995 FLC National Technology Transfer Meeting, Atlanta, Georgia. 1995.

Federal Laboratory Consortium. *Technology Transfer in a Time of Transition: A Guide to Defense Conversion*. 1994.

Interagency Committee on Federal Technology Transfer. Working Group on Technology Transfer Measurement and Evaluation. *Collective Reporting and Common Measures: Draft for Comment*. Prepared by the Oak Ridge Institute for Science and Education (ORISE) Training and Management Systems Division for the U.S. Department of Energy's Technology Utilization Office. November 1994.

McKinley, Tina. *FLC Chair, Lessons Learned in Technology Transfer: 20 Years of Federal Laboratory Consortium for Technology Transfer (FLC) Experience*. Prepared for the Committee on Science, Subcommittee on Technology and Subcommittee on Basic Research, U.S. House of Representatives. June 27, 1995.

Multi-Agency - Various Evaluators

Bozeman, Barry. "Editor's Introduction: Evaluating Technology Transfer and Diffusion." *Evaluation and Program Planning* 11: 63. 1988.

Bozeman, Barry. "Evaluating Government Technology Transfer: Early Impacts of the Cooperative Technology Paradigm." *Policy Studies Journal* 22 (2/Summer): 322-337. 1994.

Bozeman, Barry. "Evaluating Technology Transfer Success: A National Survey of Government Laboratories." *Preparing the Way: Technology Transfer in the 21st Century, Technology Transfer Society 16th Annual Meeting and International Symposium Proceedings, June 9-11, 1991, Denver, Colorado*.

Richard L. Chapman and William R. Sharp, editors: 138-153.

Bozeman, Barry. "What We Don't Know About Evaluating Technology Transfer: Some Puzzles Seeking Solutions." *Technology Transfer Metrics Summit Proceedings.* Sally A. Rood, editor. Chicago, Illinois: Technology Transfer Society. 46-53. June 1997.

Bozeman, Barry and Gordon Kingsley. "R&D Value Mapping: A New Approach to Case Study-Based Evaluation." *Journal of Technology Transfer* 22 (2/Summer): 33-42. 1997.

Bozeman, Barry and Jane Massey. "Investing in Policy Evaluation: Some Guidelines for Skeptical Public Managers." *Public Administration Review* (May/June): 264-270. 1982.

Bozeman, Barry and Karen Coker. "Assessing the Effectiveness of Technology Transfer from U.S. Government R&D Laboratories: The Impact of Market Orientation." *Technovation* 12 (4/May): 239-256. 1992.

Bozeman, Barry and Maureen Fellows. "Technology Transfer at the U.S. National Laboratories: A Framework for Evaluation." *Evaluation and Program Planning* 11: 65-75. 1988.

Bozeman, Barry and Michael Crow. "R&D Laboratories in the USA: Structure, Capacity, and Context." *Science and Public Policy* 18: 165-79. 1991.

Bozeman, Barry and Michael M. Crow. "Red Tape and Technology Transfer in the U.S. Government Laboratories." *Journal of Technology Transfer* 16 (2/Spring): 29-37. 1991.

Bozeman, Barry and Michael M. Crow. "Technology Transfer from U.S. Government and University R&D Laboratories." *Technovation* 11 (4/May): 231-245. 1991.

Bozeman, Barry and Michael M. Crow. "The Environments of U.S. R&D Laboratories: Political and Market Influences." *Policy Sciences* 23: 25-56. 1990.

Bozeman, Barry and Steve Loveless. "Sector Content and Performance: A Comparison of Industrial and Government Research Units." *Administration and Society* 19 (2/August): 197-235. 1987.

Bozeman, Barry, Maria Papadakis and Karen Coker. *Industry Perspectives on Commercial Interactions with Federal Laboratories: Does the Cooperative Technology Paradigm Really Work?* Final Report to the National Science Foundation, Research on Science and Technology Program. Atlanta, Georgia: Georgia Institute of Public Policy. Contract no. 9220125. January 1995.

Crow, Michael. "Technology and Knowledge Transfer in Energy R&D Laboratories: An Analysis of Effectiveness." *Evaluation and Program Planning* 11: 76. 1988.

Crutcher, Ronnie D. and William H. Fieselman. "Determining Metrics for Effective Technology Transfer." *Technology Transfer Partnerships: Technology Transfer Society 19th Annual Meeting Proceedings, June 22-24, 1994, Huntsville, Alabama.* Kenneth E. Harwell, Kathy Wagner and Carl Ziemke, editors. 178-184. 1994.

Geisler, Eliezer. *Why Federal Laboratories Succeed or Fail at Technology Commercialization.* Report to the National Science Foundation. 1995.

Hittle, Audie E. *Technology Transfer Through Cooperative Research and Development.* Master's Thesis, Sloan School of Management, Massachusetts Institute of Technology. June 1991.

Lee, Joseph W. *The Improvement of Technology Transfer from Government Laboratories to Industry.* A Research Project of the George Washington University's Engineering Management Department. Presented at the 15th Annual Technology Transfer Society Meeting (Dayton, Ohio, June 26-28, 1990) and incorporated in the Congressional Hearing Record (H.R. 4659) of the Committee on Small Business, September 5, 1990. 1990.

Papadakis, Maria. "Federal Laboratory Missions, Products, and Competitiveness." *Journal of Technology Transfer* (April): 54-66. 1995.

Rahm, Dianne, Barry Bozeman and Michael Crow. "Domestic Technology Transfer and Competitiveness: An Empirical Assessment of Roles of University and Governmental R&D Laboratories." *Public Administration Review* (November/ December): 969-978. 1988.

Spann, Mary S., Mel Adams and William E. Souder. "Measures of Technology Transfer Effectiveness: Key Dimensions and Differences in Their Use by Sponsors, Developers and Adopters." *IEEE Transactions on Engineering Management* 42 (1/February): 19-29. 1995.

Laboratories and Universities

Anderson, Lawrence K. and Brian D. Gurney. *Benchmarking Best Practices in Technology Transfer: Final Report.* Colorado Institute for Technology Transfer and Implementation. Colorado Springs, Colorado. Sponsored by Colorado Advanced Technology Institute and U.S. Department of Commerce. December 1993.

Roessner, J. D. and A. S. Bean. "Industry Interaction with Federal Labs Pays Off." *Research Technology Management* 36 (5): 38-40. 1993.

Roessner, J. David and Alden S. Bean. "Federal Technology Transfer: Industry Interactions With Federal Laboratories." *Journal of Technology Transfer* (Fall): 5-14. 1990.

Roessner, J. David and Alden S. Bean. "How Industry Interacts with Federal Laboratories." *Research-Technology Management* 34 (4/July-August): 22-25. 1991.

Roessner, J. David and Alden S. Bean. "Patterns of Industry Interaction with Federal Laboratories." *Journal of Technology Transfer* (December): 59 - 77. 1994.

Roessner, J. David and Anne Wise. *Patterns of Industry Interaction with Federal Laboratories: Final Report.* Georgia Institute of Technology, School of Public Policy. Martin Marietta Energy Systems, Inc., Oak Ridge National Laboratory, and U.S. Department of Energy Contract #19X-SK495C. May 1993.

University Evaluation Models

University/Industry Partnerships

Feller, Irwin and David Roessner. "What Does Industry Expect From University Partnerships?" *Issues in Science and Technology* (Fall): 80-84. 1995.

Geisler, Eliezer and Albert H. Rubenstein. "Methodology Issues in Conducting Evaluation Studies of R&D/Innovation." *Proceedings of the Symposium on Management of Technological Innovation*. Worcester Polytechnic Institute. 1983.

Gray, Denis O. and S. George Walters. *Managing the Industry-University Cooperative Research Center: A Handbook for Center Directors*. Prepared for the National Science Foundation. ISBN 0-9658444-0-4. 1997.

Industrial Research Institute, Government-University-Industry Research Roundtable, and Council on Competitiveness. *Industry-University Research Collaborations: Report of a Workshop, November 28-30, 1995, Duke University*. Washington, D.C.: National Academy Press. 1996.

National Science Foundation. *Evaluator's Handbook: NSF Industry-University Cooperative Research Centers Program*. Raleigh, NC: I/UCRC Evaluation Project. Updated 1997.

Rubenstein, Albert H. and Eliezer Geisler. "Evaluating the Outputs and Impacts of R&D/Innovation." *International Journal of Technology Management* 5 (1): 181-204. 1991.

Rubenstein, Albert H. and Eliezer Geisler. "The Use of Indicators and Measures of the R&D Process in Evaluating Science and Technology Programs." *Government Innovation Policy: Design, Evaluation, Implementation*. J. David Roessner, editor. St. Martin's Press: 185-204. 1989.

The Engineering Research Centers (ERC) Program: An Assessment of Benefits and Outcomes. National Science Foundation. Directorate for Engineering. Engineering Education and Center Division. Linda Parker, Project Director. December 1997.

University Technology Transfer

AUTM Licensing Survey: FY 1991 - FY 1995, Five-Year Survey Summary. Association of University Technology Managers (AUTM), Inc. Daniel E. Massing, editor and Chair, AUTM Survey, Statistics and Metrics Committee. 1996.

AUTM Public Benefits Survey Summary of Results. Prepared for Association of University Technology Managers (AUTM), Inc. Cranbury, New Jersey: Diane

C. Hoffman, Inc. April 1994.

BankBoston. Economics Department. *MIT: The Impact of Innovation*. 1997.

Carr, Robert K. "Doing Technology Transfer in Federal Laboratories" (Part 1). *Journal of Technology Transfer* 17 (2/3, Spring/Summer): 8-23. 1992.

Carr, Robert K. "Measurement and Evaluation of Federal Technology Transfer." *Technology Commercialization and Economic Growth: Technology Transfer Society 20th Annual Meeting Proceedings, July 16-19, 1995, Washington, D.C.*: 221-230. 1995.

Carr, Robert K. "Menu of Best Practices in Technology Transfer" (Part 2). *Journal of Technology Transfer* 17 (2/3, Spring/Summer): 24-33. 1992.

Odza, Michael. "What the AUTM Licensing Survey Statistics Mean for Federal Labs." *Technology Transfer Metrics Summit Proceedings*. Sally A. Rood, editor. Chicago, Illinois: Technology Transfer Society. 231-235. June 1997.

Pressman, Lori D., Sonia K. Guterman, Irene Abrams, David E. Geist and Lita L. Nelson. "Pre-Production Investment and Jobs Induced by Massachusetts Institute of Technology Exclusive Patent Licenses: A Preliminary Model to Measure the Economic Impact of University Licensing." *Journal of the Association of University Technology Managers* 7: 49-81. 1995.

Tornatzky, Louis G. and Joel S. Bauman. *Outlaws or Heroes? Issues of Faculty Rewards, Organizational Culture, and University-Industry Technology Transfer*. A Benchmarking Report of the Southern Technology Council, Southern Growth Policies Board. July 1997.

Tornatzky, Louis G., Paul G. Waugaman and Lucinda Casson. *Benchmarking Best Practices for University-Industry Technology Transfer: Working with Start-Up Companies*. A Report of the Southern Technology Council, Southern Growth Policies Board. Research Triangle Park, North Carolina: Southern Technology Council. October 20, 1995.

Tornatzky, Louis G., Paul G. Waugaman and Joel S. Bauman. *Benchmarking University-Industry Technology Transfer in the South: 1995-1996 Data*. Research Triangle Park, North Carolina: Southern Technology Council, Southern Growth Policies Board. July 1997.

Laboratory Economic Development Projects, Incubators, Intermediaries

Bearse, Peter. *The Evaluation of Business Incubation Projects: Comprehensive Manual*. National Business Incubation Association, for the U.S. Economic Development Administration. ISBN 1-887183-19-1. December 31, 1993.

Campbell, Candace and David N. Allen. "The Small Business Incubator Industry: Micro-Level Economic Development." *Economic Development Quarterly* 1: 178-191. 1987.

Hatry, Harry P., Mark Fall, Thomas O. Singer and E. Blaine Liner. *Monitoring the Outcomes of Economic Development Programs: A Manual*. Washington, D.C.:

Urban Institute Press. ISBN 0-87766-488-9. 1990.

Markesen, Ann and Michael Oden. "National Laboratories as Business Incubators and Region Builders." *Journal of Technology Transfer* 21 (1-2/Spring-Summer): 93-108. 1996.

Markley, Deborah M. and Kevin T. McNamara. "Local Economic and State Fiscal Impacts of Business Incubators." *State and Local Government Review* 28 (1/Winter): 17-27. 1995.

Muir, Nan. "Measuring Technology Transfer Success: A Study of Intermediary Agency Evaluation." *Technology Commercialization and Economic Growth: Technology Transfer Society 20th Annual Meeting Proceedings, July 16-19, 1995, Washington, D.C.*: 17-26. 1995.

Nexus Associates, Inc. *Guide to Economic Development Program Evaluation.* Belmont, Massachusetts. 1996.

Schroer, Bernard J., Phillip A. Farrington, Sherri L. Messimer and J. Ronald Thornton. "Measuring Technology Transfer Performance: A Case Study." *Journal of Technology Transfer* 20 (2/September): 39-47. 1995.

Tornatzky, Dr. Louis G., Yolanda Batts, Nancy E. McCrea, Marsha L. Shook and Louisa M. Quittman. *The Art and Craft of Technology Business Incubation: Best Practices, Srategies, and Tools from 50 Programs*. Southern Technology Council, National Business Incubation Association, and Institute for Local Government Administration and Rural Development. ISBN 0-927364-04-2. 1995.

Legislation

Bayh-Dole Act. Public Law 96-517. 1980.

Federal Technology Transfer Act. Public Law 99-502. 1986.

Government Performance and Results Act. Public Law 103-62. 1993.

National Competitiveness Act. Public Law 101-189. 1989.

National Cooperative Research Act. Public Law 98-462. 1984.

National Cooperative Research and Production Act. Public Law 103-42. 1993.

National Technology Transfer and Advancement Act. Public Law 104-113. 1996.

Omnibus Trade and Competitiveness Act. Public Law 100-418. 1988.

Small Business Innovation Development Act. Public Law 97-219. 1982.

Stevenson-Wydler Technology Innovation Act. Public Law 96-480. 1980.

Trademark Clarification Act. Public 98-620. 1984.

Qualitative Research Methodology

Denzin, Norman K. *Interpretive Ethnography: Ethnographic Practices for the 21st Century*. Newbury Park, California: Sage Publications, Inc. 1996.

Denzin, Norman K. and Yvonna S. Lincoln, editors. *Handbook of Qualitative Research*. Thousand Oaks, California: Sage Publications, Inc. 1994.

Glaser, Barney G. and Anselm L. Strauss. *The Discovery of Grounded Theory*. Chicago, Illinois: Aldine. 1967.

Guba, Egon G. and Yvonna S. Lincoln. *Fourth Generation Evaluation*. Newbury Park, California: Sage Publications, Inc., 1989.

Miles, Matthew B. and A. Michael Huberman. *Qualitative Data Analysis: An Expanded Sourcebook*. 2nd edition. Newbury Park: California: Sage Publications, Inc. 1994.

Qualitative Solutions and Research Pty, Ltd. QSR NUD-IST: Non-Numerical Unstructured Data Indexing Searching and Theorizing. Version 3.0. Newbury Park, California: Scolari, Software Division of Sage Publications, Inc. 1996.

Spradley, James P. *The Ethnographic Interview*. Ft. Worth, Texas: Harcourt Brace Jovanovich College Publishers. 1979.

Weitzman, Eben A. and Matthew B. Miles. *Computer Programs for Qualitative Data Analysis: A Software Sourcebook*. Thousand Oaks, California: Sage Publications, Inc. 1995.

Yin, Robert K. *Case Study Research: Design and Methods*. 2nd edition. Newbury Park, California: Sage Publications, Inc. 1994.

Index

About the Author

For 25 years, Sally A. Rood has worked in the area of technology-based economic development and other areas of concern to federal, state, and local governments. She is currently the Washington, D.C. Representative for the Federal Laboratory Consortium, a Congressionally-chartered network of federal laboratory and agency offices focused on technology transfer. She previously served for five years as Associate Director of the National Technology Transfer Center's Washington Operations, focussing on technology transfer outreach and entrepreneurial development demonstration projects for the Ballistic Missile Defense Organization. Previously, she was Manager of Technology Information Services for a government contracting firm where she implemented management support contracts for the Department of Energy, Department of Commerce, and Department of Defense. Earlier, she worked at the National Association of Counties—on technical assistance projects funded by the National Science Foundation and Department of Transportation—aimed at improving public works and other technical activities at the local level. As a university researcher and independent consultant, she has also performed work for a variety of other interest groups, including the U.S. Conference of Mayors, National Council for Urban Economic Development, Academy for State and Local Government, Coalition of Northeastern Governors, and Council for International Urban Liaison.

Dr. Rood is active with professional associations and is the Immediate Past President of the Technology Transfer Society. She has also held elected positions with the American Society for Public Administration and the American Association for the Advancement of Science.

She has written numerous articles on technology transfer and economic development, including an award-winning article in the *Journal of Technology Transfer*. Dr. Rood is also on the editorial and advisory boards for *Economic Development Quarterly* and *The Executive*, and she "founded" the newsletter *Economic Development Abroad*. Her Ph.D. is in Public Affairs and Public Administration.